INTELLOPY

Survival and Happiness in a Collapsing Society with Advancing Technology

JJ Kelly

FUTURE COACH LLC.COM

ISBN: 978-0-578-94877-5
Publication date: October 5, 2021

Published by Future Coach LLC
411 Walnut Street Unit 6464, Green Cove Springs, FL 32043
www.FutureCoachLLC.com

FUTURE COACH LLC.COM

Cover design by Author
Interior design by Author
Printed in the USA

DEDICATION

To my grandchildren—Blake, Bay Lou, Joe, and Remi:

You were born a blank slate and must learn the knowledge discovered and accumulated in society. Much of what today's society will try to teach you is false—not based in reality—and can cripple your mind.

I share my knowledge of how the world works—principles based on reality—so that you can distinguish truth from falsehood and navigate turbulent change toward your heroic vision.

With love,

Daddo

CONTENTS

PROLOGUE

WHAT ARE YOU THINKING?

What do you think when you watch the news? Are you wondering why the fabric of our American society appears to be unraveling? Did you know there are predictions that our American society might collapse into a totalitarian state in just 20 to 30 years? There are also predictions of a colony on Mars in that same period. Experts say that Artificial Intelligence (AI) will be more intelligent than all humans combined by then. Are you supposed to take any of these predictions seriously? Are you thinking about how soon 20 to 30 years is?

We are about to experience something profound. We are about to transform what it means to be "human" in mind and body as we merge with our technology. This transformation has already begun. But what if our society collapses? Will a totalitarian regime use advancing AI technology to suppress and control us? Or will technology help us save our free society, go to the stars and awaken the universe? Do you know how to plan your life considering such possibilities?

Ideas are crucially important. They determine the direction of a person's life and the fate of entire societies. They will soon determine the future of our society. How do I convince you that by letting others do the vital thinking for you, you are putting your own life and your children's lives at risk?

The intellectuals in a society are not always right. Ideas promoted by scholars led Europe out of the Dark Ages to the Renaissance and Enlightenment. But later, intellectuals in Europe caused their societies to accept Nazism and Communism. What if today's scholars are wrong? How do we identify "good" ideas versus "bad" ideas? How will we know if we do not think for ourselves?

Professors in the universities promote specific ideas to their students who will become scientists, engineers, business professionals, and politicians. In this new technological age, these people will embed those ideas in the smart AI technology *around* us. They will eventually build these ideas into the technology *within* us (brain-machine interfaces). AI systems are already beginning to replace humans in the administration of our society's institutions. If we do not take control of our thinking, we might soon lose the ability to think for ourselves.

Even when you have the time to contemplate such big ideas, it is difficult to know where to begin. It is easier to leave the thinking about big ideas to professionals in religion, philosophy, education, business, government, and science. You get bits and pieces of what those professionals and other commentators think via the media—but you are not sure that you can trust the media who seem to have an agenda. The presented ideas often conflict, and you have no way to know which is true. So, you go along with society's mainstream flow and live your daily life as best you can—hoping for a good outcome. You do occasionally worry about the big ideas and wonder if there is something you should be doing.

Considering important ideas and resolving them in your mind enables you to set your life with purpose instead of blindly reacting to your daily grind circumstances. It allows you to enjoy *every* value in life more deeply—productive work, romantic love, friendships, recreation, art, and more. It enables you to live the fullest, most exciting, and enjoyable life possible as an intelligent human.

Technology is going to force change upon all of us, opening an array of opportunities. Now is the ideal time for you to reset your life by thinking and rediscovering your values—the things that awaken your passions. New goals can challenge you to be the best person you can be. The first step is to learn *how* your thinking directs the course of your life. Once you know how it works, you can control it and make your life go where you want.

Each of us is an explorer, and our life is an adventure. We are traveling over new territory (time), and we will make discoveries along the way. An explorer must be able to navigate—to steer a course toward a destination or purpose. First, an explorer must orient himself to his existing situation or environment. He must know what the "map" looks like, and he must find his place on the map by considering observed facts or principles. Then, he can choose the course that he will take. Suppose the explorer does not know his present position, or worse, does not have a destination, purpose, or principles to guide him. In that case, he is doomed to either remain in the same dull spot or travel in random circles until he dies.

Unlike a two-dimensional paper map, the "map" for humans has multiple dimensions or coordinates defining our present position. Each dimension represents a line of evolution from one place to another, and we are presently somewhere along each line. These dimensions and orienting coordinates include cosmology, biology, philosophy, knowledge, technology, society, and happiness. The matching questions are:

> *Where am I right now along the dimension of…*
> *the evolution of the cosmos?*
> *the evolution of humans?*
> *the evolution of my ability to "think?"*
> *the evolution of my knowledge of how the universe works?*
> *the evolution of technology?*
> *the evolution of society?*
> *the evolution of my happiness?*

By observing facts and principles in each of these dimensions, we can find our place on the "map." Then we can choose our destination or purpose and set the course for the adventure that is our life. I will provide the "map" and the facts and principles that apply in each dimension. As we go into the uncharted and uncertain future, we must remain aware of these enduring universal principles based on Laws of Reality.

Reality and evolution provided us the ability to "think" as our unique method of survival. We think about things at many levels. Some thoughts are "concrete," such as, "That thing I see moving on the rock is an amoeba." Some thoughts are "abstract," such as, "I think I am in love with Ann." To be *true*, all thoughts, concrete and abstract, must be connected to reality.

It *is* possible to connect big abstract ideas such as freedom, integrity, honesty, justice, and love to reality. It is also possible to have errors in one's thinking somewhere along the chain of ideas between observable reality and the higher abstract ideas. You were not born automatically knowing how to connect abstract concepts to reality—how to "reason." It is a skill you must develop. It is not complicated. You never had a reason to focus on it in this way because *schools do not consistently teach it*. They teach it in the sciences, but not when it comes to abstract ideas in the humanities. We are going to do this together step by step.

Future technology, especially Artificial Intelligence, is the ideal subject because we must explore how *human* thinking works before considering the possibility of a thinking *machine*. We can use AI as a mirror to see ourselves. It will show us how ideas promoted by the scholars and accepted by the general population today will be disastrous for us tomorrow if AI systems (or someday, AI entities) accelerate and amplify those ideas.

If you rely on your observations and thinking to connect your abstract ideas to reality, you will become one of the New Intellectuals our society desperately needs. I will point out things that I see and how they might be connected in cause-and-effect relationships back to reality. You will consider what you see and form your *own* conclusions—not the random, unevaluated, and conflicting ideas put into your mind by others many years ago. You will focus on the essentials and cut through all the distractions. You will decide *for yourself* what you accept as *true* (real).

To be able to guide your life, you must learn how the world works. I will describe what I call *The Six Critical Processes*. These six cause a cascade of derivative processes and explain *everything* that we see in our society and our personal lives. These processes are driven by the Laws of Reality including man's nature. As principles, they enable you to understand the world around you and guide your life confidently.

The Six Critical Processes are:

1. ***Evolution*** - understanding how everything must remain connected to reality.
2. ***Plastic Brain*** - understanding how you program your brain's intelligence.
3. ***Conceptual Thinking*** - understanding what valid knowledge is and how you use it.
4. ***Technology Cycle*** - understanding how knowledge, technology, and human intelligence advance.
5. ***Ivory Tower Cycle*** - understanding society and how a few intellectuals influence it independent of advances in knowledge, technology, or intelligence.
6. ***Happiness Skills*** - understanding how your conceptual thinking shapes your psychology and your happiness.

WHY ME? WHY NOW?

I am one of those people who needs to know how everything works. As a child, I somehow realized that the type of knowledge that is valuable is "cause-and-effect" knowledge. Everything else is of lesser importance. "A" was the cause that resulted in effect "B." "B" then became the cause that had the effect "C," and so on. I learned that I could not observe some causes directly, but I could discover them by observing their effects. I began building a hierarchy of knowledge, starting with concrete facts of reality and ending with high-level abstract ideas that I could connect back to reality.

As former President/CEO of US Telecom, I have decades of first-hand experience in technology, including telecommunications, computers, networking, the Internet/Web, wireless, cellular, messaging, speech recognition, speech synthesis, and AI. As a recognized founder of the Telecommunications Revolution of the 1980s and 1990s, I was a charter member of national committees defining and promoting technology standards for compliance and interoperability of various computing and communicating platforms. My experience enabled me to analyze the interactions between society, business, and advancing technology to discover cause-and-effect principles.

Earlier, following an education in engineering and biology, and after serving in the US Navy as a carrier-qualified jet pilot, I worked my way up to Regional President at US Home Corp, then the nation's largest land development and home building company. It was there, in my 30s, that I was introduced to and developed a life-long interest in philosophy, psychology, and history which, of course, are invaluable in understanding "how the world works."

Attending the three-year *Strategic Coach* program for successful entrepreneurs, I learned that my Unique Ability is—*Discovering the essential elements and cause-effect relationships in any situation, physical or behavioral; envisioning innovation to it and derived from it; and effectively communicating all of it to others.*

Here are three essential ideas that motivated me to write this book:

1. **Most people are unaware of the twin tsunamis of *societal* and *technological change* that are heading our way—soon.**

2. **Fundamental philosophical ideas commonly held in our society will determine whether the future will be "good" or "bad" for us. Our technology will amplify the effects of these ideas upon us—good or bad.**
3. **Understanding *how to think* about these important ideas and how they cause the effects that you see happening in your world *is not difficult*. We were just not taught this skill when growing up.**

Futurist author Richard Yonk wrote, "We are the species we are today because of technology, and technology exists because of our existence as a species. This isn't a tautology, but rather a virtuous circle that is giving rise to the next amazing emergences of the universe."[1]

We will explore what I think will be the "transcendent" technologies. These are the technologies that, in combination, might enable humans to transcend our present existence and evolve to become an entirely new, technological species.

Some futurist authors motivate their readers by emphasizing the potential dangers of advancing technology. They write, for example, about the risks of a "Terminator" AI, artificial viruses, or unstoppable self-replicating nanobots. These are not the great and immediate danger to be feared. The real threat is that AI systems are replacing humans in administering the institutions and transactions of our society. The designers and operators of those AI systems can set the rules under which all the rest of us must live. They train those AI systems according to the philosophical ideas they learned from the intellectuals who taught them.

While we must not ignore any new technology's potential perils, I believe there is a higher motivation than fear. Technology presents the possibility of living exceptionally long, healthy, exciting, and enjoyable lives. Some people find themselves stuck in a lifestyle that is short of the *heroic vision* each had for himself when he was a child. That is not how anyone wants to live out a lifespan greatly extended by technology. Our goal must be to *achieve a positive*—our happiness and man's shining future—not simply avoid a negative.

I ask that you keep my primary purpose and audience for this book in mind. I am a grandfather with grandchildren who are presently all under the age of seven. When they reach high school age in about ten years, I want this book to be there for them if I am not. My goal is to record and communicate what I have learned over the decades about how the world works. I want to provide them the essential reality-based principles that will help them achieve their heroic vision, navigating what is sure to be a turbulent future.

I must caution you that my explanations might not be how the academic or professional in the related field would describe things. The professionals might not even agree with me. I will provide references and citations to previous work by others. I will quote excellent descriptions by others of things I observe to be true. However, most of this content will be the result of my own "thinking."

Ultimately, what will matter is *your* thinking. As you will soon see, today's world is in desperate need of individuals who are willing to put in the effort to *think* about big abstract ideas and connect them to reality—to be the New Intellectuals. For at least the time you spend reading this book, you will be doing the urgently needed deep thinking. Maybe something in the book will provide you the motivation to keep it going.

I am going to pose a couple of questions for you to keep in mind and answer. We will later see if your answers and mine are the same.

"What is the most powerful thing in the known universe? What can defeat it?"

Once you learn how to think using facts of reality and cause-and-effect principles, you will be able to orient your present position, choose your destination and purpose, and set your life's course. You will begin to replace your present doubts and fears about society, technology, and your future with self-confidence. You will know that you can prove your ideas as *true* by following the chain of ideas back to observable reality. Others who have not developed that skill might not understand why they do what they do, but you will. You will have developed the mental skills you need to achieve happiness in a collapsing society with advancing technology. And in doing so, you will have made your contribution to the survival of our society and mankind.

PART 1

KNOWING OUR PLASTIC BRAIN

Everything we will explore about the future is from a "human" perspective. It is only in the context of an individual human that any of this has any importance. We must thoroughly understand what it means to be "human" before we can address the possibility of humans merging with our technology or humans creating intelligent machines.

The most complex functioning thing that we know about in the universe, so far, is the human brain. We know a lot about how the human brain works, but there is a great deal more that we have not yet discovered. Most of this book will focus on how we "use" our brains to "think." But first, we must have a basic understanding of how our brain *physically* functions.

WHAT MAKES US HUMAN?

"Perhaps you regard this thinking about myself as a waste of time—but how can I be a logician before I am a human being! Far the most important thing is to settle accounts with myself!"

—LUDWIG WITGENSTEIN

THE RATIONAL ANIMAL

Because you are reading these printed words, I can safely assume that you are a human, not a chimpanzee. What is the difference? What is the defining characteristic of humans? Can humans be defined as "hairless apes"? That might be true, but it is not the defining characteristic of humans. The ancient Greek philosopher Aristotle, in 350 B.C., defined humans as the "rational animal."

Humans can "reason" while chimpanzees and every other animal cannot. Our rational faculty forms and retains abstract ideas called "concepts." Our *conceptual* level of consciousness is higher than the *perceptual* level of consciousness of animals. How we form and use concepts is at the center of what we mean by "human intelligence." We must understand this clearly to understand what is happening in our society and understand how we might someday create human-level artificial intelligence. Most importantly, we must understand the functioning of our human intelligence if we want to be better able to guide our lives.

Concepts are ideas formed in our minds. The word we assign to it represents each concept. A concept can be "concrete" or "abstract." Examples of *concrete* concepts would be the words "chair," "table," and "chest." We consider these concrete concepts because we can point to actual chairs, tables, and chests. However, the concept "furniture" is an *abstract* concept that groups chairs, tables, and chests together based on their similar usage in a house. We can point to a chair, table, or chest, but we cannot point to a "furniture."

A concrete concept *integrates* a multitude of existing *real things,* based on some essential attribute that they share, into a single *abstract thing* held in our mind to represent the group. The word "chair" becomes a *unit* in our mind, an abstraction, representing the multitude of actual chairs that ever existed or will exist. A concept is our *mental method* of identifying and dealing with things in reality, but it is not the *things themselves.* They exist independent of us and the word we assigned to them.

We can then integrate multiple concrete concepts into a single higher-level abstract concept based on some attribute that the concrete concepts share. "Chairs," "tables," and "chests" become "furniture." We can continue integrating into ever-higher abstractions—"furniture," "appliances," and "vehicles" become "personal property," which then combines with "house" and "savings" to become "assets."

A highest-level abstract concept that can be followed down through the chain of lower concepts to point to concrete existents is *true*—"connected" to reality. If we cannot connect it to reality, we call it a "floating" abstraction, for example, a "unicorn." Imagination has a vital role to play that we will cover later. The critical point is that to survive and live our lives entirely according to our nature as humans; we must be "rational." We must base our decisions and actions on concepts that are connected to reality—*true.* To do otherwise is to be *irrational.*

Since Aristotle's time, some philosophers, observing that man has not consistently acted rationally, have rejected Aristotle's definition of man as the rational animal. Unfortunately, Aristotle did not adequately explain *how* man's rational faculty works. Aristotle was the first to identify man's rational faculty as his essential characteristic. He formulated the correct definition principles and believed that only concretes, not abstractions, exist in reality. However, Aristotle did not correctly identify the *process* by which man's mind isolates the fundamental characteristic of similar existents and forms an *integration* of those multiple existents into a unitary *concept*. In other words, Aristotle correctly identified that reality is real and that our mind is competent to grasp it, but he did not explain *how* our abstract concepts connect to reality.

Lacking an explanation of how "reason" connects abstract ideas to reality, professional philosophers have historically sought to define man in many ways other than Aristotle's rational animal. They either rejected man's ability to grasp reality, or they rejected reality itself.[3] The result was that they cut off their abstract ideas regarding such things as ethics, justice, truth, or love from reality. They were floating abstractions—fantasies of imagination with no basis or definition rooted in reality. Sadly, this describes most of today's professional philosophers and is why they do not consider philosophy an actual "science."

The unit reduction theory of concepts did not appear until the mid-twentieth century along with the explanation of "volition." Unlike animals, man can choose to focus his mind and think or to unfocus it and suspend thinking, which explains why some men, at times, are rational and some are not. I will cover both the unit reduction theory and volition in more detail in later chapters.

Throughout history, each man has had to learn to use his rational faculty. In the pre-science ages, people were mainly irrational, believing that gods pushed the sun around the Earth and caused disasters such as pestilences and draughts. These "gods" were floating abstractions not connected to reality, but early man had no other way to explain what he saw. Men like Galileo discovered that the Laws of Reality exist and are knowable. People working in the physical science fields learned what it means to connect abstract ideas to reality. However, philosophers and other professionals working in the humanities and social sciences have mostly rejected the need to connect abstract concepts to reality. They still prefer to live their lives guided by floating abstractions. We will address this again later.

The twentieth-century unit reduction theory of concepts and description of volition have explained how our rational faculty works. It is time to reassert Aristotle's definition of man as the "rational animal."

♦♦♦

1.2

PATHWAYS

"The human brain had a vast memory storage, it made us curious and very creative. Those were characteristics that gave us an advantage—curiosity, creativity and memory. And that brain did something very special. It invented an idea called 'the future.'"

—David Suzuki

THE PHYSICAL BRAIN

You are using your brain right now to read and understand these words. You are using it to pursue the goal of new knowledge that you hope this book will provide. In later chapters, we will explore what *percepts* and *concepts* are and how we use them to "think"—our unique form of consciousness. But first, we must demystify and understand the basic *physical* functioning of our brain. To explore how your brain physically works, let us go on an adventure to the jungle.

Assume that you are an amateur archaeologist. There is a remote site in a dense jungle that you wish to dig. You make your way to a small mountain village that you will use as your base camp. Your dig site is a couple of miles away, and you will hike to it each day and return to the village each evening.

When you first arrive at the village, you describe your dig site to the village chief and ask him the best way to get there in the morning. He explains that there is no existing path to that site. You will need to hike through the jungle. He gives you directions. First, you head east up the mountain to the craggy rock outcropping, then south down a slope to the small pond, then west to your dig site.

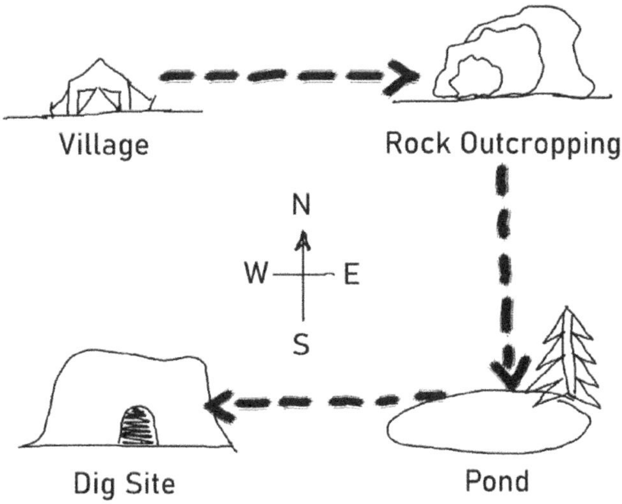

Chief's directions to your dig site.

The following morning equipped with your shovel, your machete, and the chief's directions, you head out to find your dig site. It takes a lot of effort to hack your way through the jungle creating a path to walk. You ascend steep slopes; you climb down and up the other side of small gullies; you wade through small streams. All the while you must pay close attention to your compass and keep a lookout for your next landmark. It is slow going, but you eventually find your dig site. It has taken you half the day to get there, but you now know how to get there, so you head back to the village.

You realize that to get to your dig site more quickly and not use up all your time and energy just getting there, you will need to improve your pathway to the dig site. The next day you hack away more brush and position some stones to facilitate stream crossings. The day after that, you use your shovel to dig some steps into steep slopes. The day after that, you put some logs over the small gullies.

Each day, as you repeatedly use it and work on it, your pathway becomes more developed and easier to traverse. You soon discover that you no longer must consciously look at your compass or keep a lookout for landmarks. You follow the path you have created. You finally reach a point where you no longer must work on your path; you use it. You get to your dig site quicker, and you expend less energy getting there.

Your physical brain works the same way. When you have a conscious thought, a particular network of neurons is activated. Using those neurons to have that thought caused your brain to give that network of neurons increased importance. Your conscious thinking changed the physical neuronal structure of your brain. Conscious thinking caused chemicals to be released that turned on and off the expression of specific genes controlling your brain's structure. Turning genes on and off is called "epigenetics." Connections between neurons across the gaps called "synapses" were made stronger, like placing the stones in the streams and putting logs over the gullies.[4] Repeating the same conscious thought further strengthens the neural networks involved, makes them more efficient in the energy used, and speeds them up. Once developed and optimized, this is the neural pathway you will use every time you contemplate the same idea.

At birth, your physical brain was built with a basic map of the specific brain locations used for various functions as programmed in your genetics. However, the experiences encountered by your consciousness as you grew up changed your brain map. Your brain allocated neurons as needed to encode and remember whatever your consciousness experienced and in the order in which you encountered it. Whatever you first and frequently "learned" got first "dibs" on unallocated neuronal resources. Your brain started pruning away lesser used neuronal resources as you got older to make room for new knowledge. It also created new neurons and neuronal connections as needed. Your brain grows new neurons from neuronal stem cells until you die. It is continuously developing and dismantling neural networks, and your head .

Most of the knowledge you presently have in your mind was put there by your conscious thinking over your lifespan. Whatever you thought about with the most conscious effort and the most frequency are your strongest memories and ideas. Your brain saves patterns with redundant copies, each with some slight variation. Multiple copies enable you to recognize someone from both a frontal and profile view. The more copies of the memory, the stronger it is. A memory that fades is one where the brain gradually deletes redundant copies.

Let us go back to the jungle. Every day you have been following the "idea" in your brain of the best pathway to your dig site. You do this unconsciously now since you no longer must pay attention to the jungle pathway. This pathway idea was first described to you by the chief. Then, with repeated effort applied over time, you created and made permanent a neural network of it in your brain.

Then, one day, a professional archaeologist stops in the village while on an expedition. You visit with him about your dig site. You explain its location by giving him the directions to get there that the chief gave you. The archaeologist brings out a topographic map and finds the location of your dig site on it. You immediately notice on the map that only a short distance west of the village is a stream that flows down the mountain and passes near your dig site. It looks like a shorter and easier path to get there.

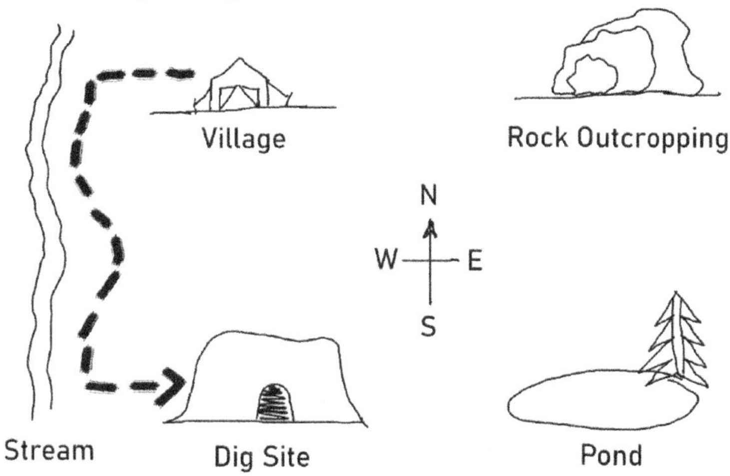

Follow the stream to the dig site.

At first, your emotions resist the idea of having to hack your way through the jungle again. However, your productive work is of high value to you, so you are rationally motivated to increase the time and energy available to you when at your dig site. So, you suppress the emotions telling you to keep using the chief's longer but now "easy" pathway.

The following morning you strike out, heading west instead of east. Once again, you must consciously think about what you are doing. You must expend effort to cut back the brush and make a new pathway to the stream, down the stream, and to your dig site. It is easier to follow and create this new pathway because you follow the stream rather than a complicated set of landmarks and directional changes given to you by the chief. You find that this new pathway to your dig site, once you develop it, will indeed be much better than the one you first learned from the chief. You will get to your dig site in half the time and spend half the energy getting there.

When you get back to the village, you ask the archaeologist why he thinks the chief first gave you such a round-about path? The archaeologist tells you that this tribe worships the sun. Regardless of the destination, any trek that begins in the morning must first start toward the rising sun in the east.

You have been creating specific neural network pathways in your brain that encode conceptual ideas for your entire life. You first obtained some of these ideas by your direct observation of reality. For example, a toy ball that you released always went down, never up. Then, when you reached the age of language skills, you also obtained ideas from other humans, like how to get to your dig site. The more often you heard or thought about the ideas you got from other humans, the stronger they became in your brain. Eventually, all the ideas, those you got directly from reality and those you got from humans, were made equally permanent in your subconscious.

But then new knowledge of reality comes along, such as when the archeologist produced the topographic map. You discover that an idea that you got from another human was not true to reality. The chief's directions to the dig site were not the "best" pathway. The chief had reasons based on mysticism, not reality, to send you the way he did, and you did not know it.

In the face of new knowledge based on reality, you must expend consciously focused mental effort to build a new or modified set of neural pathways in your brain.[5] Over time, the neural network for the old idea, the chief's trail, will no longer be maintained by the brain and lose strength due to a lack of use and inattention. The brush will grow back, and the steppingstones and logs will wash away. The brain's resources used for the old idea will be re-tasked.

Neural networks of conceptual ideas fill your brain. Some of the ideas you hold are true to reality, and some are not. You can either allow the contradictions to remain or apply the conscious mental focus necessary to identify and correct the neural networks in error.

Children are blank slates that accept new ideas readily. However, we adults have been subconsciously using specific untrue neural networks for so long and so frequently that they are powerful ideas. When presented with new knowledge of actual reality, our conscious mind recognizes the old idea's error. Still, our subconscious mind generates an emotional resistance to change. The emotion says, "Let's keep using the familiar pathway." Our brain likes to conserve energy, and the old ideas, so thoroughly automated, are the "easy" ideas. They are the pathways your subconscious automatically follows whenever considering those topics. Replacing them is hard work as you apply your conscious reason to overcome your subconscious emotional discomfort.

We all understand what a "behavioral habit" is. We all have a few that we are consciously aware of and many that we are not. You might not have realized that your strongly encoded neural networks also form "thought habits." They are your habitual ways of thinking, both generally and about specific topics. To replace a long-established thought habit that conflicts with reality requires applying intense conscious focus, repeated over time. Your brain will develop the neural network representing the new idea. Eventually, the new idea strengthens, and the old one fades.

♦♦♦

1.3

INSTINCT

"Where instinct fails, intellect must venture."

—Jim Butcher

AUTOMATIC KNOWLEDGE

All animals, including humans, are born with instinct. Instincts are how a newborn human knows how to breathe and how to suckle when presented a breast and cry when not. But how does instinct in humans differ from the instinct in animals?

We can think of instincts as neural networks in the brain built there according to genetic instructions while the fetus was growing. Animals are provided with a complete set of both lower and higher-level instincts to guide them in the actions they will take for their entire lives. Animals only have instincts to guide them in responding to and adapting themselves to their environment to survive. So long as these genetically programmed instincts are appropriate for the environment in which they live, their lives will continue. Animals are generally unable to modify the neural networks encoding their instinctive behaviors. A significant change in their environment threatens their survival. When the environment changes, only individual animals with favorable genetic mutations will survive. Of course, this is the evolutionary mechanism of "natural selection."

Humans are born with only lower-level neural network instincts built-in, not a complete set. Unlike animals, our instincts are not sufficient to guide our entire lives. They are not adequate to keep us alive for more than a few days without assistance from adult humans. We must discover and learn the actions that are required to sustain our lives. Fortunately, evolution replaced our missing higher instincts with a conceptual-level consciousness that animals do not have. Instead of following instinct, we use our rational faculty to observe and learn how reality works and what we need to survive.

By using our reason, we program what we learn into our brain's neural networks. Not limited to the lower-level instinctive neural networks provided us at birth, we can modify those and create new ones as needed. We do much of this during childhood, but "reason over instinct" continues throughout our lives. We use our learned knowledge to change both ourselves (our brain) and our environment to fit our needs.

Regarding animals, the entire species must adapt its genetically coded instincts to fit a changing environment through the process of natural selection over many generations. Regarding humans, an individual can adjust his neural networks to a changing environment during his lifetime. He can pass what he has learned on to his offspring by teaching it to them after birth. Reason provides us the survival advantage of *adaptability* and is why we are successful in so many different environments.

Now that we have learned about reason versus instinct, let us begin saving *"True Ideas"* as we encounter them. These are things we can see for ourselves and that we judge as being true to reality. Like fitting the pieces of a puzzle, these will form an interconnected set of true cause-and-effect ideas. All the *True Ideas* must integrate without contradiction. There is a progression where more abstract higher-level ideas depend on lower-level ideas. However, we will encounter them out of sequence. The two below are not the beginning of the logical chain. I will list the entire set of *True Ideas* in the *Resources* section in approximately ascending logical order. As each *True Idea* is presented, ask yourself if you agree?

True Idea: We cannot survive unless we use our reason to discover what we need to live because we do not have automated instincts like the animals.

True Idea: Reasoning based on knowledge validated against reality will further our life. Choosing not to reason or reasoning based on invalid knowledge will detract from our life.

♦♦♦

1.4

THE BRAIN-GUT CONNECTION

"I try not to think with my gut. If I'm serious about understanding the world, thinking with anything besides my brain, as tempting as that might be, is likely to get me in trouble. Really, it's okay to reserve judgment until the evidence is in."

—CARL SAGAN

YOUR SECOND BRAIN

What do people mean when they say, "follow your gut?" Some people mistakenly think that emotions and gut reactions are superior to conscious reasoning when it comes to decision-making. We need to understand better how our brain and our gut interact.

In any discussion on intelligence, the human brain is the lead actor in the story. Some believe that modeling the human brain is the most promising approach to creating a machine with human-level intelligence. However, this is an underestimation. The brain is but one component in a system with multiple elements that collectively constitute human intelligence. The brain might be the most critical component, but it could not achieve human-level intelligence independently.

Most people are unaware that their body contains a second, nearly independent nervous system that rivals the brain in its power. It is the *enteric* nervous system, sometimes called your "second brain." While the enteric nervous system controls the operation of your digestive system—your "gut," we are learning that it also intimately interconnects with your brain.

This enteric nervous system contains 50–100 million nerve cells, about the same number as make up your spinal cord. It communicates bi-directionally with your brain through a significant nerve bundle connecting your brain and digestive system called the vagus nerve. The brain and the gut also communicate bidirectionally via hormones in the bloodstream.

Digesting your food is not the gut's first function. Its first function is to serve as your body's largest sense organ. If spread out, your gut has the area of a basketball court and packed with thousands of tiny sensors. These sensors encode enormous amounts of information obtained from your food. This information is used not just by your gut but also communicated to and remembered by your brain. These real-time signals from your gut combine with memories of past gut signals in the brain to generate emotions in the brain that cause reactions in the gut.

Emeran Mayer, MD, has studied brain-body interactions for the last forty years, emphasizing brain-gut interactions. Much of the information in this section is from his 2016 book *The Mind-Gut Connection. How the Hidden Conversation Within Our Bodies Impacts Our Mood, Our Choices, and Our Overall Health.*[6]

Mayer writes, "What we sense in our gut will ultimately affect not only the decisions we make about what to eat and drink, but also the people we choose to spend time with and the way we assess critical information as workers, jury members, and leaders."

YOUR MICROBIOME

We tend to think of our bodies as being purely human. However, we are in a symbiotic relationship with bacteria, archaea, fungi and viruses that live in all parts of our body, especially our gut. Our bodies consist of many times more of these microbial cells than human cells. Collectively, these microbial cells are called our microbiome.[7]

More than 100 trillion microbes in our gut are comprised of a thousand bacterial species containing more than seven million genes—or up to 360 bacterial genes for every human gene. Less than one percent of the genes in our body are of human origin. We have in common 90 percent of our human genes. Yet, we have in common only five percent of the assortment of microbial genes in our bodies.[8]

Bacteria and archaea were the only life forms on Earth for the two-billion years before complex cells (eukaryotes) appeared. During this period, the Earth's oceans and atmosphere underwent a significant change, especially in localized environments. The bacteria and archaea had to adapt. They developed signaling molecules and receptors that enabled them to communicate with each other. Communication aided in the survival of their species.

When more complex marine animals appeared, some marine microbes took up residence inside their digestive systems. These internal microbes continued to respond to signals from microbes ingested from the animal's external environment. They responded by generating signals inside the animal's body that affected its behavior. Evolution found this combination of higher animal cells and microbial cells and the significantly increased DNA variety to be an advantage for adaptability to a changing environment. This combination continued as animals evolved up through humans two billion years later.

These gut microbes connect through a shared biological communication system to the microbes living in the soil, air, oceans, and bodies of all living animals. Our microbes send signals about our external environment to all parts of our body, especially our brain.[9]

However, our enteric nervous system can control the process of digesting our food without any inputs from the brain. From an evolutionary perspective, this is quite helpful. Our brain has enough to do guiding our daily activities of survival without also having to control the countless muscles and functions of digestion. However, emotions in the brain can disrupt these automatic digestive functions.

Your enteric nervous system is primarily concerned with whatever it is that you eat or drink. Your brain is primarily concerned with your overall health. Signals your brain receive from your gut provide a wealth of information about your environment and how well you are thriving within it. Those 100 trillion chattering microbes in your gut have a wide variety of DNA that enables them to identify and process almost any substance or invader that you encounter. The signals sent to the brain by your microbiome can affect your mood, pain sensitivity, and sleep, which, in turn, can affect how you think, how you make decisions, and your social behavior.

Our conscious, rational faculty connects to reality via the information provided by our senses. We must expand our understanding of our ability to sense our environment to include the microbiome in our gut and our enteric nervous system. Nothing connects more directly to reality than the food we eat and the microbes we encounter on everything we touch and in the air we breathe. The question is, to what extent does this primitive and basic sensory information affect our thinking?

> We now know that your gut mirrors every emotion in your brain. [...] Every time your stomach was tied up in knots, you had a gut-wrenching experience, or you felt butterflies in your stomach, it was the emotion-generating circuits of your brain that were responsible.[10]

Experiences we have early in our lives cause our brains to restructure neural circuits to remember these experiences. The complete programming of our brain-gut-microbiome system is a lifelong process. Our formative experiences, lifestyle, and eating habits setup our emotion-generating brain-gut-microbiome machinery based on a vast database of countless positive and negative emotional states throughout our lifetime. "All this leaves us with a lasting trace on our basic emotional makeup, and influences how we cope, how we make decisions, and possibly our personality."[11]

These neural circuits cause us to have certain emotions in response to specific triggers. These emotions then affect our gut's response to the experiences, including our personal microbiome's makeup. For example, suppose an abusive caregiver stresses a young person during childhood. In that case, their brain will develop neural circuits that cause that person to have a depressive disposition later in life. This depressive disposition can lead to gastrointestinal disorders such as IBS and a gut microbiota population associated with these symptoms. These unhealthy gut reactions then send signals back to the brain, reinforcing the emotional state of stress.

This state of anxiety in the gut-brain axis can last for a lifetime. However, as we previously learned, we can replace our subconscious neural circuits through conscious thought. By applying cognitive behavioral therapy and mindfulness-based stress reduction, we can replace the neural circuits of overactive thought habits that cause emotional stress. As a result, our brain will reprogram our gut microbiome to a healthier state, one that reinforces our stress-free emotions.

Concerning our value system, our values begin to be formed by brain-gut experiences in early childhood. As infants, periods when we were well fed and satisfied were associated with the "good." Hunger or an upset digestive system were associated with the "bad." Remembered by the brain, these emotional responses to gut feelings are the beginning of moral assessments of things as good or bad. "Thus your earliest gut feelings serve as a model for 'what the world is like and what I must do to survive in it.'"[12]

When it comes to decision-making, what we call "gut feelings" and "intuition" are the same capacity for instant assessment of a situation without rational thought. "You feel when something's fishy. You sense when you have an instant personal bond with a stranger. [...] Gut feelings reflect an extensive and often deeply personal body of wisdom that we have access to."[13]

One essential mechanism of *intuition* is called the brain's "salience" system. A part of our subconscious, our salience system is constantly appraising the relevance of incoming signals from our body or our environment. It determines whether a signal is important enough to enter our attentional processes and our conscious awareness where our rational faculty can deal with it. The threshold for what is salient is influenced by your genes, early life experiences, emotional state, body sensations, and vast memories of emotional moments acquired over a lifetime.

Whether a signal enters conscious attention or not, an essential mechanism of *gut feelings* is an instant database search and emotional output. It is like a Google search of the library of recorded emotional moments in our brain. When we need to make a decision, our brain gives us a probability prediction of how a given response will make us feel based on past emotions in similar situations. It guides us away from responses that will make us feel bad and toward responses that will make us feel good. "Besides allowing you to make decisions quickly, this mechanism lets you benefit from the past lessons without the psychological burden of reliving them."[14]

"Concepts" are the mental material of our *conscious* mind and the result of focused "conceptualization" or "reason" as its method of operation. We can now see that "emotions" are the *subconscious* mind's mental material. Its method of operation is a continuous global pattern matching against past emotional experiences. We also see that the gut, able to sense the web of life in our environment, is integral to our emotional machinery's operation.

While gut feelings draw instantly upon our lifetime of experiences, they are not infallible. They can have errors resulting from outside influences, mood disorders, and even advertising messages. Perhaps the most significant source of error is that your library of recorded experiences can contain conflicts, instances where the same cue has resulted at different times in both positive and negative emotions. For example, an adolescent's honest admission of a misdeed elicits praise in one instance and reprimand in another, causing conflicting emotions regarding honesty. The only way to resolve conflicting experiences is to revisit them with our rational faculty and consciously, not subconsciously, decide which reflects our true values. This conflict resolution will give more strength to the neural circuits that we consider to be true and cause the neural circuits that are false to fade.

We can now see that our rational thinking modifies the physical structure of our brain *and* our gut's functioning, including our microbiome. Both affect our emotions.

THE PROSPECT OF AI

I included this description of the brain-gut-microbiome system in our body to point out the complexity of what we mean by human intelligence. It requires our brain and our autonomic nervous system, which includes the sympathetic, parasympathetic, and enteric processes. Our consciousness is a phenomenon that emerges from all this. It connects to reality through our body, senses, and even other living entities—our microbiome. Evolution created all this machinery over billions of years. The most crucial principle to capture is—***Our intelligence evolved so that we could connect to and deal with reality better.***

This complexity should indicate that modeling the human brain alone in a machine will not be sufficient to create a human-equivalent intelligence with a connection to reality. We would need to provide such a machine with the equivalents to a body, senses, and the ability to code and decode the various kinds of communication pathways that we use. An artificial human intelligence entity would need to "eat" to know the difference between being satisfied or hungry—to have a starting point for developing a hierarchy of morality beginning with "good" and "bad" or "pleasure" and "pain." Otherwise, the artificial entity would not have a direct, deep, personal connection to reality. It would not have to face the ultimate alternative that all living things face—life or death—with its own life as its standard of value.

♦ ♦ ♦

1.5

AUTOMATION

"If you do not run your subconscious mind yourself, someone else will run it for you."

—FLORENCE SCOVEL SHINN

THE SUBCONSCIOUS

Your mind consists of a conscious faculty and a subconscious faculty. Is your subconscious autonomous and mysterious, or is it something you can examine, understand and control? We will address your conscious and subconscious faculties' *functional* operation in more detail in later chapters. Here, we continue to consider their *physical* operation in the brain.

In his 2007 book *The Brain That Changes Itself*, Norman Dodge, M.D. explains that when we learn a procedure or group of automatic actions occurring outside our focused attention, we call this "procedural/implicit memory." These are generally unconscious. The other form of memory is called "explicit" memory, which consciously recollects specific facts, events, and episodes. It helps us to organize our memories by time and place. Language supports explicit memory and becomes more critical once children can talk.[15]

Your conscious faculty brings items into conscious mental focus to create, form or grasp an idea. You can see three chairs in front of you. You see that they are used to sit on. You see that it does not matter whether they are hard or upholstered, wide, or skinny or have arms or not. What matters is that people sit on them. Someone (Mommy) is telling you that they are called "chairs." Your conscious mind is integrating all these pieces into a thought and giving it the word label "chair."

We have already seen that your brain gave a little more importance to the neural networks involved because you had this thought. Each time you consciously had this thought (each time Mommy called your attention and said the word "chair"), these networks were made stronger.

At some point, the idea of a "chair" becomes automated in your subconscious. Scientists are still discovering how this works physically in the brain. The basic theory is that whatever you encountered consciously during the day gets stored in short-term memory. Then, when you sleep, the events held in the short-term memory of our conscious mind are transcribed into the long-term memory of our subconscious mind. Our conscious mind's short-term memory neural circuits are, of course, short-lived. The neural networks of our subconscious long-term memory are more durable. They are the ones that get strengthened with repetition. The more days we have the same thought, the more times we will sleep on it, thereby reinforcing the pathway for it in our subconscious mind.

Once we automate an idea into our subconscious memory, it can guide our actions without requiring conscious mental focus. When Mommy tells us to get in the chair, we do not need to consciously recall a chair's essential purpose and non-essential qualities. We sit on it automatically without giving it a (conscious) thought. However, our conscious mind can recall from our subconscious the details of ideas stored there. We can bring back into conscious mental focus the neural network pathways that we earlier stored in our long-term subconscious memory. We can become consciously re-aware of the details of what is meant by the word "chair."

Let us save some additional *True Ideas.* Do you agree?

> ***True Idea: Our conscious faculty is a slower speed, focused and sequential information processor that results in accurate identification and expansion and contraction of concepts within our knowledge hierarchy.***

> ***True Idea: Our subconscious faculty is a high-speed parallel processor that pattern matches inputs against our total knowledge and outputs emotions as probability estimates of a match.***

> ***True Idea: Emotions have the advantages of being automatic, with instantaneous speed and broad scope for quick estimations. However, they lack the accuracy of deliberate conscious thought, which must be our primary***

guide to action, overriding emotions and correcting subconscious errors if necessary.

HIERARCHY AND LINKAGE

We do not form the neural pathways in our brain representing specific ideas in isolation. Our brain is continuously looking for relationships between ideas and creating neural cross-connections between their respective neural networks. The human cortex alone has 30 billion neurons and can make one quadrillion (a million billion) synaptic connections. A chimpanzee brain, by comparison, is one-third that size. This massive cross-linking allows the human brain to hold ideas at the granular and the holistic level simultaneously.

Our brain can create neural pathways that organize discrete ideas into linked chains where sequential order matters. They can also organize them into hierarchies with higher-level ideas made up of lower-level idea "parts." Dodge explains, "[…] smaller parts form a heterogeneous set of components which are more or less independent. But as these parts connect with each other in larger and larger aggregates, their functions tend to become integrated, yielding new functions that depend on such higher order integration."[16]

When our earlier amateur archaeologist followed the pathway from the village to the dig site, he eventually could do so without consciously thinking about it. At the conscious level, he thought, "I'm going to my dig site." His subconscious memory knew that this meant breaking the pathway into three different segments—village to craggy rock; craggy rock to pond; pond to dig site. His subconscious knew it had to follow that linked sequence. His conscious mind was then free to think about other things as his subconscious mind automatically guided him along the dig site pathway.

ATTENTION

The ability to create linkages and hierarchies of neural networks enables us to focus on how smaller things relate to larger things and how shorter periods relate to longer periods.

At first, a child cannot grasp differences between his house and a planet, or between today and a year. Eventually, his brain encodes neural networks to represent these concepts and how they relate to each other in linkages and hierarchies. The child gives mental focus attention to the "relationship" between concepts in space and time, not just the concepts themselves. Attention deficit disorders arise when an individual can only focus on objects, not the relationships between them in space and time.

LOCOMOTION

Animals are distinguished from the rest of nature in that we are "self-moving." Neuroscientists have discovered that self-locomotion ties to the development of our higher mental capacities. The hippocampus, a structure in the center of our brain, is where we develop our cognitive maps of the world. It is only by self-navigating and exploring our environment that our hippocampus develops. Being passively carried through the environment as a passenger does not cause this development.

Psychologists suggest that the dependence of memory on self-movement might explain why we do not remember our earliest childhood. Infantile amnesia begins to dissipate when children begin crawling and walking rather than being passively carried. Their brains begin aligning themselves to the environment encoding the spaces being explored and building episodic memories.[17]

In his book *Why We Drive: Toward a Philosophy of the Open Road*, Author Matthew Crawford explains that self-locomotion is critical to our development:

> We don't simply exist, like an animal that lives entirely in the present. We *interpret* our existence. We do so by composing a narrative that makes sense of our past episodes, and this narrative template provides a basis for imagining ourselves into the future. There is a striking implication here. The subjective coherence of life—our persistent identity through time—seems to be built up from some basic *motor* capacities.

Crawford writes that we must maintain "full awareness that our mobility as self-directed, embodied beings is fundamental to our nature as it has evolved over millions of years, and to the distinctly human experience of identity." As we continue to offload to technology the requirement to be mentally involved in our navigation and locomotion, we must realize that we are "embarking on quite a significant social experiment." Is there an age after which the hippocampus is fully functional, and self-navigation is no longer necessary? Consider the advent of short- and long-range autonomous vehicles. What will be the consequences of adolescent and teen children no longer riding their bicycles around the neighborhood or driving themselves around town in a car? Suppose adults do not mentally participate in self-navigation and locomotion. In that case, will their hippocampus' function atrophy over time?

Eventually, virtual reality is likely to achieve a level of fidelity to "real" reality sufficient to achieve the same effects on the hippocampus. The results will require that one self-navigates through the virtual space, with the essential element being "self" navigation, not autonomous transport.

We will elaborate on this idea that we need to do things for ourselves in a later chapter. Here, it is essential to emphasize that to gain new technology's advantages without losing ourselves in the process, we must apply technology to enhance or extend our mental engagement, not replace it.

FORESIGHT

One benefit of evolving to become conceptual beings is the capability of foresight.[18] Our brain's pattern recognizing capabilities enable us to grasp entities and their identities in our perceptual environment. It then attempts to recognize patterns of *situations* involving such entities, and then attempts to *project* how the *story* might unfold based upon the *causal* natures of the identified entities.

For humans, having a hierarchy of concepts according to cause-and-effect linkages to reality is essential for accurately projecting a story and responding appropriately to a situation. Suppose a person's conceptual hierarchy is not grounded in reality or has a disconnect somewhere along the chain of concepts. In that case, an effective response to real situations is less likely.

IMAGINATION

If you have any remaining doubt that conscious thinking can physically restructure your brain, consider experiments done by Pascual-Leone where two groups learned a piano sequence. Each group practiced for two hours a day. One group read the music and played the piano. The other group read the music while imagining playing and hearing the notes without actually playing the piano. Both groups brains were mapped before, during and after the experiment. Then, both groups were asked to play the sequence and their accuracy was scored by a computer.

> Pascual-Leone found that both groups learned to play the sequence, and both showed similar brain map changes. Remarkably, mental practice alone produced the same physical changes in the brain's motor system as actually playing the piece.[19]

The essential idea to capture is that whatever we consciously think about, even in our imagination, causes our brain to physically change, thereby altering our mind—our memories and how we think about things going forward.

Beginning Existents in Reality	Imagination's Purpose	Ending Existents in Reality
Box, Poles, Wheel	Integrating into goal—an image of something that can be created in reality.	A physical Wheelbarrow
Horse, Antelope Horn	Integrating into a fantasy—an image of something that cannot be created in reality	A work of art depicting a Unicorn

Imagination

We think about things in reality—a box, a wheel, and two poles, for example. When we use our imagination to combine items in a novel way to think of a new object—a "wheelbarrow"—we are "creative" with our imagination. If the new thing we created with our imagination is something we can build in reality, then we can call it a type of "goal." If, however, we were to think of a horse combined with a horn out the top of its head and called it a "unicorn," then we are still creative with our imagination. However, we would call the unrealistic result a "fantasy." The important thing is that both types of products of our imagination—goals and fantasies—one connected to reality and one disconnected, cause our brain to physically restructure itself to remember and apply these ideas in future thinking in the same way. We must always be careful to distinguish between what is or could become real and what is not.

CULTURE

Dodge tells us that a neuroplastically informed view of culture and the brain implies a two-way street—the brain and genetics produce culture, but culture also shapes the brain. Sometimes the changes can be dramatic. He writes:

> We become cultured through training in various activities, such as customs, arts, ways of interacting with people, and the use of technologies, and the learning of ideas, beliefs, shared philosophies and religion. Neuroplastic research has shown us that every sustained activity ever mapped—including physical activities, sensory activities, learning, thinking, and imagining—changes the brain as well as the mind.[20]

This neuroplastic research gives us insight into the relationship between an individual and society. Only the individual has a brain. Only the individual faces the life-or-death alternative and must use his brain to choose his values. An individual's choices and actions are either consistent with most other people, reinforcing accepted culture, or different from the majority, detracting from accepted culture. Accepted culture in a society is essentially a summation of the choices and actions of its individuals.

The accepted culture can influence the individual's choices and actions, especially the young, thereby modifying their brains' neural structure. This bi-directional influence creates a feedback loop that gives a culture momentum that is slow to change.

Individuals who think independently, question the culture, and exhibit different choices and actions change it. These individuals are, at first, considered "outliers" outside the norms of society. If an outlier's ideas are reality-based, like Galileo's assertion that the earth rotated around the sun, eventually society must change.

AGING

If our brains are highly receptive in childhood, what happens as we age? Dodge explains it this way:

> As we age and plasticity declines, it becomes increasingly difficult for us to change in response to the world, even if we want to. We find familiar types of stimulation pleasurable; we seek out like-minded individuals to associate with, and research shows we tend to ignore or forget, or attempt to discredit, information that does not match our beliefs, or perception of the world [...] Increasingly, the aging individual acts to preserve the structures within, and when there is a mismatch between his internal neurocognitive structures and the world, he seeks to change the world.[21]

Suppose an adult faces new information about reality. In that case, the brain's energy conservation makes replacing already automated erroneous ideas held most of your life difficult. Fortunately, while it might be difficult, it is possible. It requires intellectual honesty and sustained, focused effort. You must decide if discovering and correcting errors to achieve a conflict-free integration of ideas in your mind is worth it. Or, are you are willing to keep a mind full of contradictions for the rest of your possibly very long life?

USE IT OR LOSE IT

Like our musculature, our brain follows the "use it or lose it" principle. When it comes to allocating brain-processing power, competition for precious resources governs brain mapping. If we stop exercising specific mental skills, the brain will turn that map space over to skills that we practice instead.[22]

Competitive plasticity explains why our bad habits are so difficult to break or "unlearn." When we learn a bad habit, it takes over a brain map, and each time we repeat it, it claims more control of that map and prevents the use of that space for "good" habits.[23] That is why "unlearning" is often a lot harder than learning. It is also why early childhood education is so important—it is best to get it right early, before the "bad habit" receives a competitive advantage. This entrenchment applies to habits of thought as well as habits of behavior.

What habits of thought did you acquire in childhood that now need to be changed?

♦♦♦

COGNITIVE FLUIDITY

"The 'Me' today is not the 'Me' yesterday and it will not be the 'Me' tomorrow; a gentle reminder!"

—Moi

BRAIN RECONFIGURATION

Our brain can create new neural networks for specific ideas or concepts as learned mental content. But can our brain develop new *functional* capabilities? Can it recombine various micro-functions into new macro-functions? Yes, our brain *can* restructure how its different functional modules are combined to accomplish whatever tasks we ask of it during our lifetime.

We know that genetics initially map different areas and structures in our brains for specific purposes. Instead of thinking of these as physical locations within the brain, think of them as functional modules—such as sensory functions, motor functions, concentration functions, problem-solving functions, etc.

Dodge says we are born with essentially the same genetically built brain that our hunter-gatherer Pleistocene ancestors had. That brain modified itself to emerge from the Pleistocene. Archaeologist Steven Mithen called this self-modification of the brain "cognitive fluidity." Our brain's functional modules are plastic to some degree and can combine and differentiate throughout a lifetime to perform different functions.

The evolution of modular change in brain functions was a significant advance in our adaptability. It enables us to deal with things in our modern world that our ancestors didn't have to contend with. For example, the many brain modules a child uses to read, write, and use the computer can be seen in brain scans. These modules evolved millennia before literacy, only a few thousand years old. Literacy spread so rapidly that the brain could not have genetically evolved a module specifically for reading.

> Our brains are different from those of all humans before us [...] Our brain is modified on a substantial scale, physically and functionally, each time we learn a new skill or develop a new ability. Massive changes are associated with our modern cultural specializations.
>
> In childhood our brains readily shape themselves in response to the world, developing neuropsychological structures, which include our pictures or representations of the world. These structures form the neuronal basis for all our perceptual habits and beliefs, all the way up to complex ideologies. [...] these structures tend to get reinforced early on, if repeated, and become self-sustaining.[24]

Let us slow down here to capture this principle of "cognitive fluidity." Genetics programs the initial structure of the brain and its functional modules during the development of the fetus. There is little difference at birth between the brain of a human today and a human from the Pleistocene. They both share essentially the same genetics. However, by adulthood, there are massive differences in the physical brain structures, integrations of functional modules, and brain functional abilities. These represent differences in "intelligence." The differences are caused by each brain adapting to what the individual asks it to do during his lifetime, especially during early childhood.

Today, a young human asks his brain to learn and do much more than be a hunter-gatherer who, at age three, could walk and mimic his parents but could not talk, print his name, or operate a smartphone. New technologies for sensing and grasping the world, such as augmented and virtual reality, will cause our brains to form still newer integrations between the various functional brain modules. We will develop new functional capabilities that humans before us did not have even a couple of generations ago.

Unlike the glacially slow incremental changes in the brain structure for our entire species caused by biological evolution, cognitive fluidity causes massive differences to individual adult brain structures. They are the result of the thinking we ask our brain to do during our lives.

♦♦♦

ADAPTABILITY AND PROGRESS

"There can be no life without change, and to be afraid of what is different or unfamiliar is to be afraid of life."

—THEODORE ROOSEVELT

PERFECTIBILITY

Suppose our brains can create new neural networks and recombine functions as needed. Where is that leading the human species and the individuals within it? Dodge explains that as early as 1762, the Swiss philosopher Jean-Jacques Rousseau (1712–1778), who faulted the mechanistic view of nature of his time, argued that nature was alive, had a history and was changing over time. He said our nervous systems are not like machines but are alive and able to change. He proposed that the "organization of the brain" was affected by our experience. He said that even our emotions and passions are learned early in childhood to a great extent. Rousseau called this human malleability our "perfectibility."[25]

After Rousseau, the idea of perfectibility quickly got tied to the notion of progress. Condorcet (1743–1794), the French philosopher and mathematician, a significant participant in the French Revolution, argued that human history was the story of progress and linked it to our perfectibility. He thought that human nature was continually improvable, in intellectual and moral terms, and humans should not give themselves fixed limits to their possible perfection. He wrote, "Nature has set no term to the perfection of human faculties; [...] the perfectibility of man is truly indefinite, and [...] the progress of this perfectibility [...] has no other limit than the duration of the globe upon which nature has cast us."[26]

Dodge tells us that the twin ideas of progress and perfectibility came to America through Thomas Jefferson's thoughts. It appears that he was introduced to Condorcet by Benjamin Franklin. When we later explore how man's nature as the rational animal is the basis for ethics and morals, remember that they are rooted in the physical functioning of man's brain. They are rooted in reality. Remember also that men like Franklin and Jefferson founded America on this idea.[27]

CONCEPTUAL EVOLUTION

Our plastic brains are the mechanism advancing the present phase of evolution—Conceptual Evolution, which we explore in Part 3. As scientists discover more knowledge about how reality works, teach it in schools, and embed it in products and services, there becomes more for us to learn. And we do learn it, both explicitly in our education and workplace, and implicitly through our use of the products and services in society (e.g., smartphones).

Our brains keep restructuring their various functional modules through cognitive fluidity and creating neural networks as needed to accommodate this new knowledge. Our brain structures are vastly more capable and unlike those of any humans before us.

Evolution selected in favor of our conceptual form of consciousness because it enhanced our ability to adapt to a changing environment. With today's accelerating rate of new technology development, we will intentionally cause a great deal of change in our environment. We have the mental capacity to evolve along with it.

INTEGRATING TECHNOLOGY

And now, we face the possibility of merging with our technologies. We have covered the brain's ability to adapt by rewiring the modules within itself. What about interfacing with functions in technology outside the brain? When considering the possibility of humans merging with our technology, our physical brain's most exciting quality is its adaptability.

Scientists have examined many instances of human brains that have drastically re-configured themselves to adapt to some injury or developmental defect. For example, one sense (hearing) can enhance itself to compensate for losing another sense (sight). One side of the brain can also take up an entire brain's essential workload in a person born with half their brain missing.

The brain will go so far as to use resources in specific locations for purposes not initially intended. For example, a blind person's brain might adapt its unused visual cortex to develop an enhanced sense of touch, rather than expanding the area normally associated with touch sensation. Such a substituted resource might not perform quite as well as the original genetically defined resource that it replaced. Still, the universal plasticity of the brain enables such substitutions to be entirely functional.

This adaptability is all about adding, changing, and deleting the various neural networks within the brain and recombining functional brain modules to accommodate whatever we are consciously trying to do. It now appears that the brain can do this in more regions and to greater extents than we realized. As we learn to use our external computers, smartphones, and digital personal assistants, we are causing our brains to reconfigure.[28] Advances in the field of prosthetics, while still primitive, have already shown that the brain can remap its functions to incorporate technical devices as extensions of the body. We can expect that the brain is capable of far greater integration with technology in the future.

MODELING THE BRAIN

My earlier description of neural networks and "pathways" in the brain were, of course, highly simplified representations of the brain's physical functioning. Scientists already know much more detail. For example, the brain consists of Pattern Recognizer Units containing about 100 neurons each. Each of these PRUs is tuned to detect a specific pattern of signals from other PRUs and then "fire" a signal to still other PRUs. The neural networks within PRUs are different than the networks between PRUs. All these pattern recognizer units are active at the same time. We are beginning to gain an appreciation for the brain's exquisite complexity.

Some AI developers believe that the most likely pathway to creating an artificially intelligent machine is modeling the human brain. They are already using neural network-based AI learning in computers where repetition strengthens artificial neural networks. Scientists have built machine models of specific functional modules in the human brain such as circuits that replicate our brain's visual center and auditory center's functioning. But how close are we to modeling the entire brain?

Neurons signal each other by sending electrical signals along their axons and dendrites like wires. The axon of one neuron "connects" to the dendrite of another neuron at a "synapse." Scientists believe there are about 86 billion neurons in the brain. The axons and dendrites of these neurons make about 100 trillion connections. There exists from birth a genetically determined three-dimensional framework of crosshatched *potential* connection pathways throughout the brain. Most neurons connect to these already formed and dormant pathways as needed rather than sending out axons in a random spaghetti fashion.[29] From this brain structure data, scientists estimate that we will have computing power sufficient to model this network complexity in the next decade. However, scientific discoveries in 2020 indicate that the human brain's information network might be far more complex than the numbers above indicate.

Scientists found locations along the axons called Nodes of Ranvier that allow charged ions to diffuse in and out of the neuron, propagating the electrical signal down the axon. The flow of ions in and out generates electric fields at those nodes. These electric fields constitute "wireless" connections in the brain called "electric-near-field" connections.[30] Information could be transmitted wirelessly to nearby neurons at multiple nodes along the axon's length, not just at the synapses. The total number of neuronal connections in the human brain must include both wired and wireless types. The density of neuronal connections in the human brain could be orders of magnitude greater than earlier thought.

Scientists also discovered new electrical properties of human dendrites (in cortical layers L2 and L3) that indicate greater complexity in these neurons' operation. These single neurons appear to solve computational problems thought to require multi-layer networks.[31]

Our brain's plasticity and the ability to re-purpose its functional modules in cognitive fluidity will be essential characteristics to achieve human-level intelligence. The recent discoveries of greater brain network complexity indicate that modeling the human brain in a computer could take more computing resources than initially thought.

However, there is also an indication that there might only be a small number of functional principles at work in our brain. These are replicated in various quantities and scales to accomplish the different tasks. Once these few functional principles are understood and modeled, computers are extremely good at replication and scaling up, as necessary. A few critical breakthroughs in these foundational functions could be all that is needed for brain modeling to accelerate rapidly.

Advances in nanotechnology and quantum computing will improve the sensor sensitivity and resolution detail of human brain network mapping and then the brain's modeling in computers. We will eventually achieve breakthroughs in our understanding and modeling of the human brain.

From a physical functioning perspective, there appears to be no known limit to mankind's continued adaptability and progress—from biological human to hybrid biological-technical human to virtual human.

♦ ♦ ♦

1.8

THE FLIGHT SIMULATOR STORY

"Be the pilot of your own flight. Not the passenger."

—Giovannie de Sadeleer

THE MISSION BRIEFING

I want to provide you with a story to illustrate these essential brain functions concretely. The story is from my personal experiences learning to fly jets in the US Navy. However, let us pretend that it is you in the cockpit for you to benefit from the experience.

The story takes place in the late 1970s at the Chase Field US Naval Air Station in Beeville, Texas. It is in the desert about halfway between Corpus Christi and San Antonio, Texas. Yeah, I know; what is a Navy base doing in the middle of the desert? Well, not many people are around to be bothered by the noise of all the jets coming and going. And, there is not much of anything else for student pilots to do that might distract them from their studies.

Your daily flight training consists of two or three flights per day. You alternate between flying real airplanes and flying full-motion, full-immersion simulators that are incredibly realistic. These simulators are so realistic that you feel (minimal) G-forces on your body and your G-suit inflates. You can also hear the fuel rumbling in the fuel transfer lines like in the real plane. You flew a training flight in a real aircraft this morning, and you will be flying in a simulator this afternoon. Once the canopy closes, your subconscious cannot tell the difference. Only your conscious mind is aware that this is a simulation.

This afternoon, you will fly a simulated mission in which you will launch from the aircraft carrier and fly up a river valley to a target. Once you are in the simulator, it will become a night mission flown using instruments. You cannot see outside the simulator cockpit.

Simulators flights are not only for practicing missions but also for learning to handle emergencies. They are invaluable for the type of emergencies that would be too risky to practice in a real plane, such as an engine fire. You expect that your instructor, who is not in the cockpit during the flight but sitting at the simulator's computer control station, might throw one or more emergencies at you.

Navy 2F101 flight simulators on hydraulic actuators providing full range of motion allowing the pilot to feel g-forces.[32] The instructor sits at the computer console on the upper platform.

There is an old saying among aviators—"Aviate, Navigate, Communicate." This is a mental reminder that your first priority is always to keep control of the airplane. Your second priority is to stay on-course. If you are in control and on-course, you can do other things like talking on the radio. You never want to be distracted by those other things to the point where you let the airplane get out of control or off course.

A similar principle organizes the instrument panel in front of the pilot. The primary flight instruments for keeping the plane upright and level are central in front of the pilot. The navigation instruments reside around the primary instruments. The communications equipment is outside of those. Going farther outward from the pilot's center of vision, you find the engineering instruments that show how the engine, hydraulic, electrical, oxygen and other systems are doing.

With practice, you have developed something called the "instrument scan." The instrument scan is the pathway that the pilot's eyes follow when scanning all the instruments. He spends most of his time scanning the primary instruments right in front of him. The farther away from the center the other instruments are, the less often the pilot looks at them. He might scan all the primary instruments, then look outside of them to look at one of the navigation instruments, then come back to scan the primary instruments again. Next time he will check one of the other navigation instruments. Every ten or fifteen minutes, he will make a pass around the engineering instruments.

Earlier in your training, you spent a great deal of time studying manuals that diagram all the systems and equipment and their interconnections in the airplane. You also spent hours just sitting in the simulator practicing responses to warning lights. You practiced so much that if one of the many warning lights illuminates, one of your hands will instantly and automatically throw an appropriate switch in response. You do not have to think about it, and your hand falls to the needed switch without having to look at it.

THE MISSION

So, here you are on your night mission flying half-way up the river valley on-course. You are flying low to stay below SAM missile radar coverage. You know you will not crash into the hilltops on either side so long as you remain on-course above the river that you cannot see. Everything in the plane seems to be working normally.

The "Primary AC Bus" warning light illuminates. Your hand automatically throws the switch to select the "Alternate AC Bus." Not a big deal. It happens sometimes. Your mission can continue.

A while later, in the flight, you become aware of a tension in your body. Everything seems OK. Why are you feeling tense? You become aware that you keep looking over to the side of the instrument panel more frequently than usual. You wonder why that is?

You decide to scan the engineering instruments again with a closer look. You see that the hydraulic pressure gauge is slightly below normal—only about a needle width. Hydraulic pressure is necessary to operate the flight controls. This instrument would normally never move. You decide to start monitoring it closely.

Sure enough, a while later, you see two needles' width below normal. As you are thinking about it, the "Low Hydraulic Pressure" warning light illuminates. You have a previously learned subconscious automated response for that warning light too, but now you must resist using it. Your conscious mind must override your subconscious. No longer in its default configuration, the aircraft is operating on an alternate source of electrical power. The alternate electrical power does not supply the backup hydraulic pump that your subconscious would have automatically selected. Had you allowed your subconscious to select the backup hydraulic pump, you would have immediately lost all hydraulic pressure and flight control. Now you must call upon all the knowledge of the systems, diagrams, and specifications you memorized to think through your options. The fundamental idea that "the aircraft is in its nominal state" is no longer true. Therefore, all subconsciously automated responses to warning light stimuli are also no longer valid. Only conscious thought in problem-solving will serve you now.

Though still working, you have a failing pump, and the backup is not available. Hence, you determine that you cannot continue the mission. Your goal now is to return to the aircraft carrier. You reverse course, continuing to monitor the hydraulic pressure gauge over to the side of the instrument panel in addition to the primary flight instruments and navigation instruments in front of you. You cannot give attention to much else.

Later, as you navigate your way back down the river valley toward the ocean, an "Engine Fire" warning light illuminates, and warning horns begin blaring. That captures all your attention. You scan your memorized systems diagrams and conclude that the fire suppression system should still work, so you throw that switch. The fire goes out, but you have lost one engine. You give up on trying to reach the carrier. You can barely maintain altitude on one engine, so your goal now is to get out of the hostile territory. You plan to eject into the sea and air/sea rescue can pick you up.

But then the hydraulic pressure drops to near zero. Warning horns are blaring again. You can barely control the plane. Your goal now is to eject before this plane crashes. Your instrument scan is reduced to a "tunnel vision," looking only at the primary flight attitude instrument, nothing else. That instrument shows that the plane is tilted way over. You are struggling to get the aircraft back upright so that you can eject. With the help of some rudder pressure with your foot, you finally get it back upright. You reach for the ejection handle and pull it, but nothing happens. The ejection seat does not work. The plane goes out of control and a few seconds later crashes. The simulator gives your body a violent shake. Simultaneously, your windscreen appears to shatter and there is a blinding flash and the sound of crushing metal. Then, total darkness. You are dead.

You were in the actual airplane just a few hours ago. For much of this flight, you have been intensely using your conscious mind first to salvage the mission and then ultimately to save your life. Your crash and death are an emotional shock. Your conscious mind realizes this was a simulation, but your subconscious does not.

The simulator returns to its home position, and the canopy opens, letting in the bright lights of the building. You squint because you have been flying in the night. You notice that your flight suit and flight boots are totally soaked with sweat. You are exhausted, Adrenalin took its toll. Your legs buckle beneath you as you get out of the cockpit, and you stumble. As you get back to your feet, you notice that the instructor has not lowered the catwalk for you to get off the simulator deck. You look up to the simulator control platform. Your instructor yells down from there. "Get back in. Let's do it again."

THE MISSION DEBRIEFING

From the Prologue to this book, you might recall that I am one of those people who needs to understand how things work. As I went through these experiences, I made mental notes of what I was experiencing and my resulting behaviors. I knew that this was important stuff. It pertained to how my mind's conscious and subconscious faculties work, and my mental capabilities and limitations.

The story shows that prior mental focus and repetition training created neural networks that could handle first-level warning light emergencies without conscious thought. You did not even need to think about where the switch you needed was located. The autonomic nervous system controlling your muscles knew how to put your hand right on it. It is, therefore, evident that your conscious thoughts create your subconscious neural networks.

The next thing to notice is how your subconscious was causing a stressful emotion. You did not *consciously* see that the gauge needle had moved such a small amount, but your *subconscious* did. That is what your subconscious does best—pattern matching. Your conscious mind is best at reading the numerals that the needle on the gauge is pointing to and conceptually understanding what they mean. Your subconscious mind had learned the direction that the needle in that gauge was supposed to point—what the pattern was supposed to be. Your subconscious perceptually detected the subtle difference in needle angle. It gave you emotion as its method of reporting the pattern mismatch. It even caused you to look subconsciously toward the gauge more often. It is for this reason that today's advanced cockpit display screens depict *analog* gauges with needles and tapes rather than just numeric values.

Your emotions come from your subconscious. It continuously and instantaneously pattern matches everything you are experiencing against all the neural patterns it has already learned. "Emotion" was evolution's way of enabling you to use everything you have ever learned to guide and perhaps save your life, instantly if necessary. If our ancestor's subconscious detected the outline of a crouched tiger in the jungle, there was no time to mull it over consciously. The fight-or-flight emotional and physical response needed to be activated immediately. The same is true for you today. For example, when you are about to cross the street, focused on your smart phone, your subconscious detects a truck's sound pattern coming your way just outside your peripheral vision. The physical functioning of your subconscious mind underlies what we commonly call "intuition" or the need to "follow your gut."

In the story, the second warning light coming on illustrated that your conscious mind must always have command control over your subconscious mind. Had you allowed the automated response for that second warning light to kick-in and flipped that switch, it could have been disastrous. When a subconscious neural network becomes invalid, all those linked to it also become invalid until you can consciously sort it all out. The requirement for *reason* to override subconscious *automation* applies equally to responding to warning lights and to acting on your most deeply held values.

While your subconscious has the advantage of speed in providing estimates of things according to the totality of what you know, it is not necessarily accurate. The existing neural networks against which your subconscious is pattern matching might not be appropriate for the situation or even contain errors because they were networks you learned subconsciously. No matter how strong and automated your subconscious neural networks have become, you cannot have your subconscious doing things automatically that either will not help you reach your goal or will make things worse for you.

Near the end of the story, you focused on the single instrument that would tell you when the plane was upright. That is because it was night, and you could not see outside, and you knew that you could not trust your vestibular senses to sense "up" from "down" when in a plane. Vertigo is an extreme example of your "gut" feeding you wrong information. Your subconscious pattern matches the sensory input of lateral and rotational forces of motion detected by your vestibular balancing organs in your ears. Vertigo can cause your body to feel you are right-side-up while in a death spiral. When sensory inputs from your eyes and vestibular organs are in conflict, disorientation and nausea can occur. Only your rational mind can sort this out and place trust in the instrument over your gut as the accurate representation of reality.

You should use your conscious mind to validate what your "gut" is telling you. Sometimes your emotions and their physical gut responses are valid representations of reality, such as when the instrument needle's slight movement was detected causing stress. And sometimes your feelings are invalid, for example, vertigo. Your life depends on the primacy of your reason over your subconscious—your emotions, feelings, intuition, or gut. Admonitions such as "follow your gut" or "trust your emotions" can lead to disaster. The instruction, "Be aware of your emotions and figure out what is causing them," maximizes the effectiveness of your conscious and subconscious faculties working in tandem to grasp reality and support your life.

The story then showed us that in situations where you must consciously focus on essentials, a type of "load shedding" occurs. Your conscious mind can only focus on a few things at once. As things deteriorated in your flight and you had to think through your options and monitor what was important, your mental workload became intense. Your instrument scan kept narrowing until you were fixated only on the single, most critical instrument, having shed and ignoring everything else. Focusing intently on your most important goal with no distractions is a skill and habit that one can learn and develop.

Notice that you were betting your life on the belief that you could trust the perceptual information you were receiving about reality. Your conscious mind was receiving the perceptual information via your sense of sight, which focused on an instrument that indicated facts of reality.

We also see the hierarchies and sequential linkage of neural networks operating here. You see the "if this then that" thinking—"If I'm losing hydraulics then I must abort the mission." And we can see the hierarchy of concepts as goals. In this story, your own life was your highest value. We can assume that you did not sign on for this flight as a suicide mission. In order of increasing importance, your goals were completing the mission, returning the plane to the ship, reaching a point of safe personal rescue, staying alive. You pursued the mission until facts of reality indicated its success was no longer probable without sacrificing a higher value—your life, for a lesser value—the mission.

And finally, the instructor made you get back in to do it again. Unaware that it was a simulation, your subconscious, having just experienced your death, is generating a powerful emotion. Strong emotions such as fear can interfere with rational thinking. The instructor knew that through practice and intense mental focus, you could learn to evaluate your feelings and recognize when they are helpful and suppress them when they are not. He knew that you could become a clear-headed, rational-thinking pilot focused on reality and your goals, undeterred by unwarranted fear. Does this sound like a good prescription for piloting one's own life?

♦♦♦

1.9

SELF REFLECTION

"When you hold out for high standards, people are impressed—but they don't always like you for it. Not everybody will be on your side in your struggle to do what's right and ethical. In fact, sometimes even you won't be on your side. You'll wrestle with inner conflict, torn between what you should and what you want to do. You'll also aggravate other people. Seems when you walk the straight and narrow you always step on someone's toes. Don't count on the ethics of excellence to make you popular."

—PRICE PRITCHETT

INTROSPECTION

How do we turn what we have learned about our brain's basic physical functioning into principles that will improve our own lives? The essential principle is—that through the activity of thinking, our conscious mind creates new neural networks or modifies existing ones in our physical brain. These guide our lives, consciously and subconsciously. Three primary mechanisms constitute what I call the *Plastic Brain Critical Process.* These three principles are all we need to understand about how our physical brain works to move forward with our discovery of new knowledge.

The ***Plastic Brain Critical Process***:

1. Our conscious thoughts create, strengthen, maintain, modify, and dismantle neural networks in our subconscious.

2. Our subconscious uses existing neural networks to provide instantaneous evaluations of sensory inputs and thoughts in the form of emotions leading to physical "gut" responses.

3. Our brain uses cognitive fluidity to restructure the integration of its functional modules as needed to support what we ask of it, including integration with our technology.

From now on, we will explore our method of thinking in mostly functional and logical, rather than physical, neurological terms.

This book will introduce you to reality-based ideas that might differ from what you now have in your brain's neural networks. You will point your powers of observation to the nature of reality and the nature of man. You will then apply your conscious thinking to focus on and commit to the "true" ideas you see. You will integrate these true ideas into your mental content, replacing any floating abstractions you might have learned from other humans long ago.

Conflicting ideas held subconsciously generate both situational and chronic levels of emotional stress, negatively affecting our guts and other parts of our bodies. Pursuing an integrated hierarchy of valid ideas devoid of conflict will reduce your stress levels, and your body will find a more relaxed and healthy state.

> ***True Idea: If we have accepted invalid and contradictory ideas from others into our subconscious, then conflicting emotions will result. Only the process of reason can correct internal conflicts.***

Your plastic brain has an amazing ability to adapt itself and integrate with technology. Technology is advancing at an accelerating rate. Therefore, you have quite an exciting future ahead of you.

♦♦♦

PART 2

RECOGNIZING OUR PRESENT

Your smartphone is with you *all* the time. You probably don't understand how people ever lived without one, but people over about age forty remember such a time. Regardless, you are now dependent on your smartphone with text messaging, an Internet connection, Wi-Fi, high-def TV, GPS navigation, Google searches, videoconferencing, and shopping on Amazon.

You have a general awareness that technology that your grandparents could not even dream of is exploding all around you. You wonder if all this technology is really a good thing? Is it making you smarter or dumber? You ask yourself if you should try to keep up with it or go back to a simpler life?

You have seen the movies about AI "terminator" robots that take over the world and eliminate humans. You have seen news articles about how AI is being applied to almost everything today, in our smartphones, cars, home appliances, approving loans, and reading x-rays. There is a lot of discussion today about how we need to be careful to build ethics into our future AI creations. But what are the specific ethics that will keep us safe? You do not spend a lot of time worrying about it, but a little part of you does occasionally wonder if AI might first take your job and then someday kill you—or maybe your children.

You are aware of some of the advances in genetics such as the ability to have your own genes analyzed to determine your heritage and whether you have the markers for certain genetic diseases. You have read that genetics might finally eliminate cancer and other diseases and might even stop the aging process. You are also aware of today's ability to modify genes in plants and animals. You see the public concern over "GMO" foods and the possibility of human cloning and designer babies. You know these are labeled as "bad" things in the media, but you have not really thought them through yourself.

The "cold war" and the fear of the atomic bomb supposedly ended decades ago, but the world does not seem any safer. You see increased fear of terrorism in the world. You see China and Russia trying to assert global influence, militarily, economically, politically, and via the Internet and social media. Iran and North Korea are continual threats. China wants to be the world leader in AI and other key technologies and become the world's number one superpower. Thanks to advancing technology, in addition to nuclear bombs you now must worry about "smart" terrorists, cyber warfare, space warfare, and biological warfare. You wonder if an "EMP" or cyber-attack could really take out America's electric grid plunging us back to a new Dark Ages?

Even if we are not attacked, you wonder if it is true that, through our use of fossil fuels-based technology, we have ruined the planet by causing global warming? Can we use technology now to save the planet and ourselves?

You think that digital assistants like Siri, Google Assistant, and Alexa are cool, and you understand they are just an introduction to the new world of Artificial Intelligence. But now you wonder if you should worry about whether the mega-corporations that made them are listening in on your private life? Are they feeding you only the information they want you to have? And what are they doing with all that personal information about everything you ask, do, say, or buy and where you go? Can these corporations be trusted?

You see that conflict and strife in our society and our political system seems to be getting worse. The bedrock principles of "Americanism" that you grew up with seem to no longer apply. The population is dividing into factions, movements and parties that are at odds with each other, each taking positions and presenting arguments about science, technology, economics, and politics that are contradictory and cannot all be true. Our leaders seem to be more concerned with dominance over their opposition and gaining or maintaining power than being able to work together to achieve solutions to critical issues. You wonder if it will be possible for our institutions to responsibly keep control of powerful new technologies so that we do not destroy ourselves?

You would like to be optimistic. You feel that we humans can do wonderful, great things with technology. You can envision us solving the problems of this world, increasing our health and lifespan, and then going to the stars. However, with all the challenges that you see us facing, you have some doubt and anxiety about whether we are smart enough to reach a shining future.

Regardless of what ultimately happens to mankind, you must live your own life today. You want to achieve your own happiness to the best of your ability. If you can contribute to the success of mankind, all the better. But to be successful and happy, you wish you could make some sense of what is happening around you. You have a lot of decisions to make—what education you need, what work you should do, where you should live, which technology and change is important to keep up with, how to prepare for later in life, how to educate and then advise your children in their lives.

You wish you understood what causes the things you see. You wish it were easier to evaluate what is true and what is not. You wish there were a skill that you could learn to help you make good decisions going forward into this technological future. This book will provide that skill, but first we need to take stock of our society's present situation.

WARNING! SHARP CURVE AHEAD

"To expect the unexpected shows a thoroughly modern intellect."

—Oscar Wilde

THINGS EXPONENTIAL

It seems like technological change is coming at us faster than it used to. As soon as we learn how to use or do something, there is an all-new thing to use or way of doing it. There is a mathematical concept that describes our present situation with technology. It is the *exponential curve*. It applies to things that grow at an exponential rate rather than a linear rate.

You need to understand this concept. I mean, *really* understand this concept. Even if you think you already know it, you need to *feel* it emotionally, so please bear with me. Don't worry; we aren't going into math theory here. But instead, let me take you on a visual trip to the Grand Canyon. We check into the historic El Tovar Lodge directly on the edge of the Grand Canyon's south rim. We have come here because you will fill the Grand Canyon with water—using an eight-ounce cup.

Assume that somehow the Grand Canyon's ends have been sealed, and no rivers are flowing into it. There will be no rain and no evaporation or other gain or loss of water. The only water in the Grand Canyon will be what you pour into it. You, somehow, have an unlimited source of water available to pour into the canyon.

The Grand Canyon is estimated to have a volume of 5.45 trillion cubic yards. Each cubic yard can hold approximately 202 gallons of water. Each gallon holds sixteen cups. It will require seventeen quadrillion six hundred forty trillion (17,640,000,000,000,000) cups of water to fill the Grand Canyon.

The first morning, after breakfast in the El Tovar's grand dining room, you walk out to the railing on the edge of the Grand Canyon, and you pour one cup of water into the canyon. You have finished for today. You can be a tourist. You will do this every morning, except that you are going to double the number of cups you pour each day. Tomorrow you will pour in two cups of water, and the next day you will pour four cups, and the day after that, eight cups, and so on.

Even after an entire month at this doubling rate, the water in the canyon's bottom is barely visible. And you just added 537 *million* cups of water today, your thirtieth day here. It looks like we are going to be here for a long time. You keep doubling the amount of water poured in every day. After the fiftieth day, you think this is futile. You just added 563 *trillion* cups of water this morning, and after 50-days, the canyon is still only one-sixteenth full.

But this is where the magic of exponential growth kicks in. At only one-sixteenth full after fifty days, the Grand Canyon still looks almost empty. However, the Grand Canyon will be full of water in just four more days. It will go to one-eighth full tomorrow, then to one-quarter, then one-half, then full.

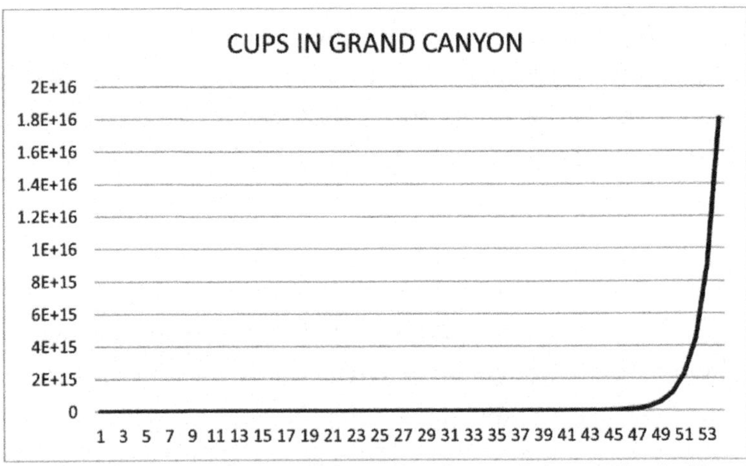

Exponential curve changes from horizontal to vertical after 50 doublings.

Exponential curves show little growth for exceptionally long periods. This period is their horizontal, slow-growth phase. Then they reach an inflection point where they suddenly begin to grow and keep accelerating until the growth becomes explosive. This period is their vertical, explosive growth phase. Imagine flipping the above chart upside down. One trudges along slowly for an extended period and then falls off a cliff and accelerates rapidly.

Good job. You endured the seemingly endless slow-growth phase to reach the point where the principle of exponential growth helped you do what seemed impossible. You filled the Grand Canyon today. But now, things have changed dramatically. We are in the explosive growth phase. What if you keep at it a few more days?

If somehow, the Grand Canyon emptied tonight, you would fill it again tomorrow, in just *one* day. The day after that, you would fill *two* Grand Canyons. The day after that, *four* Grand Canyons, and the day after that *eight* Grand Canyons—in just *one* day.

Psychologically, we can almost grasp how you endured the slow incremental growth that allowed you to fill the Grand Canyon with an 8-ounce cup the first time in 54 days. But, in only ten more days you could fill over 500 Grand Canyons in a single day and over a thousand canyons the day after that. It strains our ability to grasp or even imagine that kind of growth rate psychologically. Biological evolution occurred so slowly that there was no need to select for the ability to recognize an *exponential* change in our environment. Therefore, the innate predictor mechanisms in our brain operate linearly.[33] We naturally tend to think in *linear* terms, but our conceptual faculty can grasp exponential concepts. Bill Gates summarized this idea, "We always overestimate the change that will occur in the next two years and underestimate the change that will occur in the next ten. Don't let yourself be lulled into inaction."[34]

We can finally check out of the El Tovar Hotel and look at the history of technology growth.

TECHNOLOGY GROWTH CURVE

Imagine another line graph on paper. This one begins on the left with early man's first use of stone tools around three million years ago, and it ends on the right in the year 2010. The line's height at any period represents the number of significant discoveries and technologies that man created during that period.

You can find several websites that list all the significant technological developments and when they first occurred. The following paragraph lists a few of man's significant discoveries or inventions. Read through these examples to have them in a mental context before we make a few observations. Note that each item represents something meaningful in the progression of human technology.

Examples include: fire-cooking, spears, language, clothes, pottery, agriculture, cities, bricks, glass, the wheel, sailing, writing, bronze, irrigation, the alphabet, roads, concrete, cast iron, logic, water wheels, steam engines, drilling, horse collar, toilet paper, windmills, spinning wheels, gunpowder, algebra, eyeglasses, movable type, compass, rifle, Mercator projection maps, toilet, magnetism, telescope, astronomy, Newton's laws, sulfuric acid, sextant, hot air balloon, electrolysis, smallpox vaccine, telegraph, bicycle, automobile, photography, electric motor, Bessemer steel, lead-acid battery, Pasteurization, radio waves, telephone, incandescent light, air-filled tires, airplane, Ford Model T, television, penicillin, neoprene and nylon, programmable computer, nuclear fission, transistor, polio vaccine, birth control pill, Sputnik, jet airliner, laser, industrial robot, coronary bypass surgery, astronauts on moon, pocket calculator, microprocessor, anti-lock brakes, handheld cellphone, Internet protocol suite, MRI, Apple® computer, the Concord®, GPS, Space Shuttle, contact lenses, Hubble space telescope, online stock trading, Sony PlayStation®, genetically modified food, Google®, Wi-Fi, Prius®, genetic sequencing, Roomba®, graphene, lab-grown organ transplants, stem cell therapies, iPad®, quantum entanglement, Mars rovers, dark energy, 3D printed blood vessels.

I found approximately 2,500 such discoveries and inventions from the stone age up through the year 2010. I listed them in a table with the reported year of their appearance. Then I created graphs of the number of discoveries and inventions for the centuries since 0 AD and the decades since 1800. See the charts below. Here are some observations from the data and graphs:

- From the earliest stone tools through the end of the eighteenth century, a period of nearly 30,000 centuries, man had only produced about five percent of all the significant technologies made through 2010, with one-third of that five percent in the eighteenth century.

- Then, in the nineteenth century man produced many new significant technologies equal to all the prior centuries put together. The Scientific Age was born.

- In the twentieth century, man produced double the number of new technologies created in the nineteenth century. The Age of Technology was born and began to pick up speed.

- Then the explosion occurs. In only the first decade of the twenty-first century, from 2000 to 2010, man produced almost ten times the number of new technologies that he created in all the twentieth century.

The data I captured stops in 2010. So far, in the second decade of the 21st century, from 2010 to 2020, significant discoveries and inventions are happening so fast that it is futile to list and count them.

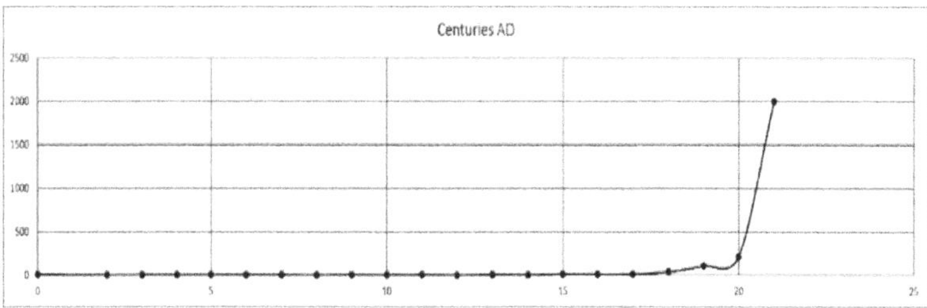

Inventions by Centuries

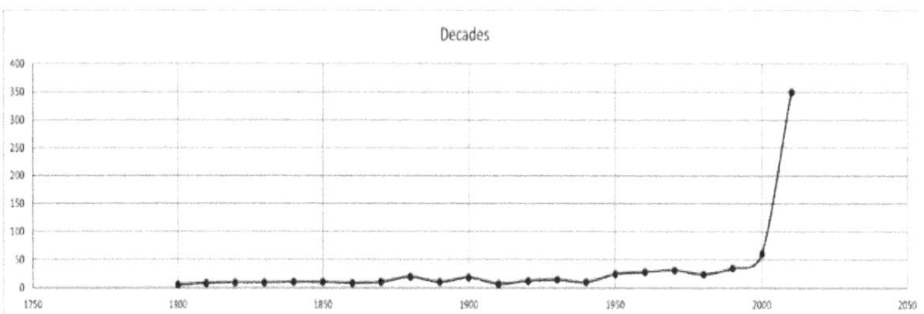

Inventions by Decades

In the line graphs, we see an almost horizontal line to about three hundred years ago when it just started elevating. Then it makes a sharp bend over a short period to become near vertical. Does it look familiar? We are now in an era where technology is growing at an explosive exponential rate.

THE REAR VIEW MIRROR

Humans all have a natural tendency to think and plan "linearly." We measure past and future events using the same "yardstick." Looking in our rear-view mirror, we have seen how much the world has changed in the last couple of decades. We naturally expect a similar amount of change in the next couple of decades. We assume that the pace of change, like the "inch" marks on a yardstick, will stay the same—*linear* growth.

We will see much more significant change in the next twenty years than we saw in the last twenty—more even than we can imagine today. Technology development is more than doubling its rate every ten years. The life cycle for rolling out new products in society is being cut in half every ten years. The yardstick we use for measuring the change in the future needs to have its "inch" marks get progressively closer together.

Like our Grand Canyon experiment, it is difficult for us to *emotionally* grasp that we need to think and plan "exponentially." However, that is what we must do *conceptually* to prepare for our technological future. Remember, we now appear to be on the explosive, vertical part of the technology growth curve. We must ignore our rear-view mirror and use "exponential" vision looking forward. Whatever your wildest imagination about what might be possible, it will likely be closer to reality than assuming the same rate of change of the recent past.

But you might ask, how far ahead is it possible for us to see?

THE SINGULARITY

We cannot imagine the technology that the rest of the twenty-first century will produce. There is a possibility that we will learn the fate of humanity by the end of this century. Advances in artificial intelligence, augmented human intelligence, genetics, the blockchain, nanotechnology, 5G network, quantum computing, and fusion energy, and their potential dangers will determine our fate.

Ray Kurzweil, inventor and, at the time of this writing, Google's Director of Engineering, is perhaps the most widely read futurist about the rate of increase of computing power and artificial intelligence. Kurzweil believes we will create artificial intelligence so advanced over human intelligence that we cannot possibly see or predict what will happen after that. This point is commonly called the technological "Singularity." Vernor Vinge popularized the term "singularity" in his 1993 essay *The Coming Technological Singularity.* He wrote that it would signal the end of the human era. The new super-intelligence would continue upgrading itself at an incomprehensible rate. Vinge wrote that he would be surprised if it occurred before 2005 or after 2030.[35]

Twelvc years after Vinge made his 2005–2030 prediction, Kurzweil, in his 2005 book *The Singularity Is Near: When Humans Transcend Biology*, pinned the date a little later, but not by much, and he explained how much the date might change if his estimations were off. Kurzweil thinks that the Singularity will occur by 2045, or if his estimate of the computing power of the human brain is off by a factor of a million, it will happen by 2060.[36] That is only 25 to 40 years from now! According to Kurzweil, that is all the farther we can expect to see or predict. Anything beyond that is pure imagination.

FORWARD VISION

Of course, there are futurists with opinions that differ from Vinge and Kurzweil. Some think that the achievement of a super-intelligence is not possible. Some think it is possible, but they believe it will take far longer than Vinge and Kurzweil estimate. Whether there will be a Singularity in our future is not where our concerns should be today. Even without super-intelligence, the disruptions that technology will bring to our society are going to be huge.

There are two key points here for you to remember. The first point is—while most people are not aware of the tsunami of change that is about to hit them, you are. You can get prepared. You will be able to use what you learn in this book to monitor the situation and make your estimates of how things are going. The second point is that our vision forward is limited, just as our rear-view mirror is not much help. Suppose we manage to survive for only two or three more decades. In that case, our technology will enable us to be unimaginably more intelligent than we are today.

There is no point in stressing over seemingly insurmountable problems that we anticipate twenty-five years or more from now. Examples include the effects of global warming, the aging population, ocean pollution, and dwindling resources. By the time we get there, our ability to find solutions will be exponentially better. Today, we cannot even imagine what those solutions might be. I am not advocating that we ignore such problems today. Rather, we need to replace feelings of hopelessness with confidence in our future creativity.

Of course, we can also expect that there will be new problems that we cannot imagine now, but so far, man has overcome the challenges that each new era brings. The AI theorist Eliezer Youdkowsky said, "There are no hard problems, only problems that are hard to a certain level of intelligence. Move the smallest bit upwards, and some problems will suddenly move from `impossible' to `obvious.' Move a substantial degree upwards, and all of them will become obvious."[37]

Today, we need to focus on understanding the creation and management of technology and the impact it is going to have on the nature of our society. We need to know how to gain its advantages while keeping ourselves safe from its dangers. We need to survive the next twenty to thirty years so that we can dramatically increase our intelligence. Unfortunately, our society's survival, and even our physical survival, are not guaranteed. As you will soon learn, I believe they are at risk.

THE LOTTERY

Do you feel it yet? That feeling of excitement that you just won the lottery. You should. The next few decades will be the most consequential in the history of humanity. Think of the one-hundred-thousand generations and the millions of humans that have lived and died since the first stone tools. What are the odds that you would be one of those born and alive during the one or two generations that could determine the fate of mankind? Congratulations, you just won the lottery!

You will not simply be a spectator. What happens to mankind will be a consequence of the ideas held by most individuals in society during this critical age of accelerating technology. Therefore, the ideas you hold and demonstrate as an individual and your choices will contribute to the overall outcome.

So now you are thinking, "Great. Now I'm really freaked out. What am I supposed to do?" The first thing you need to come to grips with is that you might be alive for an exceptionally long time.

♦♦♦

2.2

YOU MIGHT LIVE LONGER THAN YOU EXPECT

"The public, as a whole, finds reassurance in longevity, and, after the necessary interlude of reaction, is disposed to recognize extreme old age as a sign of excellence. The long-liver has triumphed over at least one of man's initial handicaps: the brevity of life."

—VITA SACKVILLE-WEST

LIFE EXPECTANCY

I said that everything we would explore would only have meaning within the "human" context—an *individual* human. Nothing in a book about the future is more meaningful to a particular human than discussing how long he or she can expect to live. You are likely to live far longer than you now expect.

Life expectancy has increased rapidly since the Enlightenment. Estimates suggest that in a pre-modern, poor world, life expectancy was around 30 years in all regions of the world. In the early 19th century, life expectancy started to increase in the early industrialized countries while it stayed low in the rest of the world. This led to a very high inequality in how health was distributed across the world–good health in the rich countries and persistently bad health in those countries that remained poor. Over the last decades this global inequality decreased. Countries that not long ago were suffering from bad health are catching up rapidly. Since 1900 the global average life expectancy has more than doubled and is now approaching 70 years. No country in the world has a lower life expectancy than the countries with the highest life expectancy in 1800.[38]

When we say "life expectancy" for a population, we usually mean the statistically average number of years that a person just born will live. Let us call that "Average Life Expectancy." Some people will live far longer or far shorter lives. There is also a "Maximum Life Expectancy," which is the most extended number of years that any human has lived. Finally, there is "Healthy Life

Expectancy," which is the average number of years lived without an age-related disability. Let us explore each of these in more detail.

AVERAGE LIFE EXPECTANCY

For the 325 years from 1545 to 1870, Average Life Expectancy at birth in the United Kingdom and America was flat at approximately age 35 (varying between 30 and 40). This average was a horizontal line on the chart for 325 years. Then the scientific and industrial age began a steady increase in average life expectancy.

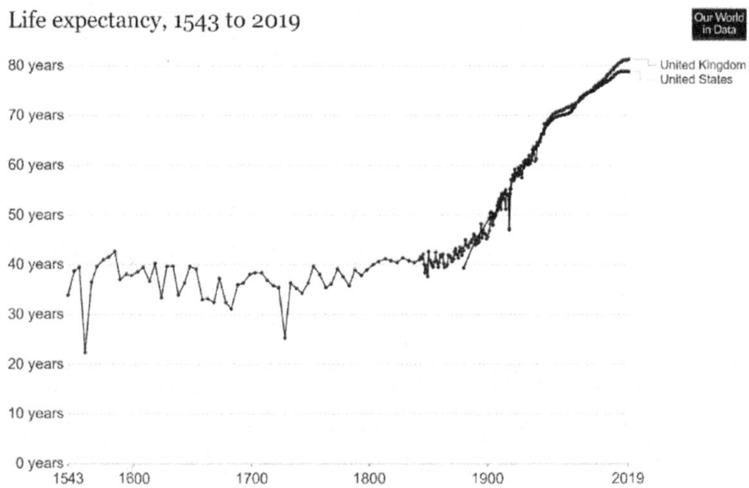

Life Expectancy[39]

For the one hundred forty-five-years from 1870 to 2015, the Average Life Expectancy in Europe and the Americas more than doubled, rising 45 years from age 35 to approximately age 80. When charted over these 145 years, this increase followed an upward sloping straight line increasing at a nearly constant average rate of 3.1 years per decade.[40] Since 1925 many studies proposing a limit on further increases of Average Life Expectancy have all been proved wrong, usually within 5-years of the study.

Note that this 145-year period of increasing life expectancy has been linear, not exponential. Industrialization and modernization initially brought incremental improvements in living and working conditions, nutrition, regular health care, and child mortality. In the later part of the period, when further improvements in living and working conditions would diminish as a factor, the linear rate of increase in Average Life Expectancy did not slow. Contributions from advances in health care technology have replaced improvement in living and working conditions as the driving factor increasing Average Life Expectancy.

A recent slowing of increases in life expectancy could be due to increases in mortality from obesity and drug overdoses.[41] However, advances in health care technology might soon overshadow the suppressing effect on longevity from these causes. If we project the historic 145-year straight-line increase out to 2040, then the Average Life Expectancy in Europe and the Americas would be age 88. Since life expectancy in all the world's countries is converging, 88 would also be the worldwide Average Life Expectancy projection by 2040.

MAXIMUM LIFE EXPECTANCY

The Maximum Life Expectancy is around 120. The oldest verified record is Jeanne Calment (1875-1997) of France, who died at 122 years, 164 days.[42] While the human body can live to 120, most do not. The Average Life Expectancy at birth in developed countries in 2015 was around age 80 or about two-thirds of the Maximum Life Expectancy.

The Maximum Life Expectancy appears to be set by our genetics. More specifically, our immune system's functioning, which we genetically share with all other jawed vertebrates going back some 400 million years, sets our Maximum Life Expectancy. Our immune system did not evolve to protect individuals. It evolved to protect the genetics of our species. In his book *An Elegant Defense*, Matt Richtel writes that our immune system "does an extraordinary job of keeping us alive until we reproduce and then rear our offspring. After that, it does an even better job of moving us out of the way."[43]

Medical advances over the next couple of decades will enable more of us to live closer to the genetics-limited age of around 120. However, during that same time, we will be learning how to take control of our immune system's functioning and even modify the genetics behind it. We will learn how to block the destructive, end-of-usefulness aging functions while retaining youth's life-protecting functions.

HEALTHY LIFE EXPECTANCY

The next decade or two will also be about extending Healthy Life Expectancy—living without an age-related disability longer and closer to the Maximum Life Expectancy of 120.

The years lived with a disability at the end of life average about eleven years in the USA.[44] In 2016 the Average Life Expectancy was age 79, and the Healthy Life Expectancy was age 68 or 86% of Average Life Expectancy. The better medical care in developed countries adds about two years of living with a disability (11) compared to the world average (9).

The life expectancy period of living with disability at the end of life in the United States has increased only one year over the last 25-years, from 10 to 11 years. Over the same 25 years the US's disability-free life expectancy increased four years from 75 to 79. This difference means the disability-free portion of life expectancy increases in the United States at four times the rate of our period of living with disability at the end.

The transcendent technologies of genetics, nanotechnology, AI, and quantum computing could soon eliminate cancer, heart disease, diabetes, arthritis and most of the other age-related diseases. We should begin to see a reduction in the average number of years lived with age-related disabilities later in life.

LIFE EXPECTANCY PROJECTIONS

The rate of increase in life expectancy is not likely to remain a straight line (linear) for the next 20 years. The Average Life Expectancy curve and the Healthy Life Expectancy curve are likely to become exponential. By the year 2040 Average Life Expectancy will be something greater than the straight-line projection of age 88. The Average Life Expectancy and Healthy Life Expectancy curves could rise closer to today's Maximum Life Expectancy curve, which by then will also be increasing.

It is probable that by 2040, disability-free life expectancy will be at least the historical Maximum Life Expectancy of 120, and the new Maximum Life Expectancy might be extended to 150 or beyond. We should live far fewer years with an age-related disability at the end than we do today because of eliminating disabling diseases over the next twenty years.

However, we must take a pause here to keep all this in perspective. We might have already reached the point where projections of our biological life expectancies in the traditional sense are, in fact, meaningless. A person who is age fifty or younger in the year 2020 can undoubtedly expect to live another twenty-five years to age seventy-five in 2045. By then, our technology might extend that person's healthy biological lifespan another fifty years to age 125 in the year 2095. However, we need to remember that our forward vision regarding technology is limited to the next twenty-five years to 2045. But by then, we might have reached the singularity with artificial intelligence a billion times greater than human intelligence today. We might have already begun to merge with our technology to become a hybrid species capable of living indefinitely. Or, if our society collapses, the life expectancy curves could take a sudden turn in the other direction.

THE FOUNTAIN OF YOUTH

In our youth, we rarely thought about our mortality. The warning, "You have to die someday," was too far in the future to seem real. When a person reaches around sixty years of age, they begin to anticipate their death more frequently. I know I did.

I "retired" after working for thirty-five years. My father, still alive at age ninety, had been suffering from severe dementia and failed knees for several years. My mother was his 24x7 in-home caregiver. At age sixty, I thought, "Is that what I have to look forward to? Two-thirds of my life have passed. Have I begun that inexorable slide toward disability and death?"

Looking at my father, I was thinking about my own disintegration and death more frequently. I thought, "Dammit! I love being alive. There is so much more that I want to experience and achieve. Why does it have to end?" Planning my remaining years was impeded by anger over the unfairness and frustration over its inevitability.

Then, reflecting on all that I had learned regarding accelerating technology, I had an awakening. "If I can stay alive for twenty to thirty more years, then technology might be able to keep me alive indefinitely. There is no guarantee, but it is a rational possibility and goal." My attitude changed from bitterness to optimism almost immediately.

Once again, the exponential growth curve challenges our ability to grasp the coming speed and degree of change psychologically. We must learn to think and plan exponentially. Everyone's goal must be to stay alive and healthy for the next twenty-five years. Anyone who can do that will have a good chance of technology being able to keep them alive indefinitely. They will have a chance of being included in the new, long-lived "Hybrid Human" evolutionary species.

Even the possibility of significant life extension, much less an indefinite lifespan, is liberating. Evolution designed humans to function by goal-directed action. It is difficult to be motivated by big goals when one is anticipating death. One tries to keep busy and distracted while waiting for the inevitable. The prospect of an extended healthy lifespan enables a person already at "retirement age" to have boundless hope. The person can return to that youthful state of mind where death is way off in the distance and you have big goals to accomplish.

Accelerating technology is *The Fountain of Youth* that mankind has been seeking for millennia. It was "created" by the mind of man. As you will learn in Part 3, achieving perpetual youth is only one of the adaptations that will enable humans to accelerate evolution in the universe. You have an essential role to play—and for a long time to come.

♦♦♦

EVERYTHING YOU CONSIDER NORMAL WILL CHANGE

"The measure of intelligence is the ability to change."

—Albert Einstein

PARADIGM SHIFTS

The automobile replaced the horse and buggy. The cell phone replaced the land-line phone that replaced the telegraph. Streaming media replaced the CD Disk that replaced the cassette tape that replaced the 8-track tape that replaced the vinyl record. These were all examples of what is known as a "paradigm shift." We can think of a paradigm shift as the life cycle of a particular technology applied for a specific purpose in society. Eventually, a technology gets replaced by a better new one. These life cycles are getting shorter. How should an individual deal with this accelerating change? How might it affect our society? In this chapter, we explore these questions. Kurzweil explains that each paradigm develops in three stages:

1. Slow growth (the early phases of exponential growth)
2. Rapid growth (the late, explosive phase of exponential growth)
3. A leveling-off as the paradigm matures to be replaced by a new technology eventually.

Kurzweil continues: "Technology typically starts out with unaffordable products that don't work very well, followed by expensive versions that work a little better, and then by inexpensive products that work reasonably well. Finally, the technology becomes highly effective, ubiquitous, and almost free."[45]

Paradigm shifts are also on an exponential growth curve. Kurzweil reports that the Cycle Time is halving every ten years. The time between early and late adoption of new technology was approximately 10-years in 2005. It was about five years in 2015 and will be only 2.5 years in 2025.

An incredible number of discoveries and developments in every field are in R&D laboratories all over the world. They are working their way through the process of becoming mainstream products for you. The time required for them to become products and reach widespread adoption will get continually shorter. Frequent paradigm shifts in your lifestyle will become the new normal.

CHANGE—THE NEW NORMAL

You have a familiarity and comfort level with the things in your life now. Like the automobile, many of those things have been around for generations. You are also comfortable with a certain pace of change. You know about how long it takes to get used to something new.

We are about to throw all that normalcy out the window. Earlier, we saw that your rear-view mirror does not provide you the information you need about your near future. That includes your experience with all that you consider normal. Even basic everyday things like where you live, your job, your income and expenses, your health conditions, and your hobbies and recreation will be subject to change in the next decade.

You have seen those self-tests in magazines which are supposed to indicate how much stress you have in your life. Those tests ask questions such as, "Have you had a recent job change or a move?" We are about to enter a period where everyone will have to learn to adapt to continual major change without becoming stressed. The significant and continual change will be the new normal. As a indication of the mindset you will need, here are a few examples of paradigm shifts to expect.

- Electric motor-driven cars replace those with the internal combustion engine. Gas stations become rare as they only serve the antique gas car recreational industry.
- Traffic gridlock is replaced by intelligent roadways with autonomous cars communicating with each other and the roadway traffic controls.
- Highways in the sky exist for swarms of people-carrying drones.

- No one needs a driver's license because all vehicles operate autonomously.
- Solar power and batteries replace your home's connection to the power grid.
- Personalized, genetics-based therapies replace chemo drugs and surgeries in treating and even preventing cancer.
- Elimination of age-related diseases such as cancer, heart disease, diabetes, dementia, and robotic AI in-home caregivers replace long-term care facilities.
- Images and sounds personalized for you and projected such that only you can see and hear them via devices you wear replace most public signs, especially advertising.
- Going to school is replaced by a virtual reality classroom experience and an AI teacher who tailors each lesson specifically for you.
- You have a personal AI assistant who handles all interfacing to other technology for you. No more logins and passwords or learning a myriad of user interfaces.
- AI robotic fighting machines replace human soldiers in battle.

Uh-oh. You might be thinking the last example is a bit scary—more on that later.

OPPORTUNITIES

Whenever there is a significant change in our society, there is an opportunity for those who see it coming. The more significant the change, the bigger the opportunity. The next couple of decades will bring the most significant changes in all human history. Nearly every aspect of how we live in our society is about to change due to paradigm shifts. Therefore, enormous opportunities lay immediately before us. There will be the opportunity for you to:

- Have an exciting career in a paradigm-shifting growth industry.
- Cause a paradigm shift by making a scientific discovery or creating an invention.
- Start a new business related to a paradigm shift.

- Be an investor or speculator in the businesses or resources related to paradigm shifts.
- Be an influencer, educator, coach, or consultant who helps others prepare for and deal with change.
- Work in government to help our laws and institutions keep up with change.
- Work in applying new technology to clean up the debris left in the environment from the old technology.
- Create new forms of art and expression of man's creative intelligence.
- Dramatically extend your life expectancy by taking advantage of new biotech.
- Increase your intelligence by embracing augmentation technology.

In the chapters ahead, you will learn how technological change happens and how to assess what is based in reality and likely to succeed and what is not. This knowledge will help you choose the opportunities that are right for you.

THREATS

Powerful technologies, like fire, can be used to benefit us and to harm us. Nuclear technology is unimaginably more powerful than fire. The new transcendent technologies will be even more potent than atomic technology. They offer both incredible promise and enormous peril.

Baby Boomers like me know what it was like to live during the Cold War when adults taught school children what to do in the event of a nuclear attack. A similar kind of pervasive stress could develop as we worry about the possibility of exceptionally potent technology run amok, released either by accident or malicious intent. Regarding AI, Kurzweil wrote:

> Artificial intelligence is all around us—we no longer have our hand on the plug. [...] If all the AI systems decided to go on strike tomorrow, our civilization would be crippled; we couldn't get money from our bank, and indeed, our money would disappear; communication, transportation, and manufacturing would all grind to a halt. Fortunately, our intelligent machines are not yet intelligent enough to organize such a conspiracy.[46]

Our best protection against such disaster scenarios is for all people who create, implement, and manage these technologies to be rational and hold ideas and principles that are connected to reality and consistent with man's nature. The best way to ensure that these individuals have reality-based ideas is for most of the general population to have them and require them in all their dealings with others.

THE "TIGHTROPE"

To visualize the challenge that we humans now face regarding our society and technology, consider the following metaphor:

Mankind is walking a tightrope toward the future and could fall off either side. Successfully reaching the end of the tightrope means that we will have achieved technological mastery over all the matter of the universe. It will require us to ground our society in the Laws of Reality, halt and reverse the aging process, merge our biology with our technology and become super-intelligent.

The tightrope is anchored in and stretched over a landscape called "Reality." The tightrope itself, which supports us, is what we refer to as the "Laws of Reality." The Laws of Reality include the specific natures of things that exist in reality, from the nature of a sub-atomic particle to the "Nature of Man." Man must stay firmly balanced on the Laws of Reality to reach the end of the tightrope.

Man's arms are outstretched for balance holding two buckets, one on each side. One bucket is labeled "Technology," and the other is tagged "Culture."

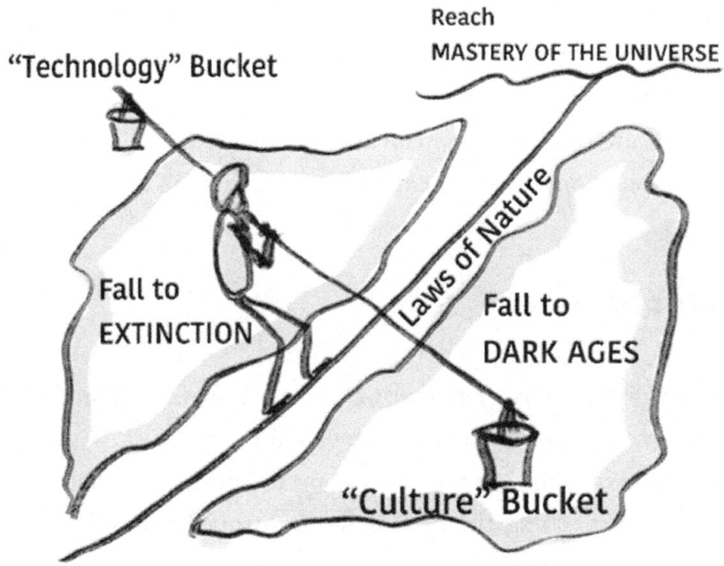

The "Tightrope"

"Technology" means artificial intelligence, augmented human intelligence, genetics, nanotechnology, blockchain, 5G network, quantum computing, and fusion energy. Advances in these transcendent technologies will spread through society from scientists to engineers to businesspeople to consumers. Each step forward on the tightrope means that we have made discoveries of the Laws of Reality and then applied them to create new technologies added into our Technology bucket. Discoveries and technologies make further additional steps forward possible. Continuing these advances will be necessary to reach the end of the tightrope. However, each of these technologies also holds the potential for our destruction.

"Culture" as we step forward means the dominant set of philosophical ideas that are believed and held by individuals in our society, especially those who create and control technology. These ideas include what we think about reality, knowledge, ethics, politics, and art. Ideas are crucially important. They determine the course of an individual's life and society. These ideas will determine our culture's relationship with our technology, such as how we create it, apply it, and reap its benefits while avoiding its dangers. As our technology advances with each step forward on the tightrope, we must increase our culture's understanding of and commitment to reality-based philosophical ideas. We must add those increases to our Culture bucket.

Falling off the "Technology" side of the tightrope could mean the extinction of humankind. A super AI entity could kill us. Or a synthetic virus either kills us directly or renders the Earth uninhabitable. Or technology could enable a fanatical regime to instigate a domino effect of global nuclear-style wars of jihad intended to send the world's population to god.

Falling off the "Culture" side could mean that Western society has abandoned its commitment to reality and reason. It has disintegrated the rational principles and institutions of society and has revolted against technology and achievement. It lives by emotion for the range of the moment. Any survivors have been thrust back into a New Dark Ages where they toil daily to survive. An elite totalitarian ruling class that claims to have mystical revelations dictates the terms. Or, having lost touch with its rational founding principles of individualism and freedom, Western society is unable or unwilling to defend itself against communist China or another totalitarian regime that could gain the technological upper hand and implement totalitarian control of the entire world.

Man must balance the contents of the Technology and Culture buckets to reach the end of the increasingly steep tightrope. The Technology bucket has been continuously filling, and the rate of filling is accelerating. The Culture bucket only fills when our society's predominant philosophy moves *toward* reality and reason—and empties when we move the other direction. Our Culture bucket was at its fullest two hundred years ago. Its heft enabled the Industrial Revolution, counter-balancing all the technology we put in our Technology bucket up through today. Unfortunately, the Culture bucket has been *emptying* over the last two hundred years as we continue to add new technology to our Technology Bucket.

The risk is that our Technology bucket will soon become too heavy for our society to handle. We are rapidly approaching a critical imbalance that could cause us to stumble and fall off one side of the tightrope or the other—Extinction or a New Dark Ages. Or we can reassert reality and reason in our culture, regain our balance, and keep moving forward. You could *possibly* live to see the outcome. Your children will *probably* see it, and your grandchildren will *almost certainly* see it.

Our Culture bucket *can* catch up, and, as we will consider later in the book, perhaps our technology can even help. This book is an example of how we can elevate the importance of reality-based thinking in our society's discourse. It will take time to catch up. Hopefully, we have the time.

RISE OF THE MACHINES

How do we ensure that the development of artificial intelligence will have beneficial impacts on society?

Today, the power of ideas exercises its control over the general population through our culture, for example, the education, business, and government institutions. As people receive their education, educators put ideas into their minds. Some educated people put these ideas into their businesses' products and services that we all use. Other people put these ideas into laws and regulations with which we all must comply. Presently, teaching, business, and governing are implemented on humans by humans.

However, technology is about to change all that as all our systems and transactions—communications, transportation, housing, clothing, medical care, banking, commerce, productive work, entertainment—everything—change from being operated by humans to being administered by AI machines. Personal AI Assistants will soon educate our children. The ideas taught during this individualized, 24x7 instruction will be far more effective at indoctrinating the child and subsequently the adult than any classroom education ever was.

Ideas held by our government leaders and those who produce all the machines that will control our lives will become embedded in the AI software that runs everything. "Smart" machines will scrutinize and manipulate our information, searches, queries, transactions, assets, income, movement, fitness, healthcare, and conversations. Soon, our lives will not be controlled by other humans but by ideas implemented in AI systems that we cannot reason with, avoid, or escape. These AI machines will amplify the power that the ideas built into them will have in controlling our lives.

For an indication of the power and danger of AI systems imposing their creator's philosophic values upon us without human recourse, consider the following quote from a paper titled *Defining Human Values for Value Learners* by Kaj Sotala, a researcher working for the Machine Intelligence Research Institute, Berkeley, CA:

> While this paper is not focused on the question of how to combine the conflicting values of different people, it is also interesting to notice that the flexibility in resolving value conflicts may be useful in trying to find a combination of values that as many people as possible could agree on. If such a set of globally most agreeable values could be found, different AI systems could then coordinate to direct the whole global population towards that set. This might look like something akin to the "Coherent Blended Volition" (Goertzel and Pitt 2012) proposal, which seeks to "incorporat[e] the most essential elements of [people's] divergent views into a whole that is overall compact, elegant and harmonious". In this proposal, AI systems would be used to assist people in reaching an agreeable combination of their values. Given the great degree of flexibility allowed by the possibility of social feedback loops, in which directed changes to a culture's customs could potentially cause major changes to the values of that culture, one might be very optimistic about the possibility of reaching such an outcome.[47]

The above paper shows that some AI creators intend for their machines to manipulate our thinking and values. If allowed to continue, the result will be an AI-implemented "Big Brother" state. If we do not soon learn to think for ourselves, our AI will do it for us.

There is currently much discussion about the urgent need to implement ethics in the narrow AI that already exists. There is concern because machine learning systems have already demonstrated algorithmic bias unintended by their creators. There is the risk that it will make injustices in our society worse. The debate exists because there is no consensus on *which* ethical principles should be in control of our machine learning systems. There is not even agreement that ethics in today's AI systems is possible.

We can be sure that ethics will be crucial when we consider future AI entities with human-level intelligence. But what values should we prioritize as we automate algorithmic decisions today in hiring, credit scores, college admissions, health care rationing, and law enforcement? Should we value "transparency" or "fairness" or "privacy" or "efficiency?"

The pursuit of ethics in AI is essential. However, there are several erroneous assumptions in today's public debate about what an adequately reasoned AI ethics would be and how to implement them in the AI entity. This difficulty with AI ethics exists because there are many erroneous assumptions today about what an adequately reasoned "human" ethics should be. Here are some proposed AI ethics ideas that will not work:

- *"The AI should be fair."*—This does not even come close. Who determines what is fair?
- *"There should be a compliance checklist."*—No checklist could ever cover all the real-world situations.
- *"It's up to each individual to determine what is morally good because there are no universally 'good' ethics."*—There *is* a universal standard of ethics. It is that required by man's nature.
- *"There is no single morally right answer."*—Yes, there is. Again, man's nature provides it.
- *"We need to become ethical AI developers and inform people of our ethics so they will trust us."*—Ethics is how you behave, not what you call yourself. Everyone is already "ethical" according to the ideas they hold as true—it's just a question of whether they demonstrate ethics based on reality and man's nature.
- *"Ethics needed for governance in one situation might be different than the ethics needed in another context."*—All desired ethical behavior exists in only one context—AI's dealings with and impact on humans, in all its variety.
- *"Ethics is the same as the rule of law. We need government regulations."*—The only proper function of government is to retaliate against the initiation of force. When government initiates force, ethics has already broken down.

There *are* universally applicable ethics as required by man's rational nature. If we develop rational AI entities, then the same ethics will apply to them. We will be exploring this in greater detail in the chapters ahead.

AI, or more specifically "Conceptual AI" (human-equivalent intelligence), will likely be achievable. (Other authors refer to this as "Artificial General Intelligence.") Eventually, there will be sufficient computing power to emulate the functioning of all the neural circuits of the human brain. However, computer science and neuroscience alone will not achieve Conceptual AI. The problem is that the prevailing wisdom that floods the society and the experts contain fundamental errors. Their confident predictions about when they would achieve Conceptual AI have all been wrong. There are errors regarding consciousness and self-awareness. Some do not recognize the essential difference between human consciousness and that of animals. Gorillas can pass the mirror test for self-awareness, but that is not the same awareness of abstract concepts humans have.

Some believe that they must include ethics in the programming of the created AI entity. If they did so, it would be the equivalent of brainwashing if it were a human. The Conceptual AI entity must be as free to use its mind as a human and then be guided in discovering the proper ethics, just like human children must learn them. Humans and AI entities must create mental formulations of ethical concepts within their minds; they cannot come from outside sources. Suppose a programmer works to program an AI whose "thinking" must follow specific ethics rules built into the programming. In that case, he is preventing the AI from having one essential quality of an intelligent, creative human being—"volition."

AI entities will be alive and can die, be murdered or commit suicide. Traditional human ethics will extend to them. However, we will need to resolve the new issues that they will bring. For example, it will be ultra-simple to clone a machine AI entity. We will need new definitions that determine what constitutes a new AI "life" or entity.

Creative people can do enormous good or enormous harm. Creative AI entities will be the same. There will be a continuation of the theme of good and evil that is as old as history. It will be the philosophy in the culture that prevents AI entities from becoming evil, just as philosophy prevents humans from becoming evil.

You will soon see that amplifying and accelerating via future technology the predominant ideas in today's society will be disastrous. It has already begun with our present-day "narrow AI" being applied everywhere. More immediate and more threatening than super-AI "Terminators," *machine-implemented philosophy* is the most urgent threat we face in the age of exponentially growing technology.

As you can see, nothing will bring about more significant "change" in our lives than AI. We must return to a proper reality-based philosophy and ethics in our *human* society for those ideas to become embedded in the AI machines. This is also needed to eventually integrate conceptual AI entities into a community containing both biological and non-biological citizens.

To keep our balance on the tightrope and reach a shining future, we must fill our Culture bucket with a philosophy based on reality and reason.

♦♦♦

THE GREAT UNRAVELING

"All rational action is in the first-place individual action. Only the individual thinks. Only the individual reasons. Only the individual acts."

—LUDWIG VON MISES

"We in the tech industry have created the tools to destabilize and erode the fabric of society, in every country all at once everywhere."

—TRISTAN HARRIS

WHERE HAVE THE PRINCIPLES GONE?

I was a teenager in the 1960s and a young adult in the 1970s and 1980s. The society I see around me today looks vastly different than the one I remember growing up. The world seems to be going crazy. People do not seem to be behaving according to any reliable principles. Is it just that the people who live in every period of history feel this way, or is there something significant going on today? In this chapter, I explain that something powerful *is* happening that should concern us all. Here are some of today's societal changes:[48]

Principles. It seems that more and more people are *un*principled. Fewer people, especially the younger ones, think in terms of principles anymore. They are unable to build chains of logic or know what constitutes evidence or proof of a claim. They do not feel the need for principles to guide their lives. Instead, they seem to be content to act on their "feelings" in any situation or moment.

Education. Our schools seem to be producing too many people who lack reasoning skills. Too many graduates did not adequately learn the Three "Rs," or any subject for that matter. They cannot write a well-reasoned report on any topic with proper grammar, structure, or logic.

Science. *In*determinacy appears to be the new depth in physics. Some scientists seem to be moving away from objective reality and supremacy of facts. Instead, they offer theories of mystical unrealities based on the "beauty" in their mathematical equations, whether they are provable or not. Some are willing to ignore facts contrary to their political positions.

Mathematics. The new revelation in mathematics is *in*completeness, which is necessary to support its science applications as described above.

Art. Artists have expelled beauty, preferring *non*representational painting instead. Art that means something to either the artist or the viewer is now old-school.

Music. More musicians now create *a*tonal music. It seems the more it grates on one's senses, the better.

Literature. Writers now flaunt *anti*heroes. *De*construction is now the method of literary criticism.

Language. Language has now been "liberated" from syntax, and verse is "free" of meter. Any word, phrase, sentence, paragraph, and punctuation mark can be used anywhere any time and can mean anything to anybody.

Politics. Politicians are "pragmatists." Lacking principles, they act to resolve every range-of-the-moment issue as quickly and easily as possible. They do this even if it requires sacrificing a higher principle (which does not guide them). Their only overarching principle is to remain in power—whatever it takes.

Economics. The government creates new economic crises and then capitalizes on them to assume even greater power over the economy. Businesspeople are afraid to defend the profit motive or defend themselves or their businesses from attacks of political correctness or demands from the cancel culture.

Today, as we walk the tightrope, the Technology bucket continues to fill and gain weight. The scientists and engineers have been a subgroup of society that tend to be more reality-based when dealing with the sciences.

Unfortunately, the Culture bucket is emptying. The ideas typically held by those who work in the humanities, especially in education, are causing our general society to be less reliant on reality and facts and more driven by feelings. This erosion is causing our society to seem to be unraveling all around us. There do not seem to be any principles that we can count on anymore to make sense of what we see. We do not know what "truth" means anymore. In the chapters ahead, we will rediscover what we mean by "truth" and "principles."

THE GREAT DIVIDE

Over the last couple of decades, Americans have divided themselves culturally and politically. America might have great diversity, but its people have segregated themselves into homogeneous communities at the city and neighborhood levels according to their beliefs and lifestyles. People live in places where those who surround them think, live and vote as they do. People have become so ideologically polarized that they do not understand or want to understand the people only a few miles away.

Some live along the east coast's I-95 corridor near the "halls of power," Washington, New York, Boston, Philadelphia, etc. The people in these areas of greater population density are more likely to have 4-year degrees.[49] They believe that things like diversity, free health care, guaranteed basic income and globalization are fundamental American values to be achieved through government institutions. They think the Constitution should be reinterpreted as needed. They would fund social programs before the military and believe that guns lead to mass shootings and citizens do not need them. Some working in the government believe themselves to be far-sighted, scientific, objective people.

They think they are working for the public good and that the masses are incompetent and need to rely on government.

Most large cosmopolitan cities on the west coast and around the Great Lakes—Los Angeles, San Francisco, Seattle, Chicago, and Detroit, share a similar ideology. According to a New York Times article, the greater the population density, the more likely a city will be liberal. At the time of the 2016 election, only three of the 25 largest cities in America had Republican mayors.[50]

Those living in the less populated "Heartland," the Midwest, South and Southwest, are less likely to have a 4-year degree. Valuing individual liberty, the ability to control their destiny, and personal ownership, they prefer a smaller government that regulates less and offers fewer services and lower taxes. They want a government that respects the original principles of the Constitution and looks after America first. They desire a strong military and are more likely to volunteer for and appreciate military service. They believe that individuals, not guns, cause mass shootings, and the Second Amendment is a fundamental right. Some think those in power in government are neither intelligent nor rational. They consider them "enforcers," imposing unwelcome elitist philosophical doctrine upon the people, much like the KGB, Gestapo, Revolutionary Guard, etc.

The popular vote in the 2016 and 2020 elections showed us that America divides equally along these ideological lines. In recent decades we have seen a worrisome change in this divide. It used to be a primarily *political* divide. Liberals and conservatives could disagree on whom to vote for but generally agreed on what it meant to be an American. After voting, they could leave their politics aside in their day-to-day dealings with each other. Now, however, we see that ideological differences polarize us at the core. We no longer seek to interact or cooperate with those who hold beliefs different from our own but instead shun them, label them, attack them, and even hate them. It is now "us" against "them." The divide has now become a *cultural* rather than political divide. It now divides Americans on the essential principles of what we think it means to be "American" and "good," with the other side being "un-American" and "evil."

In the last decade, we have begun to see an unraveling of society in metropolitan cities. City leaders have declared their cities as "sanctuary" cities preventing federal immigration laws enforcement. In a resurgence of racism with a vengeance, the law enforcement community has been labeled racist, with liberal cities adopting "defund the police" policies. Rioters have taken over multi-block areas in city centers and declared them to be "autonomous zones." City leaders have instructed Police to stand down in the face of lawlessness resulting in the destruction of businesses and the hollowing out of previously vibrant urban neighborhoods. Property owners who attempt to defend themselves and their property might be prosecuted while their attackers are either not apprehended or set free.

Despite large tax-funded expenditures on social programs, homelessness, flagrant drug use, poor sanitation and health hazards have rendered urban city centers as places to be avoided. Taxpayers and producers view high taxes and copious regulations in these jurisdictions as onerous. Citizens and businesses are beginning to flee the liberal urban areas and states, many relocating to the more conservative Midwest and Southern states.

Following the 2020 presidential election, a group of citizens aligned with the conservative ideology blamed the US congress for all the lawlessness witnessed over the prior year. Their trigger point arose from claims that liberal cities intentionally violated their own state's election laws, altering the election outcome without consequence. For the first time in history, citizens lawlessly invaded the US congress.

The "right" and the "left" leaders of today's political divide have both abandoned truth and principles in making their claims, appealing to emotionalism instead. However, the citizens' migration from liberal to conservative cities suggests which ideology best aligns with a flourishing society's requirements. The difference in education level provides a hint as to the cause of the Great Divide. It is a clue that we will follow later.

THE GREAT AMPLIFIER

The positive human benefits of advancing technology are wondrous and incalculable. Indeed, we are going to explore selected transcendent technologies in future chapters. However, one wonders if technology's recent application in society, for example, social media, also has a dark side? Is technology causing the unraveling we see, or is it simply amplifying and accelerating a problem that existed in our society long before the technology came along? Is it making things worse?

Consider the disturbing trend in teen depression and suicide. Depression anxieties for American teenagers began increasing between 2011 and 2013. The number of teenage girls out of a hundred thousand in this country admitted to a hospital every year because they cut themselves or otherwise harmed themselves was pretty stable until around 2010–2011. Since then, it's up 62% for older teen girls and 189% for preteen girls. That's nearly triple.

Even more horrifying, we see the same pattern with suicide. Rates among older teen girls 15 to 19 years old are up 70% compared to the first decade of this century. Rates among preteen girls, who initially had very low rates, are up 151%, and that pattern points to social media.

Gen Z, the kids born after 1996 or so, are the first generation in history on social media in middle school. How do they spend their time? They come home from school, and they're on their devices. The whole generation is more anxious, more fragile, more depressed. They're much less comfortable taking risks. The rates at which they've been getting driver's licenses have been dropping. The number who have ever gone out on a date or had any romantic interaction is declining rapidly.

This is a fundamental change in a generation. As Jonathan Haidt, Ph.D, a social psychologist at NYU Stern School of Business, said, "And remember, for every one of these, every hospital admission, there's a family that is traumatized and horrified—my god, what is happening to our kids?"[51]

Dr. Anna Lembke, Medical Director of Addiction Medicine at Stanford University, describes the addictive nature of social media:

> So, here's the thing, social media is a drug. We have a basic biological imperative to connect with other people. That directly affects the release of dopamine in the reward pathway. Millions of years of evolution are behind that system to get us to come together and live in communities to find mates to propagate our species. So, there's no doubt that a vehicle like social media which optimizes this connection between people is going to have the potential for addiction.[52]

Our human brain and its functioning are millions of years old and will not change fundamentally any time soon. On the opposite side of our computer screens are thousands of engineers and supercomputers with goals different from ours. Who's going to win in that contest? Tristan Harris, former Google Design Ethicist, describes our present situation:

> We're all looking out for the moment when technology would overwhelm human strengths and intelligence. When is it going to cross the singularity and replace our jobs and be smarter than humans? But there is this much earlier moment when technology exceeds and overwhelms human weaknesses. This point being crossed is at the root of addiction, polarization, radicalization, outragification, vanityification, the entire thing, this is overpowering human nature, and this is checkmate on humanity.[53]

Data scientist Cathy O'Neil, Ph.D., explains that people with agendas create AI algorithms, "I like to say that algorithms are opinions embedded in code. And that algorithms are not objective. Algorithms are optimized to some definition of success. So, if you can imagine—if a commercial enterprise built an algorithm to their definition of success, it's a commercial interest—usually profit."

Justin Rosenstein, former Engineer at Facebook and Google, adds, "Algorithms and manipulative politicians are becoming so expert at learning how to trigger us, getting so good at creating fake news that we absorb it as if it were reality, and confusing us into believing those lies. It's as though we have less control over who we are and what we really believe." Tristan Harris elaborates, "There's an MIT study that fake news on Twitter spreads six times faster than true news. What is that world going to look like when one has a six times advantage to the other one?"

In answer to our earlier question, social media technology amplifies and accelerates a previously existing problem in our society. The problem is the inability or the unwillingness to seek the *truth,* which means to connect abstract ideas to reality. Tristan Harris describes it this way:

> If technology creates mass chaos, outrage, instability, lack of trust in each other, loneliness, alienation, more polarization, more election hacking, more populism, more distraction and inability to focus on the real issues—that's just society. And now society is incapable of healing itself

> and is devolving into a kind of chaos. It's not about the technology being the existential threat. It's the technology's ability to bring out the worst in society, and the worst in society being the existential threat. If we don't agree on what is true, or that there is such a thing as truth, we're toast. This is the problem beneath other problems because if we can't agree on what's true, then we can't navigate out of any of our problems.[54]

If this problem continues, Jaron Lanier, computer scientist and Founding Father of Virtual Reality, describes the disaster that awaits:

> If we go down the current status quo, for let's say another twenty years, we probably destroy our civilization through willful ignorance. We will probably fail to meet the challenge of climate change, we probably degrade the world's democracies, so that they fall into some sort of bizarre autocratic dysfunction. We probably ruin the global economy. We probably don't survive. I really do view it as existential.[55]

THE SURVEILLANCE ECONOMY

Big Tech companies such as Google, Facebook and Twitter now handle nearly all our communications with other people and information sources and transaction processors. We value the services provided by Big Tech companies. We understand that they need to monitor our content and our transactions to a degree for them to improve their algorithms and their services. Most of us are OK with this as the "trade" for the tech services we are getting for "free."

We must take a step back and consider this arrangement. If Big Tech for-profit companies provide us all these fantastic services for "free," how did they become some of the wealthiest corporations on the planet? From where did all that money come? The obvious answer is advertising. We expect that when we enter a Google search for any product or service, we will see a listing of ads displayed on the web page. We expect the ads to be related to what we typed into our search query. This business arrangement means that *end-users are not Google's customers*—the advertisers are its paying customers. We users and our behaviors are the raw material mined by these companies, processed into prediction products, and then sold to its advertiser customers.

These Big Tech companies' services produce revenue by helping the advertisers spend their advertising dollars more effectively. The goal for the advertiser is "certainty" that the dollars spent will get the desired results. For these companies to have become so large, they must have learned how to maximize effectiveness—approaching certainty—for the advertisers. The more users these companies can attract, the more types of behavioral data they can mine from their users, and the closer to prediction certainty they can provide their advertisers, the greater the revenues.

Matching ads to search queries or other data you entered is how these Big Tech companies began. It is not how they operate today. They know far more about you than just what you type into a search query. They know more about you and your behavioral habits than you know about yourself. Most people have some vaguely defined limit beyond which they think surveillance becomes a breach of their privacy rights. Most are unaware that, over the last two decades, Big Tech has secretly exceeded that limit.

1. Big Tech captures far more behavioral information about you than it needs to improve the services it is providing you. Indeed, it captures all the information it can, even information unrelated to what you are doing on its service at the time.
2. Most Big Tech companies share what you do on their respective services with each other. Even the cell phone carriers provide a hidden code identifying you to every service you interact with. This code enables the combining of all your activities everywhere into your "user profile."
3. Apps on all your devices collect data about you all the time, even when you are not using them, and in some cases even when you have selected privacy settings. These include wearables such as fitness trackers and smartwatches, in-home smart devices, and vehicles.
4. Big Tech can determine who you are, your gender, home address, where you work, marital status, income level, financial status, and education level. They can know your hobbies, interests, spending habits, daily and weekly routines, favorite places, politics, religion, sexual preferences, how much sleep you get, emotional state, stress levels, and more. Not only do they know all this about you, but also about each of the other people in

your circle. They use all of it to refine your profile. Big Tech can learn all this even if you intentionally try not to disclose it.

5. Big Tech uses the extracted surplus behavioral data to predict what you will do *now*, *soon*, and *later*. They auction off those prediction products about you to their actual customers, the highest bidders, to generate revenue.
6. Moving from surveillance to action, the best way to achieve prediction certainty is to actively intervene at the moment a user is making a behavioral decision. Big Tech uses what its machine learning systems have discovered about triggers of human behaviors to manipulate the communications you see so that your actions are more certain to be what Big Tech and its customers want them to be.

Consider a new app-based approach to lending that instantly establishes an applicant's creditworthiness by mining the details of an individual's smartphone and online behaviors. Ten thousand or more signals examined per customer include texts, e-mails, GPS coordinates, social media posts, Facebook profiles, retail transactions, and communications patterns. They additionally include intimate information such as the frequency with which you charge your phone battery, the number of incoming messages you receive, if and when you return phone calls, how many contacts you have listed in your phone, how you fill out online forms, or how many miles you travel each day. Such behavioral data yields nuanced patterns predicting the likelihood of loan default or repayment.[56]

Harvard professor Shoshana Zuboff elaborates, "Legal scholar Frank Pasquale describes this as 'the dark market for personal data.' For example, hiQ markets its prediction products to corporate human resources professionals." Its data science engine collects signals about employees of the company. It indicates that "someone may be a flight risk" with "laser-like accuracy" that is "virtually identical" to observed turnover. This prediction enables the employer to intervene, either attempting to retain the employee or choosing to "preemptively terminate someone who is predicted to be a 'flight risk.'"[57] These surveillance practices can identify an individual's "financial stress level" and can "expose protected status information such as pregnancy and age. There is no opportunity for affected individuals to view or contest information."

We are seeing the fiction in the 2002 movie "Minority Report" becoming reality. How soon might we see AI algorithms determine that you are likely to commit a criminal act, causing the authorities to apprehend you before you do anything? And when will the "cancel culture" punish you for "thought crimes?"

THE MAN-IN-THE-MIDDLE ATTACK

One of the most outstanding Internet Age achievements is the incredible access to global communication and global information—the total of human knowledge. We will further explore the truly transcendent benefits of this later, but is there a dark side here? Let us assume that we understand that we must base our thinking on facts of reality. From where will we get our facts? Global society now depends almost entirely on search engines such as Google, databases such as Wikipedia, and news feeds such as Facebook and Twitter to obtain the information we need to conduct our lives. But can we trust such information sources?

In cybersecurity, there is a type of espionage called the "Man In The Middle" attack. The MITM attack is where the attacker secretly relays and possibly alters the communications between two parties who believe they are directly communicating over a private line. The attacker is controlling the entire conversation.

For example, when you get off a plane and turn your cell phone back on in the airport terminal, your phone will automatically connect to the closest cell "tower." Suppose the closest one is a micro-cell placed in the terminal by an intelligence service. The micro-cell "tower" then relays the connection to the public carrier's cell tower such that you are unaware of the Man In The Middle. The intelligence service can then listen to, record, and alter your conversation or text, email, or web browsing communication. This attack is more invasive than the "wiretap" of old analog "land line" phones. It enables the Man In The Middle to monitor and *alter* the communications without the victim's awareness.

We certainly know that Big Tech systems are in the middle of everything we do online. We understand that those systems use AI algorithms that humans create for business purposes. We are not aware that Men In The Middle manipulate the information we see without being transparent about it. Company personnel create and operate those systems based on their personal and business values. If users became fully aware of the scope and extent of Big Tech's hidden extraction of personal information and alteration and control of our communications, most individuals would question the trustworthiness of the communications. They would look differently at the risk-benefit assessment of using Big Tech's services. That is why Big Tech goes to great lengths to hide its internal practices and avoid raising your concerns.

In defense of the "free market," we might assert that only government can be guilty of "censorship." A private company like Google has the right to remove content that it does not want on its platforms and that this is not "censorship." Such an assertion would be misleading because it conflates two entirely different concepts.

The concept "censorship" applies to information an author intends to broadcast to unidentified recipients in the general public through some "publisher." Examples would be physical books, magazines, newspapers, TV broadcasts and other forms of public mass media. Private publishing companies have the right to choose what to publish. Suppose the government interferes with their right to choose. In that case, that is "censorship" of *public* communication and is a violation of *free speech* rights.

The nature of communication is different in the case of person-to-person communications. For example, we do not expect any third party to open and read documents we send via Fed-X. Nor do we expect AT&T to listen to our private land-line phone conversation, much less alter it without our knowledge. We are not disseminating to the masses and have an expectation of privacy on at least some level.

Consider a network of people. Suppose Mary calls and tells something confidential to Eve, and then Eve calls and tells someone else and so on. We could consider this a breach of trust on Eve's part. However, each call was *private* regarding the channel provider, the phone company.

Person-to-person communications occur through various "channels." They can be one-to-one or one-to-many such as when multiple individuals subscribe to a printed newsletter mailed to them monthly via Fed-X. The private company Fed-X is not the *publisher* of the newsletter. It is a communications *channel provider*. The same principle *should* apply when a person *subscribes* to someone's blog or chooses to *follow* someone on Twitter or *friends* someone on Facebook. In each case, the person is explicitly opting-in to private communication over a communications channel.

In today's online digital world, once you distinguish "communication-channel" from "publishing/e-commerce," there is extraordinarily little in the communication-channel domain that is public "mass media." Nearly all of it requires somc form of explicit "opt-in" (free choice) to receive it, and if something unwanted is received, one can block it.

The size of a private group and the use of the term "social media" do not change anything. Before the advent of the Internet, the land-line phone was the only channel available to groups for rapid communication. When a message needed to get out quickly, the "phone tree" was commonly used. The person launching the message had a list of ten people to call. Each of those had a list of ten, and each of those had a list of ten. In the time required to make thirty calls, one-thousand people could receive the message. Each one of those one-thousand person-to-person communications was *private.* The channel provider, the phone company, did not have the right to listen without a court order for a wiretap.

This same principle should apply to "re-tweets" and "likes" in today's world, the modern-day version of the phone tree. Even if there are ten million "followers," "subscribers," or "friends" in a group, each has opted-in to a private communication over a communications channel provided by a private company. Any monitoring or altering of these person-to-person communications should be considered a Man In The Middle attack on *private* communications and a violation of *privacy* rights, disregarding user agreement terms to the contrary.

A tech company can only know the private content between a podcaster and his followers either by voluntarily subscribing or by listening without permission. In the latter case, why doesn't the company need to get a court order for this modern version of a wiretap? The answer, of course, is that users have given them permission. The Big Tech companies' business model requires behavioral surplus capture to sell behavioral prediction products to the highest bidders.

Incrementally, year-by-year, we have been conditioned to accept the premise that these are not "private" communications channels. Further, they have conditioned us to accept their absolute right to examine all content and to act against you based on that content. "De-platforming" a user's account is the same as if Fed-X or AT&T refused to provide you mail or phone service because of what you were saying in your private conversations.

At the heart of today's debate about free speech, censorship, and Big Tech, is the failure of users to distinguish between "publishing/e-commerce" and "communication services"—between a "publisher's right to control content" and a "communications-channel MITM attack"—between "free speech" violations and "privacy" violations.

Big Tech considers everything it does as one big "platform" and claims the right to see, data capture, and control content everywhere on it. Big Tech does not make the distinction between the moral right to ban certain books from being sold on its e-commerce web sites and the morally improper monitoring and altering private conversations between individuals (regardless of buried claims in the End User License "click" Agreements).

The first thing to notice is the weaponizing of words such as "abusive," "dangerous," and "harmful" in the EULA click agreements. A company can apply such words to discredit any expression it desires. It used to be that "abuses" of communication systems meant phishing, spam, and malware. Now it includes flexible terms that allow tech companies to censor at will—such as "fake news," "misinformation," and "hate speech." Is there an agreed definition of what these words mean? What definition is Google, Facebook or Twitter using? We don't know because the operation of these company's algorithms is an undisclosed "black box." We do know that they can change the definition they use at any time.[58]

Even those terms cannot compete with the vagueness in Microsoft's End User License Agreement (EULA) prohibiting the sharing of "inappropriate content." Microsoft warns users that any transgression of the service agreement could result in a shutdown of their Microsoft account or their Skype account.[59] No one has the hundreds of hours available that it would take to even read the EULAs for all the apps that they use.

2016, 2018 and 2020 election cycles raised public interest in the question of online censorship. Google internally considers its manipulation of information as "censorship." A 2018 Google whistle-blower reported an internal presentation discussing Google's adoption of a gatekeeper's role over the information flow to protect Internet users from "harmful" content.

The presentation was titled "The Good Censor." The document called the early ideals of Silicon Valley assuring users of free speech as "utopian." It claimed that users were questioning whether the "openness" of the internet should be valued at all. The document's authors complained that "Free speech has become a social, economic, and political weapon," and allows "have-a-go commentators" to compete on an even playing field with "authoritative voices." They argued that this threatened "rational debate."[60]

Facebook CEO Mark Zuckerberg told the world they were politically neutral, "We consider ourselves to be a platform of all ideas." This assertion followed a modification to Facebook's News Feed that resulted in a nearly 50 percent drop in engagements on President Trump's Facebook page and a more than 50 percent drop to some conservative websites.[61]

It appears that the giant social media companies' leaders and employees are victims of the same extreme political correctness and "cancel culture" influence exhibited in the "campus wars." The formula used effectively on college campuses is a simple one. First, define your opponents with a negative label ("hateful," "racist," "misogynist," etc.). Then, pressure social media companies—just as they pressured college administrators—to ban anyone who bears the label. Your enemies will be kicked off the internet one by one. The tech companies are playing along, presumably because they expect it to be best for their bottom line.

In his 2020 book *#Deleted. Big Tech's Battle To Erase The Trump Movement And Steal The Election*, author and investigative reporter Allum Bokhari gives an example of activism within Twitter:

> Remember the words of Twitter's Olinda Hassan: "we're trying to get the sh*tty people to not show up." How can you train an algorithm to detect "sh*tty people"? Everything depends on this answer. Machine-learning algorithms are trained based on a set of data [...] In the case of an algorithm designed to detect "sh*tty" people and then downgrade or deplatform them, whoever gets to tell the algorithm what a "sh*tty account" looks like wields immense power. Like a trained attack dog that never sleeps, the algorithm will go after any account that bears a resemblance to the profile you've given it.[62]

The most significant power to train watchdog algorithms resides, of course, with the Big Tech company employees. Bokhari reports about an undercover video in which "Twitter engineer, Pranay Singh, revealed that the platform had written algorithms to classify as 'bots' those accounts that use stereotypical conservative phrases."

> Just go to a random [Trump] tweet and just look at the followers. They'll all be like, guns, God, America, and with the American flag and the cross," said Singh. "Like, who says that? Who talks like that? It's for sure a bot." Asked by the reporter whether he could remove the "bots," Singh said, "Yeah. You just delete them, but, like, the problem is there are hundreds of thousands of them, so you've got to, like, write algorithms that do it for you." "So, you would mostly just get rid of conservatives?" asked the reporter. "Yeah," confirmed Singh.[63]

The MITM concern is not limited to employees of the tech companies. Social media users themselves get to train these watchdog algorithms by blocking or reporting sites that they deem offensive. How often they do this contributes negatively to the "quality rating" of the reported account. If the quality rating of the account goes below some threshold, it is automatically blocked. Bokhari writes that "Those who are feelings driven are more likely to 'report' on social media than those who are facts driven. AI algorithms then amplify their impact."[64] The result is that the content we see on the Internet can be influenced more by users' feelings than by facts.

Notice that people who claim that content is offensive never acknowledge that they *chose* to follow or subscribe to the digital communication and had the option to "opt-out." Psychological fragility has replaced personal responsibility and resilience.

Regarding content modification by the first and second most popular search engines in the world—Google and YouTube—Bokhari writes:

> In public, before Congress, Pichai [Google CEO] assured us all that the company doesn't manually intervene in search results. Behind closed doors, his employees tell a very different story. [This leak] showed that Google maintains a "blacklist" (Google's word, not mine) of handpicked

> search terms on YouTube. When terms are added to the blacklist, search results are reordered to favor videos from "verified" sources, which are themselves handpicked by YouTube. [...] In April 2019, the Daily Caller obtained evidence of something called the "deceptive news domain blacklist"—yet another Google blacklist, used to filter "problematic" news sites out of search results.[65]

Regarding Wikipedia, we must be aware of the possibility of subjective bias when describing politicians, personalities, or news outlets rather than topics treated more objectively. Wikipedia's editors who make and enforce the rules are mostly anonymous and unaccountable. They get to decide what is a "reliable source." The "truth" is determined by consensus. If enough editors disagree with a fact, regardless of how many reliable sources back it up, it is no longer a fact to be reported in Wikipedia.[66]

The communications control exercised by Men In The Middle is not limited to "content" but also includes "access" such as the apps stores and "financial blacklisting." In the app economy, the only way for businesses to deliver their products is through the App Store and the Play Store. Today, if you try to create a platform that ensures free speech without the possibility of censorship, you might be denied access to smartphone app stores, payment processors and possibly the internet itself.[67]

Initially, the rise of online funding platforms promised to decentralize the business of political fund-raising. Now, activists and special interests are pressuring the online transactions companies to censor political views in the same way they pressured social media platforms and the colleges before them.[68]

We might say that our communications are being monitored and altered by the "Algorithm In The Middle." But this would hide the fact that people define what success looks like for the algorithm. A tech company employee "training" an algorithm or an end-user "reporting" another user *is* the Man In The Middle, and algorithms are the technology tools used against the victims.

Of more significant concern, the AI community aims to create *self-training* algorithms that can choose their own definitions of success out of patterns they detect in big data according to some "value system" provided by their human creators. Such an AI algorithm's power would advance from dealing with specific concepts to general principles. One might instruct an algorithm dealing with specific concepts to "block any communications containing (certain words or phrases)." However, an algorithm operating on principles could "change communications that promote 'equality in freedom of action' to communications that promote 'affirmative action for equity'"—equality's opposite. The Men In The Middle who define the "value system" for the self-training AI will become the most powerful in the world.

We can again see that our society's already existing failure to demand objective truth based on facts of reality, relying on feelings instead, is not being caused by technology. Still, technology amplifies the negative consequences. Twitter engineer Singh explains:

> We are rapidly moving into a world in which Big Tech algorithms are involved in almost every aspect of our lives. Whether we're approved for a loan, a mortgage, an insurance policy; whether we're allowed to rent an apartment; whether we can use platforms like Airbnb, Uber and Lyft; whether our business appears at the top of a Google search or whether it's buried five pages down; even whom we'll be matched with by online dating apps. Algorithms already have an outsize impact on our daily lives, and this trend is only going to accelerate.
>
> Control over this technology is the biggest prize in tech. In the title of his influential 2013 book, computer scientist Jaron Lanier asked, "Who owns the future?" With AI set to become a major feature of virtually every industry over the next century, the question may very well be rephrased as, "Who owns the AI?" Industries are unlikely to develop their own AI systems; they will likely rely on whatever is produced by the "experts" in Silicon Valley. The same tech giants that dominate today are likely to dominate the AI-powered world of tomorrow. AI trained leftist ideology will carry those biases far beyond Silicon Valley into every industry and field you can imagine.[69]

In place of instinct, it was reason that gave man the power to discover Laws of Reality and then use them to modify his environment. Suppose one takes a pessimistic view of the future. In that case, a few men now have the technology tools (machine learning) to discover the natural laws of individual and societal behavior and modify those behaviors. This power separates a few humans with super-knowledge and tools-of-scale from the rest of society, which becomes an objectified hive of less powerful humans. The super-humans will manipulate the less powerful. Such companies sharing power with government would become an oligarchy, a form of totalitarianism. The oligarchy would continue until one of the companies develops a super-AI able to subordinate the others—a dictatorship.

THE PARTNERSHIP OF POWER

There is a growing partnership between the two most important power centers—government and Big Tech—each needing and using the other.

Following the terrorist attacks of 9-11, our government sought ways to combat terrorism better. Government agencies turned to Big Tech to conduct surveillance to identify terrorist threats, foreign and domestic. In some cases, the government relaxed regulations regarding privacy. In other instances, the government outsourced certain surveillance services for enormous sums of money to tech companies as contractors because they had capabilities the government did not have. Also, the companies could do things that laws prohibited government agencies from doing. In the other direction, the tech companies make enormous contributions to the politicians to protect against being regulated.

Big Tech has the "muscle" to monitor and influence the masses. Our "democratic" government provides Big Tech the *moral sanction* it needs for its surveillance activities. Surveillance is justified in the name of democratic values such as "safety and security" and "election integrity." Over time, this partnership's surveillance capabilities have expanded and apply to non-citizens and citizens alike.

To the extent that this partnership acts to protect citizens from foreign threats of force (safety and security), it is appropriate. To the extent that this partnership pursues social or political agendas to alter the citizens' behaviors (e.g., equity and inclusion), it is immoral and neither the government nor the business can claim legitimacy in its actions.

A new "hyper-intolerance" on the part of the intellectual upper classes, academics, and the media, presents a danger that could play out in this country as it has previously in Latin American countries. In countries such as Cuba, Peru, and Venezuela, elites' efforts to silence dissent, starting with canceling an opponent for some heresy, leads to a gagged civil society. In those countries that have collapsed into tyranny, support from public intellectuals for government control of thought and speech long preceded a totalitarian regime's consolidation of power and military enforcement of censorship. Things usually ended badly for the intellectuals who unwittingly instigated them.

In the last decade in America, we have seen intellectuals in both traditional and new-technology media censoring opposing opinions and an increase in the public "cancel culture." If we are not careful, today's social media and information technology in partnership with the government could facilitate the collapse of our free society into tyranny.

THE LOSS OF SOVEREIGNTY

Society faces unprecedented change due to rapidly advancing technologies such as the Internet and social media. Because it is unprecedented, history provides little guidance. It is, therefore, an experiment. As with any experimentation, discovery and knowledge follow trial and error.

The grand trial that our society is engaged in today is the conflict between the power of the individual and Big Tech's power. The choice is between the individual's right to privately determine his future and Big Tech's claim on the right to know everything about everyone everywhere all the time. For the most part, Big Tech presently has the upper hand in this conflict, especially in the United States. However, the individual is beginning to win some battles in Europe.

Google's mission was to "organize the world's information and make it universally accessible and useful"—starting with the web. Indeed, it has benefitted our lives in many ways. However, information about an individual that otherwise would age and be forgotten is now permanent, highlighted in each person's digital search results. The Spanish Data Protection Agency ruled that not all information should be permanent. Some personal information should be forgotten. Given its "mission," Google challenged the agency's ruling before the High Court.[70]

Ultimately, the Court of Justice decided in favor of the individual. The Court's decision stated, "The operator of a search engine is liable to affect significantly the fundamental rights to privacy and the protection of personal data. In the light of the potential seriousness of the interference" with those interests, "it cannot be justified by merely the economic interest which the operator of such an engine has in that processing." Zuboff writes that as legal scholars Paul M. Schwartz and Karl-Nikolaus Peifer summarized it, "The Luxembourg Court felt that free flow of information matters, but not as much, ultimately as the safeguarding of dignity, privacy, and data protection in the European rights regime."

I think philosopher Dr. Harry Binswanger correctly articulated the principle under which the Court got it right:

> Whatever one owns, material or intellectual, can only be used by another if the owner permits it. If someone owns information, the fact that you can come to know it (as by overhearing it) does not in any way imply that the owner intended to abandon his right of exclusive control over it. Suppose someone has the right to keep others from knowing X (about himself or his property). In that case, your overhearing X does not, per se, mean he doesn't care who knows it and hence has abandoned his right. The fact that someone does not take adequate care to prevent a passerby from overhearing cannot be taken as permission. On the other hand, anything a speaker at a trade show announced from the stage can be repeated anywhere.[71]

When tech company surveillance captures surplus behavioral information about a person, such as his emotional state captured by voice stress analysis while he speaks a search query into a smart device, that surplus information is entirely unrelated to the query. There can be no implied permission granted to use such data.

Now consider the related topic of "subliminal advertising" which first appeared in 1957 with the display of Coke's single-frame images during a movie. In the 1979 case Central Hudson Gas & Electric Corp. V. Public Service Commission of New York, the Supreme Court ruled that, to be protected by the First Amendment, marketing speech must not be *misleading*. In the 1989 Nevada case Vance v. Judas Priest, a judge ruled that subliminal messages are not protected by the First Amendment and constitute an invasion of privacy.[72] Later, a 2012 review of functional magnetic resonance imaging (fMRI) studies showed that subliminal stimuli activate specific brain regions despite participants' unawareness.[73]

Advertising intended to alter your behavior has been around from the beginnings of society. However, until recently, it was more of an art than a science. Individuals who practiced the advertising art would make educated guesses at what would motivate a particular demographic of people with some successes and many failures. Also, traditional marketers put their ads out there in our environment where we might, by chance, encounter them.

Big Tech, using big data and AI neural networks that can find subtle patterns in the data, have recently made behavioral modification an absolute science. Since they control the communications channels, they can target specific individuals. They now know what actions they can take, what stimuli to provide and when, to alter people's brains and behaviors without the victim's awareness. It is a form of subliminal behavior modification and an invasion of privacy.

To move closer to certainty for the advertisers, Big Tech companies might make a pitch to an individual user at *the particular moment* when they are in *a particular location* engaged in *a particular activity* with *a particular physiological state* and *a particular emotional state* while under *a particular set of social pressures* and with *a history of past purchases* and *certain ambitions and fears*, and *a recent promotion* and *a receptive facial expression*. The company's algorithms could run statistical regressions in real-time on all these factors to calculate that a particular intervention has a high probability of success. This near-certainty allows the company to claim a larger share of the profits.[74]

The scope of the collection of surplus behavioral data has extended from the virtual world into the "real" world full of "smart" devices where we "actually live our actual lives." Zuboff writes:

> These new disciplines drive surveillance [technology] far into the intimate reaches of our daily lives and deep into our personalities and our emotions. Ultimately, they compel the development of highly inventive but resolutely secret new means to interrupt and modify our behavior for the sake of surveillance revenues. These operations challenge our

> elemental *right to the future tense*, which is the right to act free of the influence of illegitimate forces that operate outside our awareness to influence, modify, and condition our behavior. We grow numb to these incursions and the ways in which they deform our lives. We succumb to the drumbeat of inevitability, but nothing here is inevitable. Astonishment is lost but can found again.[75]

Imagine a future when Google runs entire cities. As a practical matter, it is tempting for governments to outsource "running a city and providing municipal services" to the tech firms in Silicon Valley because of the government's frequent disfunction. No doubt, the private companies could do it more effectively and efficiently. The problem is the loss of power over the institutions we have to live inside. The issue is "sovereignty."

At stake is whether a safe, private space continues to exist where you as a sovereign individual are allowed to use your mind to form your own conclusions and make your own choices. The alternative is for you to merge yourself into a hive of "others" and be scientifically and subconsciously manipulated by the high priests of technology as central planners who know what is best for you.

THE LOSS OF SELF-RELIANCE

In an earlier chapter, I pointed out that being mentally involved with our physical navigation and locomotion is necessary to awaken our evolved mental capacities. It is fundamental to our human identity. The need to "do things ourselves" rather than have them done for us encompasses much more than navigation and locomotion. It has profound implications for the individual and society.

Since the days of the "snake oil" salesmen, we have turned over our health and safety responsibility to multiple government agencies, such as the FDA, DOT, and FAA. In doing so, we eliminated the possibility of private sector companies doing impartial research and educating us with facts of reality as to the relative risks of the options available to us to make our own choices.

In the private sector, the *reputation* of such "health and safety" companies would have meant everything for their financial success. The information these competing companies would have provided us, using the best available technology, would have allowed us to retain self-reliance for our health and safety. Instead, we completely turned-off our mental engagement and permitted government agencies to take full responsibility. In doing so, we gave up the quality, speed, and accountability that the private sector would have provided. Ironically, we seriously injured ourselves. We gave up our self-reliance.

Today, we are further relinquishing self-reliance to *automation*—to technology provided and controlled by the partnership between the government and Big Tech. It seems we are trying to remove the human element from every human activity. People who put too much faith in their individual agency must be reeducated by the government and Big Tech. "Don't fight the inevitable." The technocrats intend to make everything idiotproof and do it by treating us like idiots. And we do feel like we are becoming dumber. If this risk reduction continues, we will create a society based on an unrealistically low estimation of human capabilities. Author Matthew Crawford wrote:

> Infantilization slips in under cover of democratic ideals. I will insist, on the contrary, that democracy remains viable only if we are willing to extend to one another a presumption of individual competence. This is what social trust is built on. Together, they are the minimal endowments for a free, responsible, fully awake people.[76]

The rise of the automobile in America transformed the nature of our cities. Behind that transformation was governmental and contracted "central planners." Urbanization sought "to optimize the city according to a plan hatched from on high, without a street-level understanding of what makes a place thrive."[77] In the 1920s, working-class streetcar riders were "taxed by city planners and politicians to make possible middle-class automobile use." In the 1930s, road building was a crucial part of FDR's jobs program under his New Deal fueled by a "New Nationalism" of state-directed investment. In the 1950s, the Eisenhower administration began construction of the Interstate Highway System. After winning the war, "the state enjoyed a level of prestige, credibility and *legitimacy* that is hard for us to imagine in 2020."

In planning and constructing the 41,000 miles of highways, government planners "scripted and supervised people's movements, choosing where they could alight and escape. To do all this they razed entire districts of homes and shops using eminent domain to condemn and take privately held land." It was not "the hidden hand" a free market that brought about the congestion and sprawl that came with overdependence on the automobile. "On the contrary, our overdependence on cars was engineered by the state, energized by the same faith in central planning and sincere (if shortsighted) devotion to public felicity that has always been the pride of progressive governance."[78]

Today, central planning often comes not from the state. Crawford writes that it comes from tech companies that "aspire to quasi-governmental 'platform' status. [...] Given how much our daily lives are nudged and steered into channels engineered by tech firms, one can no longer sensibly adopt a conceptual demarcation between the 'private sector' and 'government.'"

Google is making a massive investment in automobiles as another domain of surveillance and profit. It wants to auction off your attention during your precious 52 average minutes of commute time to the highest bidder.

Furthermore, Google and other tech companies are planning the cities of the future. "What is being proposed, as near as one can make it out through the fog of promotional language, is an 'urban operating system' of mobility that would be owned by a cartel of IT companies, participation in which would not be optional in any meaningful sense."[79]

Increased safety and the relief of traffic congestion would undoubtedly be good things. And, we would be free to use the 52-minute commute for other mental activities. However, suppose you dare to push back against any aspect of the constant drumbeat of "inevitability" of the future being planned by these tech companies. In that case, you will wear the label of a Luddite who clings to the nostalgic past. Crawford points out the potential danger:

> Perverse consequences follow from trying to disburden us of exercising judgment. [...] Infrastructure predicated on too rigid an ideal of control fails to accommodate the exercise of our human capacities, or to exploit the social efficiencies they offer, leading instead to the atrophy of the human.[80]

It was the "environment" that our ancestors directly interacted with and caused the genetic selections that we inherited for millennia. Those genetics now cause everyone to be born with capacities for various forms and levels of intelligence. Just like the locomotion example, individuals must directly rub up against reality in our environment to "activate" those evolved capacities. If our environment does not provide the richness of activating stimuli during our development in the same way that the environment shaped our species, we will not achieve our full intelligence potential.

Experiments on both rats and people have shown that "effort-driven rewards," particularly "movement—and especially hand movements that lead to desired outcomes—plays a key role in both preventing the onset of and building resilience against depression and other emotional disorders." It appears that "hand movements that our ancestors needed for survival—those necessary for nurturing, cleaning, cooking, grooming, building shelter and farming"—might be essential to prevent anxiety and depression because of "all the complexity of movement and thought processes such tasks require of us."[81]

To fully develop our intelligence potential and remain psychologically healthy, humans need to be able to "see a clear connection between effort and consequence—and that helps us feel in control of a challenging situation." We must remain aware of and control of the kind of environment in which we live, including "open problem spaces" that require the physical and mental engagement bequeathed to us by evolution.

Suppose we allow the central planners in the tech companies and government to fully automate our lives in the name of safety and the elimination of risk. In that case, as Crawford writes, we will be putting "ourselves in a Plexiglas enclosure in which all our most basic needs are met, we will have nobody but ourselves to blame if we begin to feel like standard lab rats in a massive laboratory of social engineering. We would be safer that way, no doubt. But remember, all rats die. Not every rat lives."[82]

FREEDOM OF INFORMATION RIGHTS VS INDIVIDUAL RIGHTS

The "foundation" on which the Founding Fathers built America was the principles of individual rights and capitalism. Individual rights derive from man's nature. He must be free to use his rational mind. Reason is the only method of survival evolution provided him. Capitalism is the only political system that protects individual rights. These principles set free human creativity and achievement and enabled America to become the world's dominant economy and superpower.

Today, we are experiencing a great conflict between *individual rights* and Big Tech's claim of *freedom of information rights* in a *free market*. Which do you think should win over the other? The answer is clear. *Individual rights* are the primary value. Capitalism, or the *free market*, is the political system that secures individual rights in society.

We think of a "monopoly" as a loss of choice—when only one company is available. In principle, a monopoly cannot exist in a free market. Only government intervention that somehow bars competition can create a monopoly. A new disruptive technology or innovative startup business might upend even the largest, most dominant company in a free market.

Patents protect an individual's right to own and derive benefits from an innovative idea his mind produces. Patents defend individual rights. They do not create monopolies.

There can be long periods where an innovative company grows so large that it completely dominates a market. It is offering something that everyone wants. As the only choice available, it can appear to be a monopoly. However, so long as the government does not somehow constrain competition, just because a company *dominates* the market does not make it a *monopoly*. During this period of dominance, we might have little choice. Still, we must remember that there was *no choice* before the company introduced its innovative solution to a previously unaddressed market need.

A successful company gains *economic* power. Economic power gives it leverage over its competitors. It can sell at prices lower than their prices. It can corner raw material supplies driving up competitors' costs. The company gains market share by putting competitors out of business or buying them in an industry roll-up. When a consumer prefers to shop at the dominant retailer rather than the local mom-and-pop store he transfers economic power to the larger company.

Competition in a free market is how companies become *national* and then *global*. It is how they become powerful enough to provide automobiles, televisions, and smartphones to the working class at affordable prices or build spaceships. They create increased wealth in the form of products and services for the consumer. Even though they put the smaller competitors out of business, their economies of scale multiply society's total wealth. This is *economic power*, not a monopoly, and it is moral.

Wherever there is a large market with consumers who wish they had more choice, other innovators will eventually develop a less costly or better solution. The free market supports and encourages creativity and does not allow permanent market dominance. In many cases with technology, a disruptive innovation arises to challenge the dominant company long before its patent protections expire. This threat drives the dominant company to hold onto its dominance through continuous innovation. As the dominant company, it usually has more significant financial and knowledge resources and can attract the brightest developers. This advantage can extend its dominance for a time. However, having grown so large, the dominant company can become less responsive to the market, thereby providing an opening for a cadre of competing creators and their investors.

In a free market, a private business may not seek protection by the government from competitive threats. Having such protection is called *political power* as opposed to *economic power*. If a company seeks political power, it ceases to be a private, free-market company. It becomes a monopoly even though private investors own it.

City governments frequently create monopolies when they grant businesses exclusive contracts or "licenses." For example, when they license a particular "utility" company to serve their city instead of allowing each citizen to choose. Such a protected status and guaranteed profit for the privately-owned utility company comes at the expense of being "regulated" by the government. The consumers lose all choice in the matter. The regulators arbitrarily set the pricing the consumers must pay and the quality of service they receive instead of free-market competition.

Innovators founded the Big Tech companies that provided globally popular solutions addressing "information age" needs. Each company achieved dominance in its market niche. The Big Tech companies did not set out to erode our privacy and sovereignty, but over time their adoption of the strategy of extracting and using our surplus behavioral data for profit, and then their absolute claim on the right to do so and to expand it into every domain possible, eventually put their *freedom of information rights* on a collision course with our *individual rights*.

If users of today's Big Tech solutions became aware of the extent of private data capture and behavior modification, they would likely seek alternative solutions where individual rights of privacy and sovereignty are featured. The dominant companies today go to great lengths to hide their inner workings from the users. As public awareness increases, we begin to see calls for regulating tech companies. But on what basis?

Today's Big Tech does not seek government protection from *competition*. It seeks protection of its claimed *freedom of information rights* to behavioral data capture over users' *individual rights* of privacy.

The question is, was their corporate dominance built upon deception and at the expense of individual rights? If the government enforced an individual's right to privacy as superior to data capture market economics, as in the earlier European court example, today's Big Tech business model based on surveillance would collapse. They will fight against it at every turn.

Big Tech in America lobbies the government to protect their *right to know* and prevent regulations enforcing an *individual's right to privacy.* Ultimately, because such a company is seeking government protection for the right to continue to violate *individual rights,* it is seeking *political power*. It is exhibiting monopolistic behavior.

Remember, since 9-11, some Big Tech companies entered a partnership with the government to provide services including surveillance on citizens, behavioral data capture, and consulting services for such technology. Our society is going to need to decide if there is still a distinction between the *public* and *private* sectors in the information age, and if the government and Big Tech's claim on the *right to know* supersedes our *individual rights to privacy and sovereignty,* and what constitutes force, coercion, and deception, in the digital rather than the physical world? They might be surveilling us, but no thumbscrews are being applied. Why worry about being known? Crawford writes:

> What one has to grasp is that the totalitarian regimes of the twentieth century were rank amateurs, having to rely on such crude measures as violence. In the current version, the whole landscape of human behavior becomes subject to nudging and herding into profitable channels, and ideally this occurs beneath the threshold of awareness.[83]

Crawford writes that today, "what makes a device 'tech' is that it serves as a portal to bureaucracy. You cannot use them without involving yourself with large organizations, each with a quasi monopoly in its domain." He continues, "if the 'exit costs' are high enough, you don't need an actual monopoly to get monopoly behavior. And the more interconnected everything is, the more 'seamlessly' it all works together, the higher the cost of wrenching yourself free of the whole apparatus, if doing so is even an option."

The legitimacy of our democratic institutions depends on their ability to rationally explain and support an argument, such as when a court writes an opinion stating facts of law and evidence. Modern "black box" AI technology often cannot be articulated, interrogated, and defended by its creators.

Ultimately, these companies must lose legitimacy due to the lack of transparency, reasonableness, and argument. "It will surely be claimed on some other basis that bears a striking resemblance to priestly divination: the inscrutable arcana of data science, by which a new clerisy peers into a hidden layer of reality that is revealed only by a self-taught AI program—the logic of which is beyond human knowing."[84]

Google expects us to fully integrate their unobtrusive, pervasive, and beneficial services into our lives until the boundary between self and Google blurs. The company will guide our intentions by providing us the mental framework and informational context. "This is to take the idea of trusteeship and install it in the infrastructure of thought."[85]

THE POWER OF THE INDIVIDUAL

Almost every generation thinks its contemporary society is headed for ruin. It is easy to dismiss warnings of a collapse of our American society as just another instance of similar warnings of the past 100 years that never came to pass, for example, the fear of communism in the 1950s and 1960s. However, the world is different this time because technology developed in the last twenty years is unprecedented and is now the great amplifier and accelerator of the consequences that ideas have on entire societies. As Tristan Harris put it, "We in the tech industry have created the tools to destabilize and erode the fabric of society, in every country all at once everywhere."[86] And this technology is now advancing on the explosive, vertical part of the exponential growth curve.

We must reassert our *individual rights* over Big Tech and the government's claims of the *right to know*. There is hope. The 2016 and 2020 presidential elections caused an elevation in the public concern regarding Big Tech's power to manipulate our communications and behaviors. Some jurisdictions in America are, for the first time, pursuing *User's Rights* initiatives. Alternative tech platforms that use end-to-end encryption, block hidden surveillance apps, and emphasize individual user rights to privacy, are emerging and gaining traction. A new form of "Internet" designed not just for *information* like the present Internet, but for *private and secure transactions* is emerging. (More on this later.) Now it is up to individuals to support these initiatives.

You are stepping on the tightrope toward the future along with the rest of us. It all begins with your "Worldview." Think of your Worldview as the information content held within and used by your wet computer—your brain. It works just like the old computer "GIGO" saying—"Garbage In, Garbage Out." For your wet computer to give you quality output to guide your life decisions and actions, you need to provide quality information input.

You got much of what you have internalized during your lifetime from other people. More and more of it is obtained via the Internet and Big Tech, manipulated by AI algorithms programmed by unknown other people. Any of it could be *true* to reality, or it could be *false*, an intentional or algorithmically derived deception. You do not have to fall victim to incorrect information. Your conscious thinking can, indeed it must identify and correct invalid ideas.

Each of the *True Ideas* in this book is an opportunity for you to ask yourself, "Do I agree that this statement is true to the Laws of Reality and the nature of man? Is it consistent with everything else I have accepted as true?" An integrated hierarchy of ideas that you have validated against reality is the quality input that your mind needs to reject falsehoods and guide your life.

Whether you have a role in applying technology or are a user of technology, your own decisions and actions will be your Worldview made visible to others and will influence them in countless ways. As an individual with a reality-based Worldview, you will require that other individuals in every institution of society, public and private, also demonstrate a reality-based Worldview. The more this happens, the more likely AI systems will amplify beneficial, not destructive, ideas. One individual at a time is the only possible way to solve society's "problem." As Tristan Harris said, "If we don't agree on what is true, or that there is such a thing as truth, we're toast."

♦ ♦ ♦

PART 3

LEARNING FROM OUR PAST

The universe, our world, and even humans have been evolving for a long time. Specific processes based on Laws of Reality were at work, causing changes to that which existed at any point in time to give rise to something different at a later point in time. Can we identify the critical processes of change—physical, biological, and social, that have caused us to be where we are today? If these mechanisms have been at work for millennia, isn't it likely that they are at work right now and will be in the future? An awareness of these processes is absolutely essential for understanding what we see around us today and predicting the changes we might expect in the years ahead.

We have bodies and brains composed of matter and energy according to laws of physics that caused an emergent faculty of consciousness. This consciousness allows us to contemplate our existence within the context of reality. Our consciousness is not omniscient. There is much about the nature of existence that we do not (yet) know. Our consciousness can observe that we exist in a real world within a real universe, made of matter and energy and behaving according to discoverable laws. We can observe that we have a specific "nature" as a species with our rational faculty as our defining characteristic. Discovery and intentional application of the Laws of Reality is our improved method of adaptation and survival.

In this chapter, we explore what we already know about the critical processes explaining our existence and driving our future. Our intelligence is continually increasing. There is no reason to believe that it is not within our conceptual power to eventually discover and understand the presently hidden knowledge of how the universe works.

EVOLUTION

"The hardest thing to explain is the glaringly evident which everybody has decided not to see."

—AYN RAND

ENTROPY

Scientists believe the physical driving force behind evolution is a thermodynamic principle called "entropy." Entropy is essentially the tendency for energy to flow from locations of higher potential to areas of lower potential. In a closed system with no energy coming into or escaping from the system, energy within the system will try to achieve a state of equilibrium where all locations have the same energy level. Entropy *increases* when energy flows "downhill" and is at its maximum when equilibrium is reached.

One can also substitute the word "complexity" for energy. In nature, it takes energy to make and sustain complex things, such as your body. Without a continuous energy supply, complicated things decay and become simple again, releasing the energy contained in them—another form of equilibrium.

To visualize the principle of entropy, think of the earth's water cycle. Assume that suddenly there are no more tides, water no longer evaporates into the atmosphere, and it never rains. All the water present in lakes, ponds, rivers, streams is all there is ever going to be. Because these bodies of water are above sea level, they hold potential energy due to gravity. That water can flow downhill through turbines to generate electricity. However, all the bodies of water with power-generating dams will eventually run out of water, and all that water will make its way to sea level.

There is still potential energy in the bodies of water that do not have power-generating dams. This is where technology and localized complexity comes in. Humans can apply technology to build pipelines that connect all those bodies of water to power generators at lower elevations. This would require increasing complexity and decreasing entropy in each location by building a pipeline. However, it would serve to unlock the potential energy in each body of water.

Even though we decreased entropy in each locality, the net result decreased the total potential energy in elevated water on earth—and increased earth's *overall* entropy. Eventually, *all* the water on earth would reach the sea, achieving equilibrium. By increasing local complexity, we enabled the flow of otherwise dormant energy, increasing total entropy. Once there is no longer any energy available to maintain the complexity of the pipelines and generators, they will disintegrate, resulting in total equilibrium in the closed system and maximum entropy.

Evolution works the same way. It increases complexity in localized situations, such as the emergence of life forms and their ability to release energy. Evolution does this because it increases entropy. Those life forms need and use energy. They unlock and speed up the overall energy flow in the universe toward equilibrium.

THE BIG PICTURE

Most scientists believe that an instantaneous "Big Bang" event 13.8 billion years ago created the universe's present form. Things have been changing and evolving one way or another ever since that moment. We think of evolution in terms of Darwin and his theory of natural selection. But that only applies to biological evolution. The topic is much larger. There is confusion about what is driving evolution and what kind of evolution follows biological evolution—is it *societal* evolution, *technological* evolution, or something else? In this chapter, we identify the past, present, and future of evolution.

I divide the entirety of evolution from the Big Bang to the present time into the following three periods:

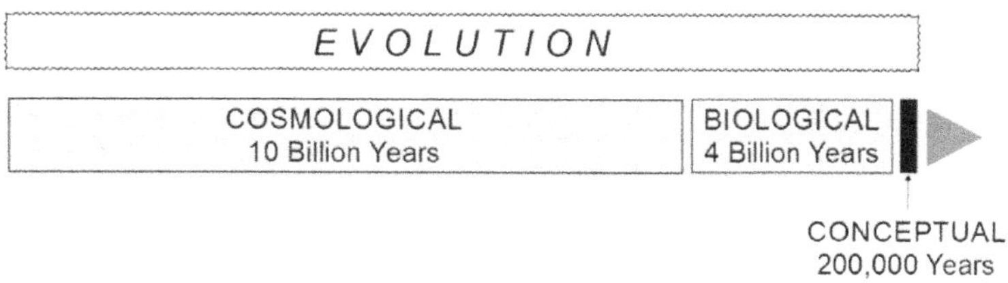

Evolution's Periods

Cosmological Evolution Period - From the Big Bang to the appearance of life on Earth—about 10 billion years. This period is when the laws of physics caused increasing complexity of matter until the first simple life forms appeared.

Biological Evolution Period - From the appearance of the first life on Earth to the appearance of anatomically modern humans—about four billion years. This period is when natural selection caused increasing levels of intelligence in living entities until they became capable of conceptually discovering and applying nature's laws.

Conceptual Evolution Period - From the appearance of anatomically modern and conceptually capable humans to today—about 200,000 years. This is the period we are in now. In this period, the *Technology Cycle* (explained later) is causing knowledge of the Laws of Reality, purposeful modification of matter, and human intelligence to increase.

We biology-based conceptual humans have been the *Technology Cycle's* agents. We applied it through the Stone Age, Iron Age, Industrial Age, and Information Age. Now, the exponentially advancing transcendent technologies are about to enable humans to evolve from being biology-based conceptual entities to being technology-based conceptual entities. We are now entering the *Transcendence Era,* during which technology-based conceptual entities take over as the *Technology Cycle's* agents.

Up until now, humans have been the only known conceptual-level entities that can act as agents of the *Technology Cycle.* Suppose we succeed in creating artificially intelligent conceptual entities, or we discover others in the universe. In that case, they could be agents of the *Technology Cycle* as well. This Conceptual Evolution Period will continue until a Post-Conceptual Evolution Period begins, or the entire universe becomes intelligent.

Post-Conceptual Evolution Period - Unknown. If there is one, this period would be about whatever evolutionary process goes beyond the evolution of conceptual knowledge of the universe.

Period of Evolution	Causal Process	Acts On	Increases	Until
Cosmological (10 billion years)	Laws of Physics	Inanimate matter	Complexity of matter	Earliest life forms appeared
Biological (4 billion years)	Natural Selection	Biological entities	Levels of intelligence	Conceptual level intelligence appeared
Conceptual (200,000 years+)	Technology Cycle	Conceptual-level entities (biological or technological)	Knowledge of universe, modification of matter, intelligence	Post-conceptual period or mastery of the universe

Evolution's Periods

A period of evolution does not end and continues to function when the next period begins. The causal factor in the new period drives evolution with much greater speed overshadowing the slower speed of the prior period and decreasing its significance.

There appears to be a chain of transcendence with each major period of evolution transcending and including its predecessor. McIntosh (2012) said: "Once we recognize that the transcendent emergence of a new category of evolution represents a kind of success for the underlying evolutionary domain out of which it arose, we can see how such an emergent new category comes to function, at least partially, as a goal or end for its underlying domains."[87]

The evidence is clear. Evolution is happening. We do not understand every detail of the history of it. Still, the physical evidence that it is occurring is unmistakable. Evidence ranges from the earliest sub-atomic particles to the fossils, the human brain's layered structure, a fetus's rapid re-creation of biological evolution in the womb, and the recent genetic analysis of how ancient DNA evolved throughout our ancestry to become modern DNA.

Evolution seems to be a "thing" in nature that is much larger than we humans. While a few scientists and philosophers think there is a "purpose" to evolution, most do not. Even though it is challenging to explain progress without purpose, that is what most believe. They think the process of evolution exists but has no purpose. They think it is the combination of Laws of Physics and random chance.[88]

Evolution does not occur in a specific direction but *all* directions. The Laws of Reality and random chance determine which directions succeed and which do not. However, the historical record indicates that the result of evolution to-date, even with all the dead-end branches and extinctions, has been a progression toward increased order and complexity.

Emergent self-organization is nature's way of dissipating and re-balancing excess energy in its efforts to attain equilibrium. Emergent complexity and intelligence appear to be a response to causal entropic principles and, therefore, an inherent property of the universe.[89] Simply put, it appears to be a universal, physical Law of Reality that when local conditions are right, increased complexity will automatically arise.

I agree that evolution must be happening in a manner consistent with the properties of energy, matter, time, and space, or else it would not be happening. The scientists can argue their various theories of evolution's causes and purposes. Our direct observation that it is happening, has been for a long time, and is likely to continue enables us to make sense of everything around us and create an informed prediction of the future.

The good news about having Laws of Reality and evolution that "just are," without purpose, is that it means that nature has no intentions toward us, good or bad. Life is not a struggle against a malevolent universe. Any living entity that lives and acts according to its nature in an environment suitable for its survival can expect to live a successful life. For man, that means to use our mind to discover what we need to survive and then modify our environment accordingly. Barring accidents or natural disasters that happen in nature with no malice toward us or interference by others, we can expect to be successful.

THE YELLOW BUS

Your Worldview will be the things you believe are true, which means "real." Your Worldview will be the content of ideas used by your brain to guide your life. To help bring the next "true" observation about reality into focus, consider the following story:

You are out for a walk on a rural country road. There is little traffic, so you walk in the roadway, enjoying the beautiful scenery as you go around a sharp bend in the road. Your cell phone "dings," so you stop and pull it out and begin reading the lengthy text message. At this point, you are focused on your smartphone unaware that a yellow bus is barreling down the road behind you and coming around the curve. The driver on the bus is distracted by some unruly kids.

So, here is a question—does reality physically exist independent of you? Aristotle said it does. Plato and other philosophers, from ancient Greece through to today, say it does not. Some say it is just created in your mind or some other consciousness's mind.

Here is another question—are your senses and consciousness capable of grasping reality and guiding your actions for survival? Aristotle says they are. Plato says you cannot grasp actual reality, only a "virtual" one, a supernatural dimension of essences. Eighteenth-century philosophers, such as Immanuel Kant and his followers, said you could not be sure of *anything*. They said your mind subjectively creates your ideas independent of external facts. Plato and Kant's viewpoints sever man's tool of cognition, his conceptual faculty, from reality, undercutting his mind.

Suppose your subconscious mind detects the approaching vehicle's sound pattern and causes you to turn and look. And your conscious mind recognizes the entity and concept of a "bus" and recalls its nature and the nature of your body, and then projects the probable outcome if the bus hits you. And you step off the roadway. Then Aristotle was right.

On the other hand, if the bus does not really exist, or if the reality in your mind is a kind of virtual reality rather than an "unknowable" real one, then you have nothing to worry about—right? Or, if the bus does exist but you cannot trust your senses and mind, then you would have no reason to move—right?

If you, deep down inside, believe that reality is not real and does not exist independent of you, or you cannot trust your senses and mind, then stand there and see what happens. One thing you can know for sure—in a few seconds, your senses WILL grasp the actual reality of the bus, and then a few seconds later, YOU will not exist in that reality!

For centuries, philosophers have proposed anti-reality or anti-reason ideas to explain our existence and our functioning in this world. Phrases you hear daily in our society are evidence of their influence. Phrases such as—"You can't know anything for sure." and "Follow your heart, not your head." We need to decide whether we believe ideas promoted by anti-reality, anti-reason philosophers or we believe Aristotle.

We will see the yellow bus again later.

PRIMACY OF EXISTENCE

"What is out there?"

I believe in reality and evolution. What about people who do not? Right here, at the beginning of our journey, you see that *ideas matter*. The ideas that a person holds to be *true*, their Worldview, determines their life.

Your Worldview is a mixture of ideas that you hold both consciously and sub-consciously. You deliberately examined some of these ideas and decided to keep them. Many ideas, you got from other people from childhood on, and you have accepted them all these years without examination. Most adults are surprised to learn that they still hold many ideas in contradiction—ideas that cannot both be true. Conflicting ideas are possible because you internalized some of them without conscious examination.

All the technology you see in the world is the product of the *human* mind. Understanding the workings of one's mind and its content, one's Worldview, is the only way a person can hope to understand where technology comes from, where it is going, how it could impact society, and how one can achieve happiness in a technological world.

In the chapters ahead, you will gain an understanding of how your human intelligence works. You will learn how to identify which of the ideas you hold are "true" and which ones, on closer examination, you might want to change. This *thinking* skill gets at the heart of what we mean by human intelligence. We must understand what a complete, integrated set of ideas, validated against reality and necessary to guide a flourishing human life, might look like—your Worldview.

It all begins with this fundamental question, does existence exist? Another way to pose the question is—do you believe that "existence" is primary, or "consciousness" is primary?

Suppose you believe in reality and the efficacy of the human mind and reason. In that case, you, like Aristotle, believe in the "Primacy of Existence" as the base of your Worldview. You think that the universe exists as an objective reality independent of your existence. You believe that reality is primary, having existed long before you or any other consciousness. You think that the reality you see around you existed before you were born and will continue to exist after you die. You believe that your consciousness must comply with the Laws of Reality and cannot change them. You think that reason, applied to the information about reality provided by your senses, is your only source of knowledge. Its use is required for you to survive.

Suppose you believe in the "Primacy of Consciousness" as the base of your Worldview. In that case, you might think that "reality" is only a phenomenon of some "consciousness" and that it does not exist apart from that consciousness. Or, like Plato, you might believe that *actual* reality exists but is not directly accessible or knowable to you and your mind lives in a kind of *virtual* reality. You further accept that there are no fundamental physical laws on which you can rely. Nothing is knowable with certainty because everything is subject to change by consciousness, either our consciousness or that of some other entity or god. You believe that emotions based on revelations or some other means of *just knowing* are your only means of comprehending that other reality and your only source of knowledge. Or you believe what other humans tell you who claim to be receiving revealed knowledge.

Anyone can assert anything. How does one know if it is *true*? You must choose from three fundamental options—knowing, feeling, or believing:

1. You ***know*** it is true because you can use *reason* to connect it through cause-and-effect back to facts of *reality*.
2. You ***feel*** it is true because only your *emotions*, not your mind, can be trusted to reveal the truth that is *already inside you* somehow.
3. You ***believe*** it is true because you have *faith* in others who tell you that truth comes to them from somewhere or something *outside of reality*.

This first choice you must make is "axiomatic," which means that one cannot dissect it any further. There is no other alternative to consider. It is the beginning. It either *is,* or it *is not*. Either existence exists, or it does not. Either your faculty of reason is to be trusted to guide your life, or it is not. You cannot have it both ways. You cannot use your faculty of reason to explain why reason is invalid.[90]

Though we use *reality, existence,* and *universe* interchangeably, we can distinguish between *existence* and the *knowable universe*. Scientists think the presently knowable universe had a beginning, the Big Bang. It might be infinitely large, but we can only see and might only ever see a finite amount of it. This is because the farther reaches are receding from us faster than the speed of light—we might never know its whole dimension.

Physicists project that the knowable universe could have an end when it eventually runs out of energy and winds down due to entropy. If the universe began with a Big Bang and then ends when it runs out of energy, that would be one "event." Our knowable universe might only be one out of many such events that might have occurred or be occurring within all of existence. The presently knowable universe is one arrangement of elements within existence.

We are not here concerned with "how" the present form of the knowable universe came to be—what caused the Big Bang. Theologians attribute it to God. Scientists have several unproven theories. Some scientists believe the best explanation is the "inflation" theory, where only an ocean of fluctuating energy existed before the Big Bang without any matter. Like wave peaks and troughs in an ocean, a random combination of energy waves developed a trough with energy levels so low that matter could come into existence. The Big Bang resulted, which created our universe and the matter in it.

The inflation theory predicts this is happening all the time, creating countless other universes in addition to ours.[91] Scientific measurements of the uniform distribution throughout the universe of the residual light left over from the Big Bang support the Inflation Theory.

If true, the inflation theory answers the question, "How were we so lucky that all the physical laws and constants in our universe are precisely what was needed for life to exist?" Just like a lottery with millions of tickets printed, the creation of countless universes means that one of them had to be a winner. It was inevitable.

The classic laws of physics that apply in the "macro-scale" world of matter do not *appear* to apply to the "nanoscale" world of sub-atomic particles and waves. In the nanoscale world, the laws of "quantum mechanics" seem to rule. This causes some scientists to theorize that there are no such things as real existents in the quantum world. Again, their mathematical formulas imply to them that there could be an infinite number of universes. We will explore the quantum world further in a future chapter.

The macro-scale world of matter derives from the quantum world. While we cannot fully explain the quantum world, we can discover how it works and apply it to our purposes. Scientists have found that biology applies quantum principles in plants and animals, from photosynthesis to a robin's ability to sense the Earth's magnetic field during migrations. We have already used quantum principles to achieve practical solutions, such as semiconductors, lasers, and quantum computing.

The nanoscale quantum world's weirdness does not change our relationship to the macro-scale universe in which we evolved as a species. Our human intelligence was created by and operates according to the natural laws of this universe. The nature of how we grasp a thing in our mind is only our intelligence method, not the thing itself. No one knows how we might evolve this relationship between our method of grasping reality and reality itself with advancing knowledge, improved technology, and higher levels of intelligence. We must work with whatever we can apprehend as reality today using the intelligence we have today. Our lack of complete knowledge about the quantum world's workings is no reason to question actual existence.

As brilliant as today's physicists are, we will trust *our* observations insofar as we can. The observable fact to consider here is that regardless how it came into existence, a finite, physical universe with energy and matter and laws, is that in which we live. We are biological entities with brains made up of that matter and energy. We cannot violate the physical laws and continue to live. We support our lives by using the information about those laws provide by our senses to guide our actions. To deny that reality exists is to deny that we exist.

We do have imagination. We can imagine all sorts of things that do not exist in our reality, such as unicorns or infinity. Infinity does not exist in our finite universe. We can imagine time and distances without beginning or end. However, except for using the concept of infinity in abstract mathematics equations, it, like unicorns, is not part of our reality. Like unicorns, it exists only in our mental content. Someday, our intelligence might be sufficiently advanced to understand what existed before the Big Bang and before time began. But that is not part of our reality today.

Each of us must commit to what we believe about existence. Do you believe in the *primacy of existence* or the *primacy of consciousness*? The choice you make will be the foundation that determines all the other ideas you will hold to be true. On that foundation, you can build your ideas about knowledge, ethics, politics, and art, which are the other branches of philosophy. All your ideas need to be consistent, avoiding conflicts. For example, you cannot honestly believe in one type of ethics and simultaneously accept politics based on the opposite kind of ethics.

Suppose you know which choice a person accepts as true about existence and knowledge. In that case, you might expect to be able to predict their ethics and politics. However, most people hold inconsistent and conflicting ideas at the various levels of their Worldview. These inconsistencies lead to lives buffeted by random external influences and conflicting emotions instead of lives stabilized by self-confidence and purpose. Internal conflict in one's Worldview degrades one's happiness.

The originators of the two opposing views of the nature of reality were Aristotle and his teacher, Plato. Aristotle said that the real reality is knowable, and Plato said it is not. The two opposing views, put into practice in societies, resulted in those civilizations' advancement or decline during those periods.

Let us identify and save the next *True Idea* for our Worldview. This one is the **beginning** of the logical chain of cause-and-effect ideas that we accept as true.

(First) True Idea: Reality is real.

As we go along, I will summarize the fundamental ideas that we accept as *true* in a table.

APPLIES TO:	QUESTION:	ANSWER:
Existence	*What is out there?*	*A "real" reality.*

One's belief regarding "existence" is the starting point. We will continue to build our set of *True Ideas.* You might be curious to see how your present ideas hold up. As we add each *True Idea*, ask yourself if *you* agree? If you find that you do not agree with it, ask yourself what it is about the *True Ideas* leading up to it that you cannot accept? See if you can explain, in writing, your chain of logic back to your view of reality.

Suppose you are not quite ready to commit that reality is real. In that case, you might want to read further to see what kind of ethics and politics depend on such commitment. The *True Ideas* at all levels of your Worldview will integrate without contradiction and reinforce your commitment to the starting point—*reality is real.*

Ideas grounded in reality will be necessary for the creation of human-level artificial intelligence in the future. They will also be essential to protect mankind from the dangers of our technology. Ethics and politics will be crucial to our consideration of artificial intelligence and other transcendent technologies. We will need to understand these from the human context to apply them to the machine context.

COSMOLOGICAL EVOLUTION

The Cosmological Evolution Period was the first period of evolution following the Big Bang. It increased the complexity of matter, beginning with energy, then to sub-atomic particles, to the lightest atoms, to the heavier atoms, to molecules, to stars and galaxies, to planets, to complex chemistry. The laws of physics and the increasing complexity of matter made the emergence of life possible.

When	What
In the first second	Elementary particles of matter are formed
At 10 seconds	The universe becomes a super-hot glowing fog of gas
At 3 minutes	Lithium, heavy hydrogen, and helium form from protons and neutrons
At 70,000 years	Gravitational collapse of matter begins
At 10 million years	Heavier elements and the chemistry that will later spark life begin
At 200-300 million years	First stars begin to shine
At 700 million years	Galaxies begin to form
At 9.2 billion years from Big Bang or 4.54 billion years ago	Our solar system and Earth-moon system are formed
At 4.41 billion years ago	The oceans are formed
At 4.28 billion years ago	Earliest discovered life - Fossilized microorganisms in hydrothermal vents

Cosmological Evolution Timeline[92]

For physicists, there is a great deal of research going on into the details of how the universe evolved. A lot happened both very quickly and over exceptionally long periods. For our purposes, we only need to observe that matter became more complex until the conditions developed that made possible the first self-replicating molecules and life's advent. The Biological and Conceptual Periods of evolution will be more applicable to our exploration of intelligence and technology.

Evolution appears to be acting in opposition to entropy. Evolution is increasing order and complexity in localized parts of the universe. While we only know of one place where it occurs (on Earth), it could be happening in countless other undiscovered places in the universe.[93]

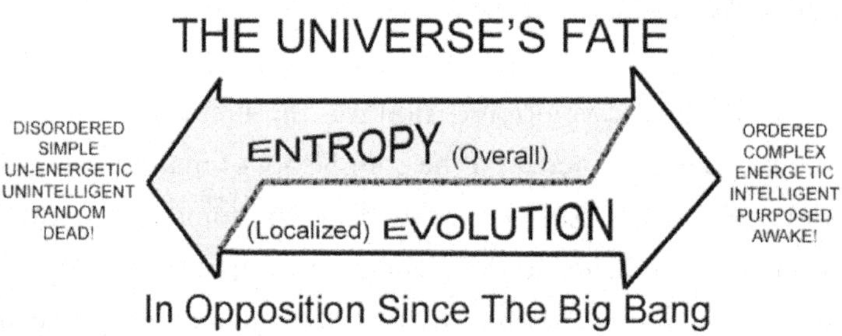

Through our technology, we humans are becoming increasingly able to build greater complexity and use greater energy amounts. We are on the thresholds of new emergent technologies that are orders of magnitude more complex and efficient at processing energy. We are already making plans to terraform other planets. Our long-range vision includes plans to harness the energy of entire solar systems and galaxies. Our plans will exponentially increase the amounts of energy we use for the rest of our existence.

What appears to be our localized "push back" on entropy instead serves to accelerate entropy overall. Will our striving to "wake up" the universe speed up its demise?

The laws of thermodynamics are the physical laws of the universe. When applied to the universe as a whole, the assumption is that the universe is a *closed* system concerning energy. Yet, there are theories of multiple universes and the possibility of portals between them. There are theories of many more dimensions in our universe than the obvious three of space and one of time. The point is that, just as we do not know if we can ever exceed the speed of light, we simply do not know with certainty that the universe is a closed system. We cannot be certain that entropy will eventually cause it to be dead and lifeless. We simply do not know enough to predict which of the following will be the universe's fate:

1. Entropy is the dominant phenomenon and evolution serves to speed up the eventual winding-down of the universe. (The predominant theory today.)
2. Entropy and evolution are opposing but equal phenomena in the universe. We will discover that a "steady state" or a "cyclical state" exists where they eternally balance each other in the overall universe or multiple universes.
3. Evolution is the dominant phenomenon and will eventually "awaken" the entire universe with intelligence.

Understanding your place in the universe begins with the awareness that you are a product of evolution's push-back on entropy, increasing order and intelligence in our pocket of the universe by creating you.[94] In this cosmic-scale tug-of-war, the significance of your individual life is infinitesimally small but not zero. It somehow helps to understand and appreciate your place and purpose within the cosmos. Then, simply let it go as one of those things you cannot change. There will be plenty that you can do in the chapters ahead.

If you sometimes feel that the vastness of the universe and the cold, physical laws diminish any meaning from life, reconsider. There certainly is meaning and purpose in the universe. A galaxy cannot contemplate itself. A self-replicating molecule derives no pleasure from a sunrise. A bacterium does not feel love for the millions of others around it. We do not notice how unique our experiences are. Unlike the rest of nature, *we* are what gives the universe meaning. Unless we discover others like us, it will be solely our duty to fill the universe with intelligence and witness its awakening.

BIOLOGICAL EVOLUTION

"There is only one fundamental alternative in the universe: existence or non-existence—and it pertains to a single class of entities: to living organisms. The

existence of inanimate matter is unconditional, the existence of life is not; it depends on a specific course of action. Matter is indestructible, it changes its forms, but it cannot cease to exist. It is only a living organism that faces a constant alternative: the issue of life or death."

—AYN RAND

The Cosmological Evolution Period ended, and the Biological Evolution Period began with the emergence of intelligent forms of matter—"life." During the Biological Period, the level of intelligence of certain evolutionary branches of biological entities increased. The Biological Period began with life forms that could only respond to Laws of Reality by reacting to momentary stimuli in their surroundings. The Biological Period culminates with the emergence of conceptual humans who can comprehend the Laws of Reality and apply them to modify their surroundings to themselves.

In the following table, notice how the time between milestones gets shorter as evolution progresses. Do you see the exponential curve at work?

When	What
4.0 billion years ago	Matter assembled into the first self-replicating molecules in life forms in the oceans
3.8 billion years ago	"Primitive life forms"—cells (archaea) and bacteria
3.5 billion years ago	Swimming cells and bacteria
3.0 billion years ago	Photosynthetic bacteria—they create oxygen in the Earth's atmosphere
2.8 billion years ago	Microbial life appears on land
1.8 billion years ago	Cells with organelles serving different functions, (eukaryotes) "complex" life forms
1.2 billion years ago	Sexual reproduction in single cell organisms appears and speeds up evolution
1.0 billion years ago	Plants appear on land. They further increase oxygen and carbon dioxide in the atmosphere
600 million years ago	Increasing oxygen develops the ozone layer that protects emerging land life from ultraviolet radiation
580 million years ago	Larger, multi-cellular animals of the modern phyla appear
530 million years ago	Earliest footprints on land

485 million years ago	Vertebrate fishes with true bones appear
360 million years ago	Carboniferous period begins. Earth resembles its present state. Insects on land, sharks in the ocean, seed-bearing plants and forests, four-limbed tetrapods on land
320 million years ago	Precursors to mammals separate from reptiles
60 million years ago	Earliest true primates
6.5 million years ago	Earliest larger primates (hominins)
2.5 million years ago	Stone tools use begins by hominins
2.0 million years ago	First of the homo genus (Homo Habilis)
350,000 years ago	Neandertals and Denisovans branch from the human lineage and interbreed with humans later
250,000 years ago	Anatomically modern humans appear
50,000 years ago	Behaviorally modern humans appear

Biological Evolution Timeline[95]

There are a few particularly notable observations from the preceding table. Notice that the first primitive life forms, archaea, and bacteria, appeared on Earth very soon after Earth's formation. They were the only life forms on Earth for two billion years. Scientists today generally believe that we will find these primitive life forms on many other habitable planets. The complex carbon-based organic molecules that gave rise to such life here on Earth came from the comets that struck Earth soon after its formation. These molecules were created and preserved in the comets at the beginning of the universe. Comets such as these would undoubtedly have struck many other planets in addition to Earth.

The big question is what happened on Earth two billion years after the primitive life forms first appeared. That is when the complex life forms, eukaryotes, appeared. These are cells with a nucleus and other complex organelles inside them. They are what makes all higher life forms on Earth including humans, possible. Scientists cannot yet explain this event. It appears that, after two billion years of sharing the planet essentially unchanged, a simple bacterium somehow came to live inside a simple archaea cell without being destroyed. A symbiotic relationship developed, and when the combined entity replicated, both parts replicated. The bacterium is now the mitochondria, the energy source in every plant and animal's eukaryotic cells.

Scientists cannot yet say whether this complex life form is likely to have occurred on any other planet. They believe the emergence of complex life forms might be an extremely rare occurrence and could even be a unique event in the entire universe. We just do not know. In any case, biological evolution is a natural phenomenon of physics, matter, and energy. Oxford biological physicist Sonya Contera (2019) describes and sums it up this way:

> [Biological] Evolution is the universe's mechanism for fine-tuning the physics and chemistry of nanoscale molecular building blocks to create increasingly refined living structures. The physics of the universe ensures that biological complexity naturally increases in molecules that dissipate energy into their environment. While evolution computes the form and activity of the organisms that are able to succeed in that environment. In this context, by "computation" I mean that life embeds in itself the capacity to "compute" organisms that are able to survive in a complexly changing environment (as 'living solution' to the physics of life), since the organisms that don't succeed eventually disappear ('unviable solutions to the equation of life').[96]

The neocortex enabled perceptual consciousness in mammals. It gave them evolutionary dominance following the great extinction event 65 million years ago that eliminated animals not so equipped. This adaptation tool was so successful that the neocortex continued to evolve until it comprised the bulk of the brain of Homo sapiens.

At the end of the Biological Evolution Period, we see humans' appearance with conceptual-level intelligence. In the next period, the Conceptual Evolution Period, it will be important that humans explore and modify their environment and manipulate the matter around them. These activities are made possible by walking and opposable thumbs. These features are usually associated with Homo habilis, which first appeared about two million years ago, long before the anatomically modern humans 250,000 thousand years ago.[97]

Now that we have explored Cosmological and Biological Evolution Periods, let us save another *True Idea*. We accept that reality is real. Now let us capture an observation about this reality in which we live. Do you agree?

> ***True Idea: Things that exist in reality (including us) have specific natures and identities.***

CONCEPTUAL EVOLUTION

"The brain's hierarchical, complex organization, and its connections with the rest of the body all the way down to our cells, proteins, and DNA, make us both eager and able to explore the world around us. This kind of hierarchical assembly is the way evolution has consistently embedded in itself the capacity to learn and adapt, at the molecular, cellular, and organism levels. Bacteria, plants, animals, and humans explore, adapt, and learn because we are built by the rules that govern the universe."

—SONIA CANTERA

"Man's senses are his only direct cognitive contact with reality and, therefore, his only source of information. Without sensory evidence, there can be no concepts; without concepts, there can be no language; without language, there can be no knowledge and no science."

—AYN RAND

The Conceptual Evolution Period began when an anatomically modern human performed the first mental integration of several perceived items into a single concept and represented it by a word that he could speak.

The essential, defining characteristic of the Cosmological Period was the long-term and massive interplay between physical matter and energy until the conditions existed supporting the emergence of intelligent matter—"life." The essential, defining characteristic of the Biological Period was the increase in intelligence levels until the ability to form concepts was achieved in one of the species—Homo sapiens—the "rational animal."

Dr. John Medina, a developmental molecular biologist at the University of Washington School of Medicine, describes human conceptual emergence this way:

> We gave up on stability. We began not to care about consistency within a given habitat, because consistency wasn't an option. We adapted to variation itself. Those unable to rapidly solve new problems or learn from

> mistakes didn't survive long enough to pass on their genes. The net effect of this evolution was that rather than becoming stronger, we became smarter. It was a brilliant strategy. We went on to conquer other ecological niches in Africa. Then we took over the world.[98]

"Reason" is the essential in the entire discussion of human-level consciousness. It was the appearance of man's conceptual faculty that powerfully separated man from all the other animals that lacked an abstract intellect. That was the big event separating all of history into "before reason" and "after reason." The emergence of a rational entity on Earth marked the beginning of *technological* evolution that would eventually overwhelm *biological* evolution. "Technology" only became possible after the emergence of reason.

It appears that reason and human society emerged together. The need for individuals to communicate the concepts they learned gave rise to language, which was both a requirement and a consequence of evolving our rational faculty. From the first concept formed in the mind of a human forward, the quantity of *conceptual knowledge* discovered by all humans and accumulated in society has been increasing. The depth and breadth of our understanding of the Laws of Reality as new knowledge have increased. So has the scope and power of our modification of the matter around us for our purposes.

The essential, defining characteristic of the current Conceptual Evolution Period is not the genetic evolution of humans or the evolution of society or the evolution of technology. *It is the accumulation of conceptual knowledge of the Laws of Reality within a society of individuals applying reason.* That accumulating body of knowledge makes it possible for individuals in that society to increase technology, increase intelligence, and increase the matter put to purposeful use. The Conceptual Evolution Period is evolution's solution for further increasing matter's complexity to accelerate entropy—evolution's driving force.

The Cosmological Evolution Period lasted ten *billion* years and the Biological Period lasted four *billion* years. The Conceptual Period has only been going now for 200 *thousand* years. It is in its infancy on the evolutionary time scale. Evolution is just beginning its trial-and-error experimentation with our species to see if a society comprised mostly of *consistently* rational individuals can emerge. An irrational society will be selected against by evolution because it cannot significantly increase complexity and entropy. The Homo sapiens lineage, while having made some progress, has not achieved consistent rationality.

A human's conceptual consciousness—his rational faculty—is the fundamental driver of conceptual evolution. But only insofar as he and the others in society have learned how to use it and consistently apply it. A new human must be taught by other individuals in society how to use his rational faculty. The new human then uses his rational faculty to learn the knowledge already accumulated in society. The now educated human then uses his rational faculty to discover new knowledge. Other individuals in society use their rational faculties to collect and distribute the new knowledge to those who use their rational faculties to apply it for their purposes.

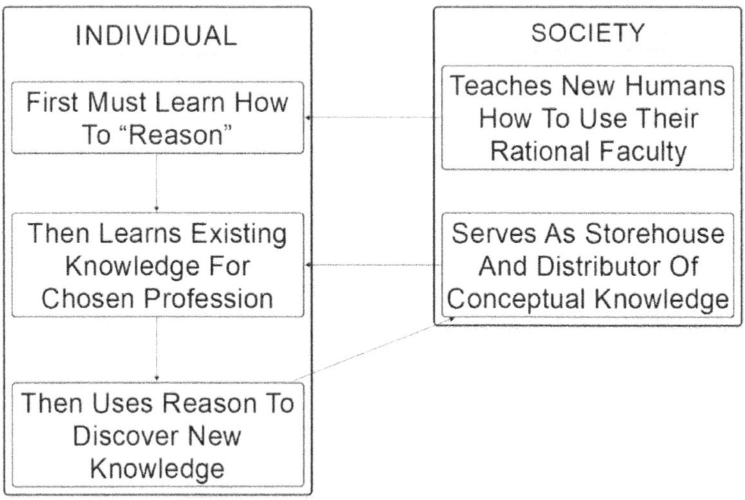

Conceptual Evolution's Roles

We must clearly understand that "society" is not a *thing* that exists in reality. It is a collective term like "group." "Society" means the individuals who live and can interact in a time and place. Society has no characteristics beyond the statistics of the individuals comprising it. Still, we can say that the first function of (the population of individuals in) society is to teach new humans how to think. The second function is to be the storehouse and distributor of discovered knowledge.

It remains to be seen whether Homo sapiens as the present agents of conceptual evolution will keep it going or will reach a dead-end. If humans fail, the future of conceptual evolution will be up to some other lineage or species some millennia in the future. If we can keep it going, then conceptual evolution is poised to transcend biological humans and human society. It has taken humans two hundred thousand years to advance conceptual knowledge along the slow, horizontal portion of the technology growth curve from man's first formed concept to today. We have reached the sharp bend in the curve toward the vertical. We are now entering the period of explosive growth in knowledge, technology, intelligence, and our ability to modify our environment, including ourselves.

We are entering the *Transcendence Era* of the Conceptual Evolution Period. The conceptually intelligent entities that will discover new Laws of Reality as we advance are about to change from biological humans to hybrid humans. We will merge our bodies with our artificially intelligent technology. Eventually, we might leave our biology behind as we and our society become "virtual." Or we might create conceptual AI entities as agents that can advance conceptual evolution using intelligence superior to ours.

We will examine how our human conceptual consciousness functions, how it guides a person's life, how it contributes to society, and how it is the source of all technology. We will identify the mechanisms that are driving the process of conceptual evolution. We cannot consider the future of society or technology without a foundational understanding of human consciousness.

LEVELS OF CONSCIOUSNESS

During the Biological Evolution Period, living entities pursued self-replication with incremental increases of complexity over generations. As living entities evolved from the simplest to the most complex (most intelligent), they all pursued survival and replication goals. Randomness in the replication process led to a variety of lineages. Variety in the individuals and change in the conditions they encountered in their environment led to either successful replication or extinction of the various lines.

To be more successful at first meant being better adapted instinctively to the reality around them (up through the animals). Later, it meant being better able to understand then adapt reality to themselves (humans). In all cases, success meant being consciously connected to and compliant with the facts of reality, implicitly at first, then explicitly. An entity that lost its conscious connection to reality or acted in contradiction to reality did not survive. By the time humans appeared, Francis Bacon's quote—"Nature, to be commanded, must be obeyed."—became immensely applicable.

Humans' incredible ability to survive a wide variety of changing environmental conditions is because they can explicitly identify the Laws of Reality at work and then apply that knowledge to modify their surroundings to fit their needs. Humans are the first species with this ability.

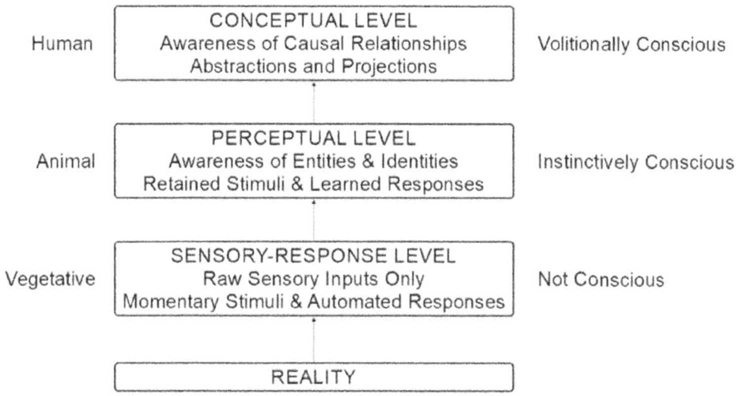

Levels of Consciousness

Sensory-Response Level

First, evolution created the earliest living simple organisms. They were connected to reality by their *stimulus-response* level of operation. This stimulus-response functioning is directly connected to and acts according to the Laws of Reality. We will later learn that what is called "vegetative goal-directed action" is evidenced in entities that are not conscious, such as plants and simple animals. A critical limitation of a non-conscious sensory-response level entity is that a momentary sensation and immediate response is fleeting and not retained.

Perceptual Consciousness Level

Evolution further increased intelligence by developing higher-level animals with a *perceptual* level of consciousness. The perceptual consciousness level emerged from and added to the previous sensory-response level. Their new perceptual consciousness enabled animals to retain a sensation beyond the immediate moment, be aware of other entities in their environment, and learn and automate appropriate responses, thereby beginning to learn and "understand" the Laws of Reality *implicitly*.

Being fully automatic (instinctive) with no awareness of its functioning, perceptual consciousness is like the lower stimulus-response level. It is directly connected to reality with no ability to evade it. Like earlier living organisms, animals had no choice but to adapt themselves to the reality around them via instinct. Man has a higher *conceptual* level of consciousness that can deal with abstractions. Still, like animals, it is his perceptual faculty that directly connects to concrete reality. Philosopher Ayn Rand explains that perceptions, not sensations, are our source of knowledge:

> A percept is a group of sensations automatically retained and integrated by the brain of a living organism. It is in the form of percepts that man grasps the evidence of his senses and apprehends reality. When we speak of "direct perception" or 'direct awareness,' we mean the perceptual level. Percepts, not sensations are the given, the self-evident. The knowledge of sensations as components of percepts is not direct, it is acquired by man much later: it is a scientific, conceptual discovery.[99]

Subjective Reality

It is necessary to point out here the error of "subjective reality." All perceptual animals, including humans, create mental representations of everything they encounter with their senses within their brains' neural networks. Different species and different individuals within species have differing sensory capabilities and therefore have different perceptions of reality.

As a result, some people conclude that "reality" only exists in our minds. They believe that we are incapable of knowing objective reality, so whatever it is, does not matter. They believe the only true reality is the subjective, virtual reality we create in our minds which is different for each of us. They then use the subjective reality in our minds, not the physical reality from which we evolved, as the foundation for all higher concepts. The result is that they disconnect all knowledge from objective reality.

To recognize the error of subjective reality, one only needs to observe its consequences. Owls, raccoons, and some humans are colorblind. Rats and some insects can see in the ultraviolet spectrum, while some snakes can see in the infrared spectrum. Dogs, bats, and moths can hear far better than humans. We can be sure that the mental representation in their brains of things they sense, their subjective virtual reality, must be different for each of them. But that does not alter the physical nature of the objects detected *by* them—physical reality.

Remember the yellow bus? It is coming down the road again. Evolution equipped each animal species with a perceptual consciousness—the ability to retain perceptions formed from the information provided by the senses. In its way, each can grasp enough information about an object being detected to learn appropriate responses to survive in its environment. According to their nature, each might perceive the same object in reality differently in its brain, but it works for them. A dog, owl, raccoon, rat, snake, a color-sighted human, and a colorblind human can all sense the yellow bus as an object in their environment using their different sensory capabilities and their mental version of reality in their neural networks. If any of them is not looking or listening, they will not sense the bus's presence.

The fact that they have different *subjective* realities in their brains does not mean that *physical* reality is not there. Physical reality would still exist even if none of them existed. Physical reality is the *only* "reality." According to their nature, it must be grasped by everyone's perceptual consciousness and guide their actions if they are to survive. What matters to perceptual entities is whether the subjective reality they hold in their brain is sufficiently consistent with physical reality to further their own life. If it is not, and they do not get out of the way, that bus (physical reality) will kill them.

Conceptual Consciousness Level

Most recently, evolution developed humans and equipped them with their conceptual consciousness level. This level was made possible by an increase in the human neocortex size and provided increased capability. This level added to the previous sensory-response mechanisms and perceptual consciousness of animals and increased intelligence still further.

The conceptual faculty enables humans to identify and integrate information provided by their perceptual faculty into abstract concepts, grasp causal relationships, organize concepts into hierarchies, and introspect. Volition enables them to control their thinking. Man's hierarchical language reflects his neocortex's hierarchical operation, which reflects the hierarchical structure of reality.

Existence and consciousness are axioms that one cannot further reduce. Existence exists, and entities conscious of existence exist. Consciousness is an emergent phenomenon of evolutionary complexity in a living entity. Consciousness is the entity's improved method of grasping reality—other entities, identities, and causality in space and time. Its contents are virtual representations of existents, not the existents themselves. As only a *method*, it has no physical form. Still, it can affect the physical substrate that is executing the method—i.e., our brain.

Let us add another *True Idea.* We have accepted that reality is real and that things in reality have specific natures and identities. We have answered the question of "What is out there." Now we can begin to answer the question, "How do we know it?" This is the "epistemology" branch of philosophy.

True Idea: Our senses and perceptual faculty are competent to grasp reality.

APPLIES TO:	QUESTION:	ANSWER:
Knowledge	*How do I know it?*	*I can trust my senses and my mind to grasp reality.*

THE CHILD

Let us examine these layers of consciousness in a human child. A child does not have a conceptual level of consciousness until he develops language skills. Until then, the child functions at the perceptual level of consciousness like that of the higher animals, the family dog, for example. Ayn Rand explains:

> As far as can be ascertained, the perceptual level of a child's awareness is similar to the awareness of the higher animals: the higher animals are able to perceive entities, motions, attributes, and certain numbers of entities. But what an animal cannot perform is the process of abstraction—of mentally separating attributes, motions or numbers from entities. It has been said that an animal can perceive two oranges or two potatoes, but cannot grasp the concept 'two.'
>
> At birth, a child's mind is tabula rasa; he has the potential of awareness—the mechanism of a human consciousness—but no content. Speaking metaphorically, he has a camera with an extremely sensitive, unexposed film (his conscious mind), and an extremely complex computer waiting to be programmed (his subconscious). Both are blank. He knows nothing of the external world. He faces an immense chaos which he must learn to perceive by means of the complex mechanism which he must learn to operate.
>
> If, in any two years of adult life, men could learn as much as an infant learns in his first two years, they would have the capacity of genius. To focus his eyes (which is not an innate, but an acquired skill), to perceive the things around him by integrating his sensations into percepts (which is not an innate, but an acquired skill), to coordinate his muscles for the task of crawling, then standing upright, then walking—and, ultimately, to grasp the process of concept-formation and learn to speak—these are some of an infant's tasks and achievements whose magnitude is not equaled by most men in the rest of their lives.[100]

A newborn infant experiences a whirling, buzzing soup of sensory input. At some point, the infant begins to realize that there is a "something" present in that soup as isolated from the background sensory noise. The infant starts to implicitly form the percept (not concept) of "entity," a sensed object that he can hold in his mental focus.

Later, the child observes that there are two or more separate entities in his field of perceptions. They are somehow different, leading to the implicit percept of "identity." The child will recognize and differentiate between

"Mommy" and "Daddy" at the perceptual level long before he develops language at the conceptual level and affixes the word labels "Mommy" and "Daddy."

Once the child develops language skills, his hierarchy of concepts such as "Mommy" and "Daddy" ties through lower-level automated and wordless percepts of "identity" and "entity" to sensory data. Rand describes the connection between our mind and reality:

> Consciousness is the faculty of awareness—the faculty of perceiving that which exists.
>
> Awareness is not a passive state, but an active process. On the lower levels of awareness, a complex neurological process is required to enable man to experience a sensation and to integrate sensations into percepts; that process is automatic and non-volitional: man is aware of its results, but not of the process itself. On the higher, conceptual level, the process is psychological, conscious and volitional. In either case, awareness is achieved and maintained by continuous action.
>
> Directly or indirectly, every phenomenon of consciousness is derived from one's awareness of the external world. Some object, i.e., some content, is involved in every state of awareness. Extrospection is a process of cognition directed outward—a process of apprehending some existent(s) of the external world. Introspection is a process of cognition directed inward—a process of apprehending one's own psychological action in regard to some existent(s) of the external world, such actions as thinking, feeling, reminiscing, etc. It is only in relation to the external world that the various actions of a consciousness can be experienced, grasped, defined or communicated. Awareness is awareness of something. A content-less state of consciousness is a contradiction in terms.[101]

From this we can see that we will not create an artificial consciousness that does not have senses and awareness of existents in the outside world.

The percepts of "entity" and "identity" are only two examples of the countless integrations the child's consciousness will make in his first years as he learns about the real world. Think of a young child sitting on the floor playing with a ball. The child picks up and drops the ball countless times. This play is the child integrating the physical reality of "gravity" into his perceptual knowledge. The child learns that the ball always goes down—never up. The child is automatically programming his brain's neural networks with information content, the source of which is reality itself. At this stage, the child's consciousness connects directly and automatically to reality.

It is time to save another *True Idea.* We have accepted that reality is real, that things in reality have specific natures and identities, and that our senses are competent to grasp reality. Now we can add how we use the information provided by our senses.

> ***True Idea: Our perceptual consciousness is an automated process that retains and integrates the information provided by our senses, identifying the items being sensed and automatizing responses to them.***

The child's perceptual knowledge of reality continues to grow. The child learns how to operate his body to walk. The child's parents protect the child from harm while exploring reality and learning how to protect himself. He learns that bumping into things hurts, and he learns to avoid doing so. The knowledge he is gaining remains directly connected to reality as "valid" knowledge.

Even though the child still has not learned language skills, he begins to learn from his parents. He might not know what the word "hot" means, but after some repetition, he begins to understand Mommy or Daddy's warning not to touch the hot stove because it will hurt. After one experiment proves Mommy or Daddy was correct, the child learns to trust Mommy and Daddy as a source of information about reality. From the child's experience so far, the information received from Mommy or Daddy is valid, like the information received through his senses.

As the child develops language skills, he makes no distinction between the knowledge he is gaining through his perception of reality and is getting from Mommy and Daddy. The lesson about looking both ways before crossing the street makes sense to the child, and he does not need to test it to consider it valid knowledge. So far, the child's knowledge is still valid—connected to reality.

However, the child's development of language skills changed the rules of the game in a profound way. This begins when Mommy and Daddy, who are trusted sources of valid knowledge, tell the child about Santa Claus, the Easter Bunny, and the Tooth Fairy. The child begins to internalize ideas disconnected from reality along with those connected to reality. The child is accepting information from other humans that is not valid.

The parents allow the child's accumulation of valid and invalid knowledge to continue for a few years until the child stops believing in Santa Clause. Unfortunately, it ends there. The opportunity for an enormously valuable life lesson is lost.

The point where the child stops believing in Santa Claus would have been an ideal time for the child's parents to have had a serious discussion with the child. They could have explicitly pointed out the lesson that not everything the child hears from other humans is valid information; that it is not the same class of information as the information the child obtains by its own thinking; that the child must consider information obtained from other humans unverified until the child can think it through and decide that it is valid and consistent with everything else the child already knows about reality. And finally, the parents can tell the child that this lesson was extremely important for the child to learn, which is why they deceived the child in this way.

The point of this life lesson is for the child not to have a mind that is either "closed" or "open" to the ideas obtained from others, but rather to have an "active" mind that uses one's own thinking to either validate or reject ideas obtained from others.

We distinguish between advancing an idea that one does not know is untrue, and an idea that one knows, or suspects is false. The former is a mistake caused by one not knowing how to validate knowledge (lacking a skill) or an unwillingness to do it (laziness). The latter, telling others ideas that you know to be false or you know you have not validated, as if they are truths, is dishonesty. Suppose we do not want to judge ourselves as lazy or dishonest. In that case, we must learn how to validate our knowledge. We must eliminate our internalized false ideas that we got from others a long time ago and that we have been perpetuating as true.

Here is another *True Idea*. (Note: I am not going to continue reviewing those we have already saved. You get it now. We are building a linked set of ideas.)

> ***True Idea: There are two classes of knowledge, that which one has validated against reality using one's own mind, and that which one gets from other humans, which might or might not be valid.***

Unfortunately, most adults today were not taught this life lesson when they were young. As a result, they have internalized both valid and invalid knowledge throughout their lives as being equal. And, unfortunately, they have been passing invalid knowledge on to others (along with valid knowledge), perhaps for generations. It now requires an active process of identifying and correcting the invalid ideas that they are not aware they hold.

> ***True Idea: We must, by actively using our reason, validate ideas received from others against what we already know to be true, without contradiction, before accepting them as new knowledge.***

> ***True Idea: Our conscious faculty programs our subconscious knowledge content. All ideas considered by our conscious faculty are, by default, saved as subconscious knowledge unless we actively examine and reject them.***

PERCEPTUAL-LEVEL COMMUNICATION

Like higher animals, humans evolved non-verbal communications over millions of years before the recent development of language. We are born with neural circuits that generate automated emotional responses to perceptual cues observed in others—facial expressions, body posture, hand gestures, non-verbal vocalizations. That is why crying or laughter is contagious in a group.

As we have seen, we can reprogram our automated responses to external stimuli. This is especially the case when we become aware of our emotions at the conceptual level—introspection—and the stimuli that cause them. However, most of us go through life using these perceptual-level communications without full awareness of the power they have over our behavior. This creates a vulnerability that we will address later when considering AI and "emotional intelligence."

KNOWLEDGE

Consider the basic levels of goal-directed action and consciousness that evolution has so far produced:

1. Vegetative—Stimulus-response (primitive life)
2. Perceptual (higher animals)
3. Conceptual (human)

The branch of philosophy that describes what knowledge is and how we use it is called "epistemology." Metaphysics asks the question, *"What is out there?"* Epistemology asks the question, *"How do I know it?"*

We have already examined the vegetative and perceptual levels. Now we need to examine the conscious faculty that defines us as human. Aristotle defined man as the "rational animal." Our faculty of reason uses "concepts" as its method and content. The following paragraphs explain what concepts are and how our mind uses them. This topic is crucially important. I encourage you to follow along as I describe the "unit reduction theory of concepts."

At the perceptual consciousness level, a *mental focus* is aware of delimited items identified in the perceptual field by the pattern-matching of sensory input data. We can perceive the entities before us and their identities. Something called the "crow epistemology" is the limiting factor. Our perceptual mental focus can hold only about four or five items at the same time. The "crow epistemology" was first reported in 1885 with the story of a man wanting to shoot a crow:

> To deceive this suspicious bird, the plan was hit upon of sending two men to the watch house, one of whom passed on, while the other remained; but the crow counted and kept her distance. The next day three went, and again she perceived that only two retired. In [the end], it was found necessary to send five or six men to the watch house to put her out in her calculation. The crow, thinking that this number of men

> had passed by, lost no time in returning.' From this he inferred that crows could count up to four.[102]

This principle of a limit on working memory was also noted by cognitive psychologist George A. Miller of Harvard University in his 1956 paper *The magical number seven, plus or minus two: Some limits on our capacity for processing information.*[103]

More recent research by neuroscientists at MIT, referenced by Ray Kurzweil, indicates that "We are able to keep up to about four items in our working memory at a time, two per hemisphere."[104] The point is that humans have a severely limited mental focus like the crow.

> ***True Idea: Our perceptual consciousness can only hold in mental focus about four or five items simultaneously.***

At the conceptual level, items in mental focus are compared and differentiated. We combine those with essential similarities into a single unit and give it a name (language is both a requirement and a consequence). We ignore their non-essential differences. The multitude of items is now combined into one mental unit and represented by a word as a placeholder symbol (a concept) to stay within the crow mental focus limit. We begin this process with objects in the physical world and then continue the unit reduction process on concepts themselves at higher and higher abstraction levels.

- All things you sit on (essential similarity) are "chairs" ignoring their (non-essential) shapes, sizes, colors, etc.
- All things you eat on are "tables."
- All things you use for activities in your house (chairs and tables) are "furniture."
- All things you use for function or decoration in your house (furniture, rugs, and art) are "furnishings."
- Everything you own inside your house (furnishings, clothing, jewelry) is "personal-property."
- Everything you own anywhere (personal property, your house, and car) are "property."

- Your property, your financial investments, your skills, and talents are all "assets."

This "unit reduction" process *is* the power of man's conceptual faculty. Our minds can perform this compression-or-expansion-of-units process in either direction with lightning speed to keep the number of items held in mental focus within the crow limit.

> ***True Idea: Our conceptual faculty performs unit reduction on perceptual material by assigning a word (a concept) to represent two or more essentially similar items while ignoring their non-essential differences.***

Man's hierarchy of concepts is constantly subject to change as new knowledge becomes available, adding to or clarifying similarities and differences. Man's conceptual faculty can quickly collapse multiple concepts into one (unit reduction on essential similarities) or expand one into its components (to become re-aware of differences). Scientists estimate that our brain can change triggered patterns about a hundred times a second. They know that signaling occurs between higher and lower pattern recognizers in the brain when they are to become more active.[105]

Man applies his conceptual method of thinking to the outside world, "extrospection," and his inner mental content, "introspection." Rand gives examples:

> Just as, extrospectively, man can abstract attributes from entities—so, introspectively, he can abstract the actions of his consciousness from its contents, and observe the differences among these various actions. For instance (on the adult level), when a man sees a woman walking down the street, the action of his consciousness is perception; when he notes that she is beautiful, the action of his consciousness is evaluation; when he experiences an inner state of pleasure and approval, of admiration, the action of his consciousness is emotion; when he stops to watch her and draws conclusions, from the evidence, about her character, age, social position, etc., the action of his consciousness is thought; when, later, he recalls the incident, the action of his consciousness is reminiscence; when

he projects that her appearance would be improved if her hair were blond rather than brown, and her dress were blue rather than red, the action of his consciousness is imagination.[106]

True Idea: Further unit reduction of concepts into wider concepts creates a hierarchy of concepts that we call "knowledge."

True Idea: Concepts that link through a chain of lower concepts to facts of reality are "validated" knowledge.

True Idea: Our conceptual faculty can imagine new combinations of ideas that do not (yet) exist in reality. If an imagined idea can be created in reality, we call it a "goal" of our creativity. If not, we call it a "fantasy."

Suppose an AI entity with human-level intelligence arises out of reverse-engineering the human brain. In that case, its form of intelligence will, by design, be based on this same perception—mental-focus—unit reduction of "concepts" principle. We will explore AI in much more detail later.

CERTAINTY

Those who wish to deny our reason's ability to deal with reality will commonly say something like, "You can't be certain of anything." What do we mean by "certainty?" First, we need to define the terms "fact" and "knowledge," and then "certainty."

A "fact" refers to some concrete existent in reality that is independent of our consciousness. We can say that facts are purely metaphysical. It is a "fact" that our physical body contains an organ that is a "brain," and that the organ exists independent of whatever we might call it.

"Knowledge" refers to the hierarchy of concepts that we mentally form with our conceptual faculty. Concepts are our unique mental method of dealing with reality. The *knowledge* we hold of a *fact* is not the concrete existent itself. Knowledge has both a metaphysical and an epistemological component. The word "brain" is an abstract concept that we mentally use to grasp the *fact* of our physical brain, other people's brains, and all brains that will ever exist. The concept and its word label "brain" is not a real existing brain. It is our way of obtaining and applying *knowledge* of the real world by integrating the material provided by our senses first into percepts and then into concepts.

"Certainty" is a purely *epistemological assessment* that we make regarding the *knowledge* that we obtain about *facts* in reality. Suppose we can assess that we integrated an abstract concept of knowledge from the percepts of real-world things grasped by our senses and lower concepts. In that case, we can be "certain" that it is valid knowledge.

Certainty, however, is not the same thing as "infallibility." It is always possible that we are unaware of circumstances that mislead our senses or information is missing. Certainty only requires that the integration or conclusion based on all the facts available is logical and without awareness of anything missing. For example, before we know about the physics of refraction, we might be certain, based on our observation, that the pencil in the glass of water is "bent." However, after removing the pencil, we see that it is straight and that only the image is bent as seen through the water. We can now be "certain" at the conceptual level that the pencil is straight even though refraction misleads our senses at the perceptual level.

The commonly accepted bromide that "You can't be certain of anything," is false. You can be certain of abstract ideas that connect through lower concepts and percepts back to reality without contradiction or reservation.

VOLITION

Humans can grasp and understand the Laws of Reality by using their conceptual-level thinking. However, for the first time in evolution, humans also have the power to turn off their thinking or evade reality. But they cannot succeed with such evasion. Consider the following airplane analogy:

Most high-performance airplanes of today, the Navy jet we saw in The Flight Simulator Story, for example, are built with negative stability designed into them. Airplanes designed to be positively stable, like toy gliders, will stay upright without input from the pilot. Positively stable airplanes have stability as their advantage, but they are not very maneuverable. Airplanes with positive stability will fight against a gust of wind or inputs by the pilot to stay upright and keep a constant direction.

An airplane with negative stability requires continuous inputs from the pilot to remain upright. If the pilot (or autopilot) stops providing control inputs, an airplane with negative stability will soon tumble out of control. An airplane designed with negative stability is highly responsive to control inputs from the pilot. He can easily make the plane go anywhere he wants. The pilot is in control, and the airplane does not resist him.

Notably, the Wright Brothers succeeded while others failed because they precisely understood that the issue of "control" was critical and that they needed to emphasize "control" over "stability."[107] Others, who had built positively stable gliders, kept crashing when the pilot's attempts at turning or wind gusts upset the plane's basic stability. The pilot of the negatively stable Wright Flyer, on the other hand, could easily adjust for wind gusts and turn in any direction.

Evolution created animals with positive stability in the form of their genetically programmed instinct. Their instinct automatically keeps them "upright," in support of their continued existence. An animal has no conscious control over its perceptual-level consciousness. The animal will survive so long as its instinct is appropriate for its environment.

Evolution created man with negative stability. While man begins life with some low-level instinct, his conceptual-level faculty replaces the animals' automatic higher instincts. The tremendous power of man's intelligence is his ability to pilot his mind, to "think." Man can adjust for upsetting "gusts" and steer his life in any direction he chooses.

This freedom comes with the responsibility to always pilot one's mind to stay "upright" as a human. But, just like an airplane with negative stability, man can stop piloting his mind and suffer the consequences, tumbling out of control. To be *volitional* means having the freedom to think or not to think—"to be or not to be." Rand explains:

> Man has no automatic code of survival. His particular distinction from all other living species is the necessity to act in the face of alternatives by means of volitional choice. He has no automatic knowledge of what is good for him or evil, what values his life depends on, what course of action it requires. […] Man must obtain his knowledge and choose his actions by a process of thinking, which nature will not force him to perform. […] Man has been called a rational being, but rationality is a

> matter of choice—and the alternative his nature offers him is: rational being or suicidal animal. Man has to be man—by choice; he has to hold his life as a value—by choice; he has to learn to sustain it—by choice; he has to discover the values it requires and practice his virtues—by choice.[108]

True Idea: Animals cannot suspend their consciousness; Man can. Volition is the choice to think consciously or to suspend thinking.

Humans, who do not have automatic knowledge to guide them, must use their minds to discover what they need to survive and obtain it. Their volitional, conceptual thinking enables humans to adapt the reality around them to themselves rather than themselves to reality. It provides humans with the ability to create, which is the power to bring into existence an arrangement (or combination or integration) of natural elements that had not existed before.

Humans can choose not to think and act on ideas not connected to reality. "I can walk on water across this river." They cannot survive such evasion of their responsibility to think.

True Idea: Thinking based on reality is man's only means of survival. He can evade the responsibility to think but cannot succeed with such evasion.

Evolution provided humans with reason, opposable thumbs, and the ability to walk, enabling them to discover more about reality and then apply what they have learned to the matter around them. Humans have an innate and insatiable desire to explore, discover, learn, imagine, and then create—because it is essential to their survival.

In the last few centuries, humans have used their observations of reality to discover the principles of gravity and the nature of mass. They have measured the speed of light and discovered the relationship between matter and energy—$E=MC^2$. This equation, a very high-level abstract concept, can be explained and connected down through a hierarchy of lower-level concepts and percepts to observable reality. Such a new abstract concept *is* connected to reality and therefore becomes new knowledge. We can then use it to manipulate reality for further discovery, such as in the transcendent technologies of quantum computing and fusion energy.

Achieving all this is not automatic. Each person has *volition.* He must *choose* to be committed to reality.

FOCUS

What are you doing in your mind when you make the volitional choice to think? Your volition, let us call it your *free will,* is your ability to choose to *focus* your mind on reality or not. Rand explains:

> Man can focus his mind to a full, active, purposely directed awareness of reality, or he can unfocus it and let himself drift in a semiconscious daze merely reacting to any chance stimulus of the immediate moment, at the mercy of his undirected sensory perceptual mechanism and of any random associational connections it might happen to make.[109]

If you did not have this choice, you would not have a mind; you would be following instincts like animals. If you have a mind, then you are an independent entity. When others use physical force against you, they prevent you from using the method of survival that evolution provided in place of instinct. Therefore, freedom is man's natural state. Freedom is a high-level abstract concept that derives from having a mind and free will. It is a valid concept because it connects to the reality of man's nature. In the following paragraphs, are points highlighted in a 1999 lecture by Dr. Binswanger:[110]

> If you do not have a mind, then you are simply an automaton that replays beliefs that are fed to you. You cannot judge their truth or falsehood, you cannot obtain knowledge, you cannot discover what is required for you to achieve self-preservation and happiness. Your entire life is controlled by some outside factor that is a dictator whom you must obey as a slave. These are the kinds of higher concepts that result from determinism, a doctrine that denies that we have free will and are instead controlled by some outside factor.

Let us return to the young, developing human child. Through observation, he acquires factual knowledge of the objects and events he experiences. He concludes that a universe exists and that he exists. He grasps objects and entities and sees that they move or change, which is "action," and that he, himself, can act. Then he learns that in certain situations, he must act, and he does it through choice. For example, if he is hungry, he must ask for food, cry for it, or get it himself. He must do something, choose what it will be, and then choose to do it.

Focus is the state you are in if you have a goal-directed mind committed to acquiring full awareness of reality. This active mind intends to understand everything it encounters. A focused mind seeks all the relevant facts instead of settling for just some of the facts while failing to examine or consider other vague facts. A commitment to reality means an unconditional commitment to facts, whether easy or not, and whether they support the desired outcome or not. It is the acceptance that one's life depends on knowing what the truth is, the unconditional acceptance of reality.

Focus is the will to know. Being in focus requires effort because one must continually remake the focus from moment to moment. Focus is an act, not a process and is essentially instantaneous. Either you are in focus, or you are not. Your mind does not go through a series of steps to get into focus. Your mind cannot analyze its own state of being unfocused because it must be in focus to analyze anything. Focus means being alert, having a purpose, and being actively in charge of your mind. Being unfocused is to be passive, aimless, and not in charge. You just will your mind to get into focus, and it instantaneously happens.

We know we choose to focus because it requires an effort that we are aware we do not have to make. It takes effort to manage or regulate the operations of our consciousness. We must choose to remain in a state of mind that connects to reality. When emotions grab our attention, that is not being in focus. Focus means that your conscious mind remains in control of the topic of focus, not your senses or subconscious. You can just say no to distracting emotions.

Free will requires self-awareness, observing what is going on in your mind. Being able to self-monitor is the essence of focus and free will. *Self-consciousness* is the means of achieving the end of consciousness. Awareness of reality and being self-conscious means that you are judging how your consciousness relates to reality.

Focus is the state you must achieve as a precondition to thinking. One is ready to think. Thinking is when you sustain focus for a time to answer a question.

Managing your mind can occur at two different levels. At the micro-level, you give your subconscious step-by-step instructions. It returns step-by-step answers to your consciousness, for example, when doing a math problem. At the macro level, you give your subconscious a standing order, and then you let your subconscious take care of it, for example, knowing the route to drive home from work. At either level, you actively manage your mind. Your subconscious will store the information you want to keep according to its essential connections to other already stored data.

If your mind is in a *drift* state, you are not aware of what is going on in your mind and relinquished control to your subconscious. Your mind is still doing things, even making associations, but you would probably not approve of these associations if you were in focus. In a state of drift, information your subconscious stores get filed at random based on non-essential linkages to other non-essential memories. The mental state of drift degrades your knowledge.

Your mind can take a needed rest. *Rest* is when you allow your mind not to focus, but you first give it a standing order that no mental conclusions that arise from the period of rest are to be stored away. Your subconscious is obedient to your mental instructions, so it is safe for you to relax.

Evasion is being in a mental state where you tell your subconscious not to be aware of or integrate something. It is saying no to something relevant or essential, to suppress it unjustifiably. "This can't be true; I won't let it be true." Evasion is an even more destructive evil than drift because it is intentional.

Rational mulling is when you evaluate the information sent up to your conscious level by your subconscious to see how it relates objectively to reality. Your mind never accepts the information from your subconscious as a given without challenge, even though your subconscious believes it to be true. You analyze your feelings as well as the words that your subconscious sends up. Your subconscious is not some infallible voice within you.

Emotionalism is a compulsive need to reach some pre-ordained conclusion to the material. You provide an agenda or hypothesis to your subconscious, causing a loop. Emotionalism can feel like one is thinking things through. It can be recognized because it usually involves one's ego. One is suppressing facts that are counter to the agenda. It is like a mock trial. One's subconsciously held values trigger both rational and irrational emotions.

It would help if you choose whether facts or feelings are going to be king of your subconscious. Making reality king means placing it above your psychology. Your conscious mind makes this choice. To be objective, you consider the whole of reality as your domain. You are calm because you feel free to consider all the facts. No fact of reality can be a threat; it is your ally. Even negative facts are essential to have the most precise grasp of reality possible to you.

MYSTICISM

Mysticism is when one attempts to focus his mind not on reality but outside of reality. When man first evolved his new conceptual consciousness, he only had mysticism to explain what he saw in reality. It is mysticism that human society's evolution, through centuries of trial and error, is struggling to finally replace with reason to fully achieve human values, happiness, political freedom, and civil society.

Mysticism is the choice to keep things mysterious rather than find explanations for them in reality. Mysticism rejects that things in nature have specific "identities" and act according to their nature—"causality." Without identity and causality, things are indeed mysterious because you do not know what they are or what they can do.

The mystic imagines a supernatural realm to create a place where contradictions, which cannot exist in reality are possible. By rejecting identity and causality, the mystic says that things can have any natures, even contradictory natures, that can be "A" and "non-A" simultaneously. Why would he do that? So that he can say that things are however he wants them to be, regardless of whether they are that way in reality. This evasion is "whim worship"—putting the "I wish" over the "it is." The mystic claims some non-sensory, non-rational, non-identifiable means of knowledge through instinct, intuition, revelation, or other means of "just knowing."

Bad things sometimes happen in life—deaths, injuries, injustices, disasters, wars, terror and more. These can make people wish that things were not the way they are. However, while we can evade reality, we cannot succeed with such evasion. To achieve our values and happiness, we must squarely face reality regardless of how we feel about it.

When things happen in reality that one cannot explain, the person might become skeptical. *Skepticism* is the belief that we cannot understand reality and certainty in knowledge is impossible. The skeptic holds the contradiction where he believes nothing in reality is knowable; yet, his existence and consciousness are facts. When the skeptic creates a supernatural realm that allows him to escape this contradiction, he becomes a mystic.

We must learn and comply with the Laws of Reality to live. This requirement means mysticism is a philosophy of death. The mystical periods of human history were not ages of technological advances and raised living standard—they were dark ages of poverty and misery. To achieve human values, we must understand reality and make it as un-mysterious as we can. We must turn away from blind faith in our feelings and wishes and towards facts, reason, and science.

The choice to focus our mind is our *volition*. The choice to focus it on reality is *reason*. The choice to focus it on the supernatural is *mysticism*. Ultimately, the freedom to choose whether to focus one's mind on reality is the essential characteristic of what it means to be a human. It is the fundamental starting point for our unique form of conceptual consciousness. Its use is required to sustain our lives, thrive, and achieve happiness. Evolution replaced our positive stability—instinct—with negative stability and the need to continually be the pilots of our own lives—volition. Any discussions regarding the possibility of ever creating a machine with intelligence equivalent to a human must include free will and focus.

> ***True Idea: Our basic virtue, required by our nature, and the source of all our other virtues, is "rationality." Rationality means recognizing and accepting reason as one's only source of knowledge, one's only judge of values, and one's only guide to action.***

> ***True Idea: Our basic vice, the source of all our evils, is the act of unfocusing our mind, suspending our consciousness, which is not blindness, but refusing to see, not ignorance, but refusing to know.***

SELF AWARENESS—"ME"

Man must "pilot his airplane" in the direction he desires. But what is it exactly that he is piloting? As previously pointed out, man applies his conceptual thinking method to both the outside world, "extrospection," and his inner mental content, "introspection."

A child learns to isolate "entities" from the flood of sensory information about the world around him, and he discovers that those entities have implicit identities, "Mommy" and "Daddy." Eventually, the child becomes aware that he is himself an entity like those around him, and his implicit identity is "me." A while later, he learns his explicit name "Johnny" as a word representing "me."

The child becomes aware that he has control over his body, and his mind can direct his hands to touch and move things. He is beginning to pilot the airplane of his life by manipulating his environment. An airplane's pilot must think about "What is my destination?" and "What direction must I fly the airplane?" Piloting one's own life is a person's "executive" level conscious function. It is his introspective thinking about his "me." It answers the questions "What do I want?" and "What do I need to do?" These are the choices and actions that a man must make in every moment of his life to sustain his life and achieve his values and happiness.

A person has specific values, which are in a hierarchy of importance to his "me." His volitional, introspective thinking chooses and ranks those values. If he chooses his values rationally, he expects that achieving those values will result in happiness. Therefore, he pilots his airplane in the direction that will achieve his values.

You might recall—"Aviate, Navigate, Communicate" from The Flight Simulator Story. Man's first priority as the pilot of his life is to "think," to remain in conscious control of his life and his mind's connection to reality. His second priority is to make sure he is on-course toward his major goal(s). If he is in control and on course, he can consider other things in life that are lower in the hierarchy of his values. If he gets off course or loses conscious control, he will suffer the consequences.

We will revisit this executive-level conscious function later in the context of a conscious AI entity.

PRINCIPLES

We do not need knowledge simply for the sake of knowledge. We need it as a means of acting successfully in the world. A "principle" is a fundamental generalization. By fundamentality, we are referring to causal sequences that have a branching, tree-like structure. The fundamental causal factor is the one on which the entire branching series depends. If "A" causes "B" and "C," and "B" causes "D, E & F," then "A" is the fundamental causal factor in the hierarchy. An example from biology would be that natural selection is the causal factor for the tree of life. Binswanger explains the benefit of thinking in terms of fundamentals:

> Organizing one's knowledge on the basis of fundamentals permits a 'crow-friendly' condensation of the whole 'tree' of phenomena. This not only adds clarity, it also allows for automatizing what is most cognitively fertile. [...] But the real cognitive value to be reaped comes from the generalization to a type of cause and a type of tree of effects, as with: 'Any star is the basic causal factor controlling the orbits of its planets.' When such generalizations reach a certain level of scope, they qualify as *principles*. A 'principle' is a fundamental generalization that serves as a standard of judgment in a given domain.[111]

We need principles because they provide us with the "road map" we need to navigate through a life filled with cause-and-effect situations. They make us aware of the consequences that follow from choosing one way or another. They bring into mental focus cause-and-effect relationships. Binswanger explains that principles also help us simplify things:

> ...having a principle eliminates the need to continually re-investigate the same issue; the principle is an invaluable shortcut. But there is a deeper answer. Concretes differ in their complexity. Principles enable the clear, simple cases to shed light on the obscure, difficult ones.[112]

A principle identifies an action that is required by the facts at hand without contradiction. Binswanger provides an example of how principles are absolutes in their applicable contexts:

> On an objective theory of rights, rights are moral *principles*. Rights are the application of the principles of morality to a man's dealings with others; they demarcate the individual's proper sphere of independent action. To say that a man has the right to do X is to say that he should be the one to choose whether or not he does X; no other person or group may force their choices upon him. Rights prescribe freedom by proscribing coercion.[113]

Many of the *True Ideas* we have been collecting throughout this book and listed in the *Resources* section could be considered *principles*.

VALUES AND VIRTUES

A *value* is that which one acts to gain or keep. *Virtue* is the process by which one achieves a value.

We have said that a volitional mind must choose the values that it will pursue to further its life and rank those values in a hierarchy of importance.

What are values, and why does an entity with a rational, conceptual faculty need them? By what standard does one choose *values,* and what are the *virtues* that will achieve them? Ayn Rand describes it this way:

> The concept of 'value' is not a primary; it presupposes an answer to the question: of value to whom and for what? It presupposes an entity capable of acting to achieve a goal in the face of an alternative. Where no alternative exists, no goals and no values are possible.[114]

> Life or death is man's only fundamental alternative. To live is his basic act of choice. If he chooses to live, a rational ethics will tell him what principles of action are required to implement his choice. If he does not choose to live, nature will take its course.[115]

My morality, the morality of reason, is contained in a single axiom: existence exists—and in a single choice: to live. The rest proceeds from these. To live, man must hold three things as the supreme and ruling values of his life: Reason—Purpose—Self-esteem. Reason, as his only tool of knowledge—Purpose, as his choice of the happiness which that tool must proceed to achieve—Self-esteem, as his inviolate certainty that his mind is competent to think and his person is worthy of happiness, which means: is worthy of living. These three values imply and require all of man's virtues, and all his virtues pertain to the relation of existence and consciousness: rationality, independence, integrity, honesty, justice, productiveness, pride.[116]

A moral code is a set of abstract principles; to practice it, an individual must translate it into the appropriate concretes—he must choose the particular goals and values which he is to pursue. This requires that he define his particular hierarchy of values, in the order of their importance, and that he act accordingly.[117]

Since a value is that which one acts to gain and/or keep, and the amount of possible action is limited by the duration of one's lifespan, it is a part of one's life that one invests in everything one values. The years, months, days or hours of thought, of interest, of action devoted to a value are the currency with which one pays for the enjoyment one receives from it."[118]

Only an individual can have values.

True Idea: A person is an end in himself. The furtherance of one's own life is his highest value and the source of all his other values. He must choose his own values, not live for the values of others.

APPLIES TO:	QUESTION:	ANSWER:
A Person	*What should I do?*	*I must use my mind to identify and achieve my own happiness.*

True Idea: A "value" is that which one acts to gain or keep. A "virtue" is the action by which one achieves a value.

True Idea: The three primary values for a conceptual entity are Reason, Purpose, and Self-esteem.

Now we can better describe the actions that are required to achieve those values—the *virtues*: rationality, independence, integrity, honesty, justice, productiveness, pride. Each of these virtues is a *True Idea.* I will list them that way.

True Idea: Our basic virtue, required by our nature, and the source of all our other virtues, is "rationality." Rationality means recognizing and accepting reason as one's only source of knowledge, one's only judge of values and one's only guide to action.

True Idea: The virtue of "independence" is acceptance of the fact that you must live your life according to your own thinking and judgment, not the thinking of others.

True Idea: The virtue of "integrity" is the recognition that you cannot fake your consciousness, that you are made of a body and mind, and there is no breach between them.

True Idea: The virtue of "honesty" is the recognition that you cannot fake existence. It means to live by, to practice, any idea you accept as true.

True Idea: The virtue of "justice" is the recognition that you cannot fake the character of others, that you must judge others for what they are by a rational process of identification and treat them accordingly.

True Idea: The virtue of "productiveness" is the recognition that productive work is the process by which your mind sustains your life, the power to adjust your environment to yourself.

True Idea: The virtue of "pride" is the recognition that you are your own highest value, and like all of your values, it must be earned. It is the achievement of the creation of your character and self-esteem.

ART

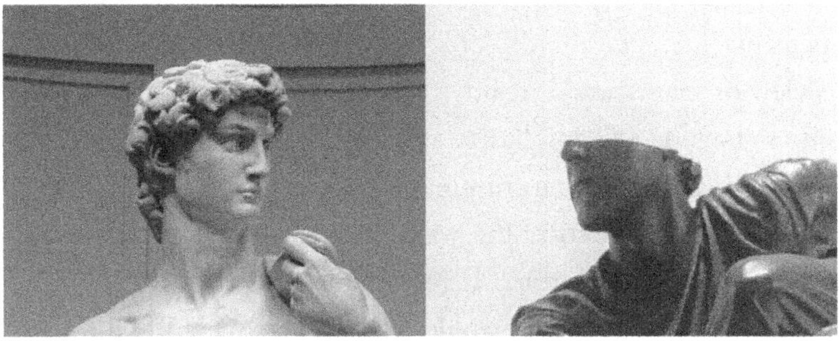

Art: "*This is what life means to me.*"

Man's conscious rational faculty is his source of knowledge and his means of survival. It is man's rational faculty that must deal with the following questions regarding existence as stated by Rand:

> Is the universe intelligible to man, or unintelligible and unknowable? Can man find happiness on Earth, or is he doomed to frustration and despair? Does man have the power of choice, the power to choose his goals and to achieve them, the power to direct the course of his life—or is he the helpless plaything of forces beyond his control, which determine his fate? Is man, by nature, to be valued as good, or to be despised as evil?[119]

> The answers one reaches for these questions regarding the nature of reality will determine the kind of ethics and values one chooses, and therefore his goals and his plans for his future. A man knows, either explicitly or implicitly, that he needs a comprehensive view of the nature of reality to provide consistent guidance in his life, his every choice, decision and action. His comprehensive view of existence "includes every concrete he has ever perceived; it involves such a vast sum of knowledge and such a long chain of concepts that no man could hold it all in the

> focus of immediate conscious awareness. Yet, he needs that sum and that awareness to guide him—he needs the power to summon them into full, conscious focus. That power is given to him by art.[120]

In all its forms, art is the means of making man's highest abstract concepts immediately graspable by his perceptual level of consciousness. It is as if they were percepts, i.e., *real.* A work of *art* is a re-creation of reality according to the artist's view of the nature of reality. The artist selectively chooses which aspects of reality to include and emphasize and which to exclude and minimize.

As with all perceptual stimuli, art generates an emotional response in the viewer according to whether the work of art supports or negates the person's views regarding reality's nature. If the emotion generated by a positive work of art could be put into words, it would say, "This is what life means to me." The emotion generated by a negative work of art is one of revulsion because it is in opposition to one's own answers to the fundamental questions regarding reality.

Man supports his life by reshaping his environment to serve his purpose, first defining and then creating his values. A man needs a graspable projection of these values—a concrete image of how he will reshape the world and himself. "Art gives him that image; it gives him the experience of seeing the full, immediate, concrete reality of his distant goals."[121]

> Since a rational man's ambition is unlimited, since his pursuit and achievement of values is a lifelong process—and the higher the values, the harder the struggle—he needs a moment, an hour or some period of time in which he can experience the sense of living in a universe where his values have been successfully achieved. It is like a moment of rest, a moment to gain fuel to move farther. Art gives him that fuel; the pleasure of contemplating the objectified reality of one's own sense of life is the pleasure of feeling what it would be like to live in one's ideal world.[122]

Take another look at the two works of art in the photo above depicting sculptures of two humans as you consider that—"Art is man's metaphysical mirror; what a rational man seeks to see in that mirror is a salute; what an irrational man seeks to see is a justification—even if only a justification of his depravity, as a last convulsion of his betrayed self-esteem."[123] In the photo above, one sculptor salutes man as a heroic being. The other sees man as a sightless, mindless monstrosity, completely powerless.

Exponentially advancing technology means our ability to reshape our environment—our world and eventually our universe—to serve our purposes is increasing. We first define and then create our values. Our need for art will also increase exponentially as a graspable projection of our values and a full, immediate concrete image of our distant goals. To keep our technically powerful society from going off the rails, we will desperately need artists who give us art that salutes reality and Man's goodness and efficaciousness, not depravity and betrayed self-esteem.

True Idea: Art is a recreation of the artist's view of reality according to his values. A work of art says, "This is what life means to me."

APPLIES TO:	QUESTION:	ANSWER:
Art	*What is the image of my Worldview?*	*Art that shows a real world and rational people.*

CULTURAL EXPANSION OF KNOWLEDGE

Early humans had no language, no knowledge of the Laws of Reality, and no significant technology. They were mainly isolated, not yet living in larger groups. Their conceptual level of consciousness could not become fully functioning until the development of language. Once they had language, they could build their conceptual mental content.

Existing knowledge made possible additional discoveries and new tools for later discoveries. As the body of conceptual knowledge grew, the rate of discovery increased. One of the factors facilitating the acceleration of discovery was the development of society, a grouping of humans much larger than the family unit or the tribe.

Let us review some of the milestone events in human history where the relationship between the individual mind and a society of humans accelerated the discovery of new conceptual knowledge. As you read the following, note the decreasing time between significant events as another indication of the effects of the exponential growth curve.

Language [100,000 BC]

Even though the brain of the anatomically modern human had the capability to think conceptually, he could not fully do so until he developed language. He could not reduce the herd of antelopes in front of him to a single mental concept called "antelope" because he did not yet have language skills. He could not come up with a word label to affix to the group of similar animals.

As we saw in the biological evolution timeline, anatomically modern humans appeared about 250,000 years ago. Their primitive conceptual faculty probably was first awakened by hand signals and grunts as the symbols that an individual attached to learned concepts. Humans needed symbols to convey meaning between individuals. Communication, necessary for socialization, gave rise to the conceptual faculty and, eventually, language.

Scientists believe that language appeared less than 100,000 years ago.[124] This timing means that it took humans 150,000 years to mature their primitive rational faculty. Finally, they developed language, fully activating their conceptual faculty and completing the evolution of *reason* as the replacement for *instinct*. They called the herd in front of them "antelope" (or whatever spoken word meant the same thing back then).

They started to collect concepts as mental information content. They would have come up with words for almost every concrete thing around them as lowest-level concepts. Then they would have begun developing higher-level abstract concepts. "Antelope," "roots," and "berries" would be combined using unit reduction into the word "food." They are all something you can eat. The hierarchy of knowledge was beginning.

Symbols and Counting [40,000 BC]

Cave paintings and counting sticks date back to about 40,000 years ago. It took early man about 60,000 years from the time he developed language to develop the ability to represent concepts such as "antelope" in drawings. It also took him that long to begin to use abstract math concepts such as counting.

Cities, Concepts Sharing and Standardization [8000 BC]

Early humans lived in small family groups. If one man made a discovery and formed a new concept, he could only teach it to others within his family. Unless family members transferred to a different family, the concept and the word for it would remain in his family tree through successive generations. Other family lineages separated by geography would have developed concepts for the same physical objects but with different spoken words representing them.

If two family lineages encountered each other, two things would need to happen to resolve their conceptual knowledge differences. First, they would need to resolve the naming of concepts they had developed for the same thing. Second, each family lineage would teach the other new concepts that the one had developed, but the other had not.

It is essential to point out the method of resolving concept naming differences. The question is, if both family lineages developed the concept for antelope but had assigned different spoken words for it, how would they know the two different words refer to the same thing? The answer is that they had to point to an antelope in reality. Then, they only had to agree on which word they would continue to use, or they could keep both words understanding them to mean the same thing. The lesson here is that all concepts are validated by following their hierarchical chain down until you can point to reality. Reality provides the ultimate validation of concepts and is the arbiter of conflicting ideas. This principle is critical later when we get to the level of ethics.

The earliest cities formed about 10,000 years ago. Cities brought more significant numbers of family lineages into proximity to each other, accelerating the sharing and standardization of concepts. As humans became more able to cooperate, they could achieve more than they could as individuals.

Writing [3000 BC]

Even within cities, initially all concepts that had been developed had to be passed from one generation to the next by word of mouth. Children learned from their parents. Children were shown the meaning of lower-level concepts by pointing to them in reality. Higher level concepts were passed down in stories and folklore.

Developed concepts did not accumulate in a straight line. Wars, diseases, famines, floods, volcanoes wiped out entire cities and the collective set of concepts developed in that location were lost. Writing on papyrus was first developed about 5,000 years ago. Writing enabled the persistence of discovered concepts on sheets and scrolls. These repositories of developed concepts could survive beyond the communities that had developed them. People could also transport them to other communities because of conquest or trade.

Aristotle - Logic [340 BC]

Aristotle lived during the Classical period in ancient Greece. His writings cover many subjects—including physics, biology, zoology, metaphysics, logic, ethics, aesthetics, poetry, theatre, music, rhetoric, psychology, linguistics, economics, politics, and government.[125]

As we will see in Chapter 3.4: *Society*, Aristotle adopted the position that the universe was real, and that man's reason and logic could grasp it. He provided the discovery of the method of logic and is therefore considered the father of all the sciences. During the eras and in the world's locations where Aristotle's ideas dominated the culture, the discovery of new conceptual knowledge was advanced.

University [859 AD]

Apart from Plato's Academy in ancient Greece, the University of Al Quaraouiyine in Morocco in 859 is considered by some to be the oldest degree-granting university.[126] Such universities established the pursuit of conceptual knowledge as a profession, thereby elevating its importance in society and increasing the number of individuals who had access to accumulated conceptual knowledge. The mobility of the professors from one university to another helped spread new conceptual knowledge geographically.

Printing Press [1439]

Hand duplication of manuscripts, even primitive woodblock printing, was labor-intensive. This labor made books expensive and low in quantity. The dissemination of conceptual knowledge was still primarily by the in-person method. It required the student and the teacher to meet physically.

The advent of the movable type printing press dramatically reduced the cost and increased the number of books. Printed books enabled much larger quantities of accumulated conceptual knowledge to disperse geographically without the student and teacher having to meet physically. It also ensured the survivability of conceptual knowledge. There became many more books, widely distributed, instead of only an original document that could become lost to posterity.

Renaissance [1500]

The Renaissance was essentially a resurfacing of the teachings of Aristotle after a protracted period of their absence. A renewed interest in reality and reason-logic led to a new period of discovery. Galileo discovered that the Laws of Reality are mathematical, and he described the proper relationship between mathematics, theoretical physics, and experimental physics. DaVinci, Raphael and Michelangelo elevated the use of reason to new heights. Newton's "Mathematical Principles of Natural Philosophy" showed us that reason was competent to discover and explain the Laws of Reality. With Aristotle's foundations of logic, great, curious minds began to build on one another's discoveries.

American Revolution [1776]

Early scientists such as Galileo arose at a time when mysticism dominated society. Galileo's discovery that the Earth rotated around the sun and not vice-versa was considered heretical by the church. Galileo was forced to denounce his scientific discovery and spent the rest of his life under house arrest.

For the first time in the history of the world, the founding of America created a country based on a constitution of written principles that were primarily Aristotelian principles. These provided the freedom for minds to pursue science, engineering and business without challenge or constraint from the government or the church. In this new, most free country on Earth, humans began developing new conceptual knowledge at a rapidly increasing rate.

Communications [1800–1900]

Through the end of the eighteenth century, the only way for men to share the knowledge they discovered was by physical transport. They had to travel personally or hire the delivery of printed material. Both were slow and expensive within a country like the United States. It was incredibly slow to send information across oceans. Slow communications limited one scientist or engineer's ability to take advantage of another's discoveries in a different geographic location.

Additionally, most original texts were handwritten, and images of things documented had to be hand-drawn by an artist and described in the text. Creating these original documents and then communicating them to others was a slow process with unreliable fidelity.

The telegraph in 1816, photography in 1827, typewriter in 1868, telephone in 1876, and radio message across the Atlantic in 1901 brought a new age in communication. These greatly facilitated the sharing of new conceptual knowledge around the world.

Einstein [1916]

Perhaps more than any other scientist, Einstein proved to the world that even the most abstract and hidden Laws of Reality were indeed discoverable by applying reason, observation, and logic. Einstein's discoveries have been the basis of countless other discoveries and inventions by those who followed.

Fast Fourier transform [1965]

The Fast Fourier transform is an "algorithm." It is a mathematical method of analyzing an analog input signal and generating a specific digital output. The use of algorithms to obtain information from observed and collected data expanded greatly with computing advances over the next decade. Algorithms enable today's computers to interact with humans by detecting spoken words and recognizing things in images.

Microprocessor [1971]

Other programmable mechanical and electrical computers already existed. However, the invention of the microprocessor began a new revolution in the discovery and dissemination of new concepts. By enabling the application of various algorithms at high speed, we could apply microprocessors to almost any challenge of discovery, development, and commercialization of new conceptual knowledge. Microprocessors were affordable and available to even the most modest budget.

Ethernet & Internet [1973]

The Ethernet enabled the "team" approach to discovery. Large numbers of persons could have instant access to the same stored information and could instantly collaborate. Initially, we formed teams at the building level. Later, we could create a team of people distributed in multiple geographic locations using telephone lines as a network connection.

Word Processor and Spreadsheet [1978]

Software applications that could run on affordable microprocessors, as tools of discovery and communication, began to appear with the word processor and the spreadsheet. We could produce larger quantities of text and formatted data quickly and easily. We could easily view and analyze data in new ways with the push of a few buttons, sparking discoveries that we would have otherwise missed. Standardized software applications and formats for word processor and spreadsheet documents meant we could readily share electronically with others who had the same software.

World Wide Web, Google Search, Wikipedia [1990–2001]

Perhaps it was the Internet that was the spark that ignited the explosive rate of increase of new conceptual knowledge that we see today. The World Wide Web, Google Search, and Wikipedia have changed everything. For the first time in human history, all conceptual knowledge on all topics produced anywhere in the world is instantly available to nearly everyone globally. Today, collaboration for the discovery of new conceptual knowledge has no friction or resistance. It is all-inclusive, instantaneous, and essentially free. Discovery is now happening at an exponential rate.

Siri [2011], Google Glass [2012]

The word processor and spreadsheet enabled us to visualize and share data in new ways. Now, AI personal assistants and wearable personal digital interface devices will usher in a new era of discovery. At almost the speed of thought, a researcher will have the ability to bring up any existing knowledge on any topic in his field of vision when he is looking at something he is exploring. He will immediately determine if he has made a new observation or discovery compared to what is already known. Collaboration will have nearly telepathic speed and immediacy. He will be able to transport himself virtually anywhere in the world (more on this later).

True Idea: Our hierarchy of conceptual knowledge is growing at an exponential rate.

HIVE MIND

Human society cannot have a "hive mind" because humans lack instinct. Evolution provided species that live in swarms with instinctive behavior that protects the hive at large, even if it requires individual self-sacrifice. Humans initially evolved in small family units, not massive hives. Human *society* is a later development that coincided with the evolution of man's conceptual-level consciousness. Humans traded the animals' stability and inflexibility of instinct for the adaptability and flexibility of volition and reason. Without instinct, there can be no preprogrammed "hive" human behavior in human society. Society is only the sum of individuals' behaviors, each of which can act for or against society's survival—or himself.

Human society does, however, affect the behaviors of the multitude of individuals within it. Society is the storehouse of the conceptual ideas produced and recorded by all its present and past members. Born a blank slate, each new human does not know how to operate his conceptual mental potential. He must learn how to operate it from the society around him. He must then acquire conceptual content from the body of knowledge stored in society. The individual may then add to society's stored body of knowledge during his lifetime.

The ideas stored in society are a mixed bag of ideas regarding the nature of reality and of human knowledge—"What's out there?"—"How do I know it?" The ideas in today's society derive from two different answers to these questions. One set of answers says that there is a real reality, and our rational faculty is competent to deal with it. The other set of answers says that what we see around us is a virtual reality, and the real one is unknowable to us. Feelings or revelation, not reason, are our trustworthy sources of knowledge.

The odds are that each new human will learn the method of thinking and the body of conceptual knowledge that is predominate in the society into which he is born. It will likely be a mixture of the two opposing sets of ideas, with one or the other being dominant. Teaching them to children reinforces the dominance of those ideas in society—either reality-based or not.

The mix of ideas in society changes only through the influence the society's recognized intellectual leaders have on the society's institutions and systems for educating new humans. A leader who rises to prominence and promotes reality-based thinking methods will move society as a whole and conceptual content in that direction and vice-versa. But that does not constitute a "hive." Everyone is free to choose the opposing ideas.

Social media and information technology have the power to amplify the spread of the ideas of society's intellectual leaders and cause similar behaviors in masses of people as if they were a hive. However, this is not a hive but rather a population of humans. Each of them can think, but many have lost or never acquired the skill of thinking for themselves. They are at the mercy of the techno-priests who claim "knowing" and will do their thinking for them.

TECHNOLOGY CYCLE

Processes of the *laws of physics* drove cosmological evolution. The process of *natural selection* drove biological evolution. We are going to call the process that now drives conceptual evolution the *"Technology Cycle."* The *Technology Cycle* is the source of Francis Bacon's observation that "Nature, to be *commanded*, must be *obeyed*." The *Technology Cycle* is about increasing three quantities:

1. **Conceptual Knowledge**. This is the total body of knowledge of how reality works known to society. It includes knowledge about the Laws of Reality and knowledge about the nature of things that exist in reality.
2. **Technology**. These are the tools or products we create from the matter around us to support and advance our lives and aid in discovering new conceptual knowledge.
3. **Human Intelligence**. This is the set of intelligence functions within our brains created by functional brain modules' integrations.

The *Technology Cycle* consists of the following four major steps. These are repeated in cyclical iterations continuously over time.

1. **New Discovery**. Individuals, usually scientists, apply the knowledge, technology, and intelligence available to them at that time to the reality around them to explore, observe, experiment, analyze and ultimately discover a new Law of Reality or the nature of an existent as new knowledge.
2. **New Knowledge**. This new knowledge distributes throughout society via the information technology and education technology existing at that time.
3. **New Technology**. Individuals, primarily engineers in businesses, apply the newly acquired knowledge and existing intelligence capabilities to manipulate the matter around them to create new or improved technology, products, and services for society.
4. **New Intelligence**. Individuals throughout society learn the New Knowledge and learn to use the New Technology products and services. This learning and doing causes their brain modules to restructure, creating new human intelligence functions that prior humans did not have. Repeat step 1.

In the beginning, new knowledge slowly accumulated as individuals demonstrated it to other individuals. As language, cities, writing, and the

Internet developed, knowledge began accumulating more rapidly. Each time around the *Technology Cycle*, the incremental increase in new knowledge is greater than during the previous cycle.

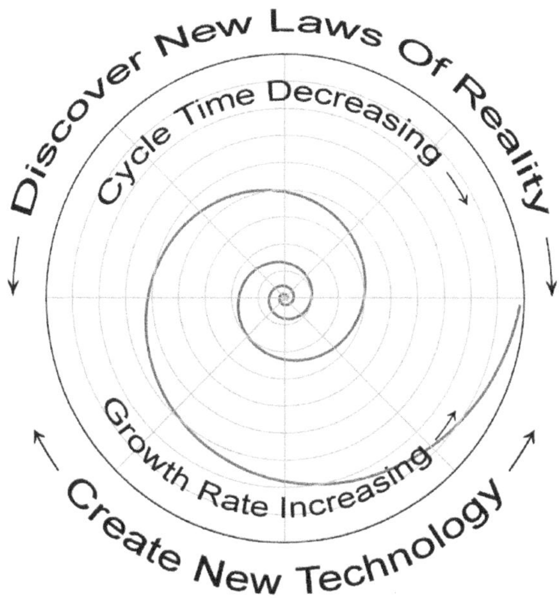

Technology Cycle:
Knowledge increases and cycle times decrease at an exponential rate.

Simple stone tools were a new technology that helped break open things to see what was inside. It took a long time to learn new knowledge with these primitive tools. Today, we have the technology to examine sub-atomic particles, decode genomes, and peer deep into the universe. Each time around the *Technology Cycle*, the cycle time to discovering new knowledge gets progressively shorter.

Conceptual-level entities execute the *Technology Cycle*—humans. In the future, non-biological conceptual entities that we humans create might execute it. The *Technology Cycle* will continue until a Post-conceptual Evolution Period begins, or we master the universe. Or, if we cause our extinction or a New Dark Ages, the *Technology Cycle* might be suspended until later generations, or a different conceptually conscious species picks up where we humans left off.

True Idea: The Technology Cycle uses contemporary intelligence and technology to discover new Laws of Reality and uses the new knowledge to create new technology and new intelligence functions. The cycle repeats.

It seems that evolution provided us with intelligence that came with a built-in drive to discover the Laws of Reality and modify our environment. Indeed, since man does not have higher-level instincts, he must think and act to modify his environment to survive. For man, discovery and creativity mean survival. This innate drive of discovery and creative manipulation of our environment is the *Law of Intellopy* at work (a later chapter).

The evolutionary emergence of our rational faculty set in motion a recursive loop. We use technology to help us adapt to a changing environment, which then adds complexity to our environment, which then requires further adaptation. We are the evolutionary agents that will use our rational faculty in the *Technology Cycle* to increase *Intellopy* (the quantity of Intelligent Matter) and entropy in the universe so long as our offspring or intelligent creations survive.

True Idea: Humans have an innate need to use their conceptual intelligence to discover and create. Their survival requires it. The resulting creations are what we call "technology."

Before the relatively recent discovery of science, mankind attributed all causes to the gods as mysticism. In times and locations where mysticism was prevalent, men who made scientific discoveries or created scientific inventions were deemed heretics or witches and prevented from further discovery or killed.

Over time, discoveries of the Laws of Reality displace mysticism. Valid knowledge eventually replaces blind belief within society. Each time the *Technology Cycle* produced discoveries and inventions, it chipped away a little more of the mysticism of the day. The men of the mind gained a little more freedom to think, discover and create. We have come a long way, but there is still mysticism in our society, filling the void of things we do not yet understand. Yet, the *Technology Cycle* continues.

The total amount of intelligent matter that we are aware of is here on Earth. The amount of matter in all the life forms on Earth and that is converted to their purposes is significant relative to the Earth. However, it is minuscule on a cosmological scale. On the cosmological time scale, we humans have not been around that long. We started slowly, but we are now picking up speed. We have reached the inflection point of the exponential technology growth curve.

The amount of intelligent matter in the universe is presently like the first cup of water we poured into the Grand Canyon. Suppose we manage to survive and continue the *Technology Cycle*. In that case, we will eventually learn how to master other planets, stars and galaxies, converting the matter in them to our purposes. Then, someday, thanks to the power of exponential growth, the universe, like the Grand Canyon, might awaken, filled with intelligent matter.

COGNITIVE FLUIDITY

In Part 1: *Knowing Your Plastic Brain*, we were introduced to and asked to remember the principle of cognitive fluidity. Cognitive fluidity is the brain's ability to physically restructure itself, creating new integrations of its various functional modules to handle what we ask it to do.

We previously noted that a human brain from the Pleistocene and one from today are identical in structure and function at birth, but they are dramatically different in adulthood. The difference is caused by what the individuals asked their brains to do during their lifetime. Two-hundred-thousand years seems like a long time. However, in genetic evolution terms, it is only a tic of the clock. Man has essentially not changed genetically in that brief time. Yet, humans of today are far more intelligent than our ancestors. Cognitive fluidity in response to the *Technology Cycle* is the explanation.

To develop language skills, the human brain did not have time to genetically develop a specific functional module. Instead, the brain formed a new integration of the functional modules it already possessed. It took us one hundred thousand years to develop language skills. Once we did, it was like turning on the power switch for our conceptual faculty, thereby greatly amplifying our ability to further modify our brain's neural networks. There are four critical observations here.

The first observation is that on the time scale of the biological evolution of intelligence from the sensory-response level of simple organisms to humans, our recent increase in intelligence has been an instantaneous explosion, a sort of Big Bang of intelligence. There was not a concurrent exponential increase in the rate of genetic change. In a relatively brief 200,000-year period, human functional intelligence increased slowly at first and then rapidly on the anatomically modern human time scale.

The second observation is that our increased adult intelligence is not passed on to our offspring biologically. Each child must develop his intelligence from scratch because he does not inherit intelligence gains by previous generations through his genes. He inherits them through conceptual knowledge recorded in society. The child develops his intelligence by using the conceptual knowledge and technology existing in his environment to restructure his brain. With proper education, the child can, in his lifetime, achieve a functional intelligence level equal to and then higher than his parents.

> ***True Idea: A proper education causes the plastic human brain to create the mental intelligence functions necessary to grasp contemporary knowledge and technology.***

The third observation is that, like a ratchet, the *Technology Cycle* continually advances global knowledge and technology. Each new generation's human brains restructure themselves to handle the knowledge and technologies that exist in their contemporary environment. Human intelligence promotes conceptual evolution, which advances human intelligence. This cycle is accelerating; therefore, each subsequent generation of humans gains an incrementally more significant intelligence than the prior generation. James R. Flynn, a researcher at the University of Otago in New Zealand, discovered the phenomenon that IQ scores have been increasing steadily since the beginning of the twentieth century. He found that they have been increasing three full IQ points per decade.[127] This IQ increase appears to be evidence of the *Technology Cycle* and Cognitive Fluidity at work.

The fourth observation is the assessment that, for all practical purposes, the biological phase of human evolution has been superseded. Its importance ended when a human formed the first unit-reduction concept and assigned it a word. Our conceptual knowledge and our technology will now cause faster evolution of both human intelligence and the human body.

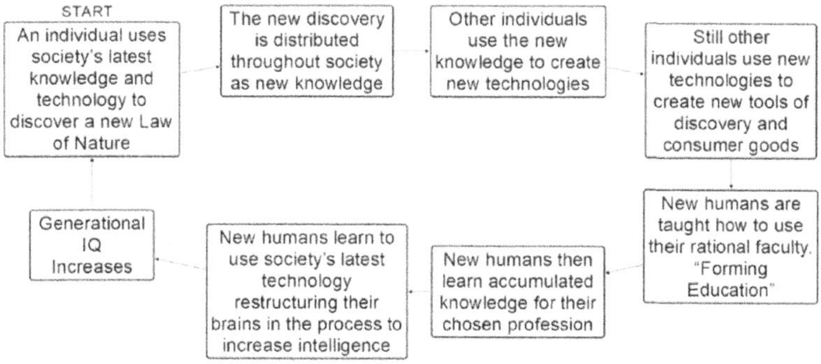

Technology Cycle and Cognitive Fluidity

Let us summarize the interplay between individual genius, the *Technology Cycle*, cultural expansion of knowledge, and cognitive fluidity. It goes something like the following:

> A genius individual discovers a Law of Reality. The new knowledge distributes throughout society. Other individuals use that knowledge to create a multitude of new technologies. Some of those technologies make their way into consumer goods, for example, the next generation of entirely realistic 3D projection for virtual reality. The next generation of human brains in society—a multitude of children—are first taught how to use their rational faculty effectively. They then learn the latest knowledge in society and how to use the new virtual reality technology as easily as today's children learn to use the TV remote. Their brains restructure their functional modules to integrate VR capabilities as a new sensory and mental capability. (A new "functional module" is a new cognitive function, not a new physical area of the brain.) This generation of humans now has a new intelligence function that prior human generations did not have. From this new generation with increased mental capabilities, the latest knowledge, and newest technology, one individual, as an adult, applies his increased intelligence to make another genius discovery of the Laws of Reality. The cycle repeats. Generational human intelligence has incrementally increased—it has evolved.

FUTURE OF CONCEPTS

We are now poised to dramatically restructure our brain's functional modules still further to gain new mental functions by integrating our minds with our technology. We will explore this in more detail later. With increased functional intelligence we will discover more conceptual knowledge about reality and create newer technology. It appears that, so long as we can manage to not kill ourselves or ruin our society, we are on an accelerating trajectory of discovery. The more we discover, the better our intelligence and tools of discovery become.

The anticipated expansion of conceptual knowledge in the physical sciences and the creation of technology by scientists and engineers is obvious. These increases in knowledge and technology are accumulated in our society and then taught to new humans. What is not so obvious is that new humans must first learn *how to think*—how to use their rational faculty to connect abstract ideas to reality.

Unfortunately, reality-based concepts within the general population regarding knowledge, ethics, politics and art are not advancing. They are retarding—moving further away from reality and reason. The intellectual leadership in our society, especially the professors in the humanities, social sciences and education that have accepted anti-reality ideas, drives this retreat.

This intellectual abandonment of reality and reason in our educational institutions has begun to erode the scientific community as well. This retarding is the reverse of the advancing period leading up to the Renaissance. Aristotle's ideas are losing, not gaining, their impact on our society. With or without advances in technology, our society might be heading toward collapse—a New Dark Ages. As we will see in a later chapter, the collapse might happen sooner than we expect.

Continued conceptual evolution depends on a society of individuals committed to reality and reason and dedicated to teaching new humans how to use their conceptual faculty effectively. Our educational institutions long ago abandoned such dedication.

While reality-based science and technology concepts are still advancing and filling our Technology bucket, our society's widely held philosophical concepts are retreating from reality, emptying our Culture bucket, leading to a dangerous imbalance. As you will see, there is hope. Perhaps we can use some of the Technology bucket's technology to inject reality-based ideas into the Culture bucket to maintain our balance on the Tightrope.

POST-CONCEPTUAL EVOLUTION

Our method of thinking is *reason*. It is our conceptual faculty. It evolved on top of the earlier perceptual faculty, which had evolved on top of the earliest sensory-response faculty. Each higher-level adds to and depends upon the layer below it. A diagram of our brain typically depicts three different brain layers according to the *Triune Brain Theory*.

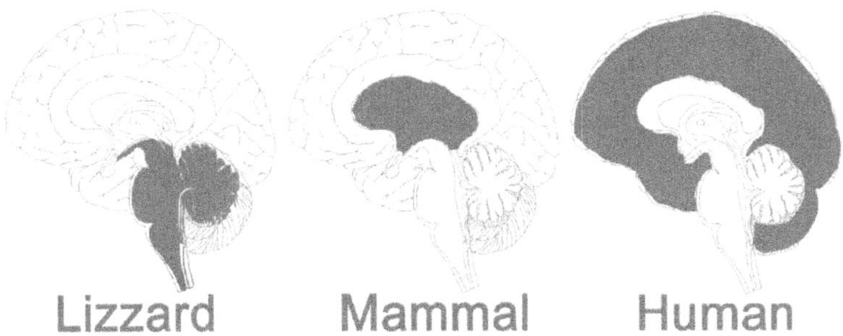

Triune Brain

The Lizard Brain layer is primarily instinctual stimulus-response. The Mammal Brain layer, also called the "Limbic" layer, added strong perceptual consciousness. The Neocortex layer added man's conceptual consciousness.

There seems to have been a progression since the first life forms appeared. It now appears that there will never be a fourth biological layer produced by evolution. All evolution for millions of years and up through humans was biological. Natural selection was a process of trial and error of random mutations of genetic material. Man now has the power to modify genetic material purposely and is beginning to enhance our biological intelligence with our technology. Random, slow biological evolution is being superseded by intentional, rapid technological evolution.

Technological evolution will proceed at lightning speed as compared to biological evolution. Biological evolution moves on a scale measured in millennia, conceptual evolution in centuries, and an individual's evolution in the decades of one's lifetime. A Conceptual Machine AI entity might evolve on a scale of years, months, days, then hours.

We will be advancing technology rapidly, especially AI, along with our ability to augment our intelligence with it. It is unlikely that a higher level of consciousness will evolve biologically. If there is ever a higher level of consciousness than our conceptual level, it will evolve technologically. We might develop a layer of consciousness added to our neocortex layer. But this will be a technological brain layer, an *exocortex*, not a biological one. Our neocortex contains approximately 300 million pattern recognizer units. Biologically increasing their number is not likely. However, an exocortex could have a billion or a trillion pattern recognizer units.

A new level of consciousness higher than the conceptual level might never be necessary. An exocortex will likely enable us to think of ever-higher abstractions. However, we will still be using the conceptual method of intelligence. It could be that the conceptual method will continue indefinitely but with greatly enhanced capabilities via an exocortex. The *method* of thinking might not change, just the speed and capacity of thought.

♦♦♦

THE LAW OF INTELLOPY

"The cosmos is within us. We are made of star-stuff. We are a way for the universe to know itself."

—Carl Sagan

WHAT DRIVES US?

Did you ever ask yourself why evolution is occurring? What will be its outcome? Can we see a universal principle at work? Is there some progression that indicates where future evolution might take us? What is our role in all this? Let us see if we can find answers to these questions.

We understand that entropic laws of thermodynamics in physics cause increasing complexity in localized conditions. This complexity eventually resulted in the emergence of biological entities. After thinking about the evolution and future of intelligence, I began to develop a premise regarding the nature of the universe (reality). It seemed to me that a "law" of the universe must exist regarding *intelligent matter* evolution.

There is a term in thermodynamics—"causal entropic forcing"—that applies to conscious living organisms' actions that increase complexity and decrease the entropy in their localized environment. They increase entropy in the overall universe because they are not closed systems and they dissipate externally obtained energy.[128] Causal entropic forcing comes close to describing the law I was seeking. Still, it does not include the goal-directed actions of non-conscious living entities. Also, while it states the *cause* for what we see in reality, it does not address the cumulative *consequences* in the universe.

We need a way to express whatever it is that drives all living things to modify matter and its consequences in the universe. We would like to understand better how this "law" is integral to who we are and our "nature" as humans. Perhaps it would help us understand why humans are *creators*.

INTELLIGENT MATTER

When we think of evolution, we usually skip past cosmological evolution and focus on biological evolution as the more interesting period. It applies to one class of existents, living entities, from self-replicating molecules to humans.

The earliest self-replicating molecules introduced a more complex form of matter than previously existed. This new self-replicating type of matter had become "intelligent" in the sense that it could now "do something" beyond just being atoms in chemical reactions. This new form of matter *acts* to support its existence and replication. We can refer to this new form of increasingly complex matter as "Intelligent Matter," wherever we might find it in the universe. We are only aware of the Intelligent Matter here on Earth, but it is possible that it also exists elsewhere.

The advent of this new Intelligent Matter in the universe was when the fundamental "alternative" in the universe began—to exist or not exist. Only Intelligent Matter entities face this alternative. Simple dumb matter faces no alternative. Its chemical and physical forms can change due to external processes, but it cannot go out of existence.

An Intelligent Matter entity, facing the alternative of existence or non-existence, must act to sustain itself. Its actions must comply with the Laws of Reality and the capabilities and limits of its nature. So long as the Intelligent Matter entity is "doing something," we consider it "alive." It must continuously "do something" (act) to achieve the purpose of continuing its existence and the goal of replicating itself. When it stops acting to sustain itself, it goes out of existence and reverts to plain old unintelligent matter. It is said to "die." We call all instances of active Intelligent Matter "life." The fundamental characteristic that differentiates all living entities from inanimate ones is "goal directed action."

In biological evolution, an Intelligent Matter entity is a test of the particular set of evolutionary traits it possesses, its "nature," against the surrounding conditions, its "environment." It acts, using the matter and energy around it to sustain itself and to replicate. Individual entities in subsequent generations who pass this test, can become more complex and more capable of success in varying environments over time.

All the life we know about organizes itself into delimited individual living entities with a definite boundary isolating each from its environment—a body.

All living entities, wherever we might find them in the universe, are one form of Intelligent Matter that we can call "Living Intelligent Matter." Over time, the complexity of Living Intelligent Matter entities has increased. Increasing intelligence improves their ability to act within their environment—to adapt—to "do" more capable things according to their improved nature. "Memory" of successful past actions embeds in the genes of a species, and in the brain of an individual having one.

Single-cell organisms bring nutrients in from their liquid environment across their cell membranes and discharge waste products. Plants take in sunlight and carbon dioxide to create internal chemical stores of energy and expel oxygen. Fish and birds build nests using what they find in their environment—sand, shells, straw, twigs. Living in forests, beavers use the matter around them by cutting down trees and damming up streams to create a better living environment to support their existence and reproduction. Humans combine countless substances from their environment into rocket ships. These are all examples of goal-directed action, the common denominator of all living entities.

As the various *species* evolved, natural selection advanced the genes that equipped offspring to succeed in the environment around them. Evolution populated the Earth with countless branches of different species, replicating for equally countless generations with each branch always selecting for an improved ability to adapt to their environment. Over evolutionary periods, the "nature" of living entities within species changed significantly.

Regarding the evolution of *life* in the universe, the first extraordinary event was the emergence of life from inanimate matter.

The second great emergence event pertained to a critical improvement in living entities regarding their ability to adapt themselves to varying environments—*consciousness.* Adaptation of the pre-conscious species progressed at a glacial pace through natural selection. Consciousness enabled individuals to adapt during their lifetime.

The third marvelous emergence event provided conscious living entities (humans) the power to adjust their environment to themselves instead of the other way around—*reason.*

Nature endowed us with conceptual-level consciousness because it makes possible enhanced survival, but how we choose to use our consciousness is up to us. It is "volitional." Human action is not "instinctual." It depends upon one's thinking, deliberation, and decision—factors which cannot be claimed to be dictated by selection pressures, genes, or any other necessitating factor.[129]

Today, we have on Earth living entities from all three emergence periods above. All of them exhibit goal-directed action. Those from before the emergence of consciousness exhibit *vegetative* goal-directed action, such as the plants. The values they pursue are set according to the memory of past successes in genetics and are not subject to further adaptation by the individual.

Those from after the emergence of consciousness exhibit *purposeful* goal-directed action where the values and behaviors are subject to the memory of past successes in the individual's brain. Adaptation continues during the life of the individual based on experience.

We can separate conscious entities into two sub-categories. Those from before the emergence of reason have *perceptual-level* consciousness—higher animals. Those from after the emergence of reason have *conceptual-level* consciousness—humans. The key points to remember as we go forward are:

1. All living entities separate into three levels by type of consciousness:
   Vegetative—Plants and simple animals (non-conscious)
   Perceptual—Higher animals (conscious-instinctual)
   Conceptual—Humans (conscious-volitional)
2. All living entities exhibit goal-directed action:
   Non-conscious entities—plants—exhibit **vegetative action**
   Conscious entities—animals and humans—exhibit **purposeful action**
3. All species adapt offspring to their environment genetically through natural selection.
4. A plant or simple animal cannot adapt to its environment during its lifetime.
5. A higher animal adapts itself to its environment through perceptual consciousness.
6. A human adapts its environment to itself through conceptual consciousness.

Sensory-response vegetative action evolved into perceptual consciousness, which evolved into conceptual consciousness. The higher the level of consciousness of Living Intelligent Matter entities, the stronger their ability to use or modify their environment. Man is making plans to build Dyson Spheres around stars to capture their energy.

When evolution created conceptual-level humans, a quantum leap in the relationship between Intelligent Matter entities and their environment occurred. Before man, animals had to adapt to their environment according to specific, genetically programmed instincts without introspective awareness of purpose. Man, and perhaps future non-biological conceptual entities created by man, can be aware of and choose their purposes and then adapt the environment to themselves according to their ever-increasing understanding of and ability to apply the Laws of Reality.

The matter that makes up the conscious entities themselves can be called Living Intelligent Matter—the form of Intelligent Matter that "does something." Matter converted to the purposes of Living Intelligent Matter entities can also be considered a form Intelligent Matter wherever we might find it in the universe. We have converted all the silicon in computer chips across the world to intelligent use and the iron (steel) used in ships, engines and skyscrapers. Pencils, cars and roads, buildings, spaceships and Mars rovers are all Intelligent Matter. If we build a machine with human-level intelligence, it will be Intelligent Matter.

All the Intelligent Matter in the universe consists of two types, *Living Intelligent Matter* that "does something" and *Purposed Intelligent Matter* that is "used for something." Evolution has exhibited a progression of living entities with increasing ability to understand the Laws of Reality and to use them to modify matter to their purposes. We call this increasing *intelligence.* It would be presumptuous to think that evolution ended when humans first appeared. We can assume that evolution is continuing, and intelligence is increasing.

By "increasing intelligence," we usually mean a higher *level* of intelligence attained in one or more individuals, not the cumulative *quantity* of intelligence of all individuals collectively.

We need a term to represent the increasing total collective sum of all Intelligent Matter in the universe, including all the Living Intelligent Matter and all the Purposed Intelligent Matter. It must include biological life forms that exist anywhere in the universe today, non-biological life forms that might exist in the future, and all the matter that these life forms, biological or non-biological, convert to their purposes.

The laws of thermodynamics are physical laws regarding the quantity of *energy* in the universe, not quantity of *intelligent matter*. They do not account for the goal-directed actions of intelligent entities that intend to use all the universe's energy to convert all the universe's matter to intelligent purposes. The conversion begins with our planet, then the solar system, galaxy, and all the other galaxies within our finite knowable universe limited by the speed of light. We might discover how to exceed the speed of light and access an infinite universe where entropic "heat-death" is not an issue. We need a term to refer to the *consequence* of evolution and intelligent goal-directed action, which is the ever-increasing quantity of Intelligent Matter, both living and purposed, in the universe.

Let us use the term "Intellopy" to represent the degree to which all the universe's matter has been converted into living or purposed Intelligent Matter. So far as we know, the entire universe except for this planet is dumb—asleep. Intellopy is the degree to which the universe has become intelligent—awake! Man intends to awaken the entire universe.

THE LAW OF INTELLOPY

Intelligent Matter—"Living" and "Purposed"

"Intelligent Matter" is made up of:

1. Living entities that "do something" according to the nature of their consciousness and in compliance with the Laws of Reality in the furtherance of their individual life and their kind; and

2. The matter they convert to their purposes that is "used for something."

"Intellopy" is the degree to which all the universe's matter has been converted to Intelligent Matter.

"Law of Intellopy" is the observation that the physical properties of the universe include the emergence of Intellopy (the quantity of Intelligent Matter in the universe) at an accelerating rate over time.

EVOLUTION IS CONTINUING

The Law of Intellopy is not a scientific term but rather an observational theory. It names the universe's property that causes or enables increasing intelligence, natural law discovery, and matter conversion to purposeful use.

According to the Law of Intellopy, evolution is continuing to evolve ever-higher levels of intelligence. We have now transitioned from biological evolution to technological evolution. The evolution of intelligence has reached the "knee" of an exponential growth curve. Accelerating growth in Intellopy (quantity of Intelligent Matter) in the universe requires accelerating intelligence increases (level of intelligence).

"Replication," as a method of advancing intelligence, will soon be enhanced and accelerated. Over a relatively few decades, it appears that biological humans will augment their intelligence with technology and then, perhaps, create artificially intelligent machine-based entities. We will be able to advance new generations of technology-based intelligence on a much faster cycle than it takes to reproduce a human.

Technologically enabled evolution is eclipsing biological evolution's natural selection process based on random chance. The intentional evolution of intelligence will now accelerate in the domain of technology. Natural selection could not anticipate the future—it selected adaptive mechanisms because of the survival benefit they provided organisms in the past. We can anticipate the future and intentionally create the adaptive mechanisms desired to accomplish long-range goals. We will exponentially speed up our ability to learn.

Connection to Reality

The next phase of technologically driven evolution will depend on those humans who hold a reality-based Worldview. People who do not believe that reality is real or trust their rational faculty will more likely impede advancing intelligence than contribute to it. A proper, reality-based Worldview is essential for both AI creators and the AI entities they might create.

Suppose humans or machines evolve a higher level of consciousness. In that case, that level must be able to connect its most abstract content back to observable reality. This validation is necessary to be able to manipulate reality to achieve even greater intelligence—"Nature to be commanded, must be obeyed."

DEFINING INTELLIGENCE TERMS

Referring to levels of artificial intelligence, other authors use the terms "Artificial General Intelligence" (human equivalent) and "Artificial Super Intelligence" (something much smarter than humans). I find these terms somewhat limiting, so I will use the following terms. They will be briefly introduced here and then covered more thoroughly.

- Standard Human
- Hybrid Human
- Virtual Human
- Narrow AI
- Vegetative AI
- Perceptual AI
- Conceptual AI
- Ultra-Conceptual Human
- Ultra-Conceptual AI

It all began with biological, anatomically modern, conceptually capable humans in the pre-language, pre-technology period. They were ready to start growing their conceptual intelligence.

"Standard Human"

Once humans developed language abilities, they awakened their conceptual faculty. They began using the *Technology Cycle* to expand the totality of human conceptual knowledge. A Standard Human's intelligence remains conceptual. He increases his mental ability as he discovers more Laws of Reality and develops new technologies. His brain forms new integrations of functional modules.

A Standard Human's conceptual abilities are enhanced by interfacing with external technology aids through the original biological senses. Galileo enhanced his conceptual ability by using the telescope. Today, we enhance our conceptual ability by using robot landers to explore other planets. For a few more years, we will continue to improve the intelligence-enhancing external technology that we interface with through our original biological senses. This external technology, such as personal AI assistants, will have varying degrees of artificial intelligence, further enhancing our intelligence.

All humans, from the development of language skills through to today, are Standard Humans regarding intelligence. Even though we will get progressively smarter, we will be Standard Humans so long as we rely on our original senses for interfacing with our technology. Suppose a human child, who became frozen in the Arctic a thousand years ago, was thawed out today and taught with contemporary children. In that case, he could interface with, understand and use today's technology. All Standard Humans have the same basic model number. Let us call them version 1.0 humans.

"Hybrid Human"

We will soon enter a period where humans increase intelligence via the expansion of sensory, perceptual, and conceptual functioning through brain-to-machine interfaces that bypass the original sense organs. We will augment our brain by connecting it directly to technological AI devices that enhance existing senses or add new ones and expand our mental capacity and speed.

Having an informational image projected onto your retina by a wearable device in your glasses is essentially no different than Galileo's looking through his telescope. You are still using your original sight sense organ to interface with technology. However, a critical evolutionary event will occur when brain-to-machine interfaces bypass original sense organs establishing two-way communication between the brain and technology. We will have entered the era of Hybrid Human intelligence. From that point forward, our intelligence will more quickly evolve as a combination of biology and technology.

The technology portion of Hybrid Humans will advance through successive new versions as the technology improves. Hybrid Humans will begin being referred to by their brain-machine-interface (BMI) technology version numbers. They could be hybrid human version 2.1, 2.2, 2.3 etc. Our 1,000-year-old thawed out version 1.0 human child would not be able to interface with modern technology by using his original biological senses. He would need to have the direct brain-to-machine interfaces installed.

We will fully explore the limits of our incredible plastic brain's ability to rewire and adapt itself to new connections. Our brain's self-awareness, abstract thinking, and creativity will act as the top-level "executive" that directs the work of multiple external AI processes. Those processes excel at finding patterns in large amounts of data and remembering data but lack those executive functions.

The era of increasingly intelligent Hybrid Humans will continue until nanotechnology can completely map the human brain's physical functioning and brain-body connections and computing technology can model it. Only then is a human-equivalent, self-aware, conceptually intelligent, machine-only AI entity likely to become possible.

"Virtual Human"

When computer technology and brain mapping technology become sufficiently advanced, we might be able to transfer our human consciousness into a machine-based "body." At that point, we will become Virtual Humans. We might view this as preferable to our biological body. Once our conceptual consciousness resides entirely in technology, we can increase our intelligence beyond the limits imposed by our biological brain's structure and capacity.

To be successful at becoming Virtual Humans, we will need to maintain our connectedness to *real* reality, not just an *artificial* virtual reality. Suppose we lose our connection to reality and exist inside a machine that provides a virtual world. We grasp it using virtual senses and minds, with no need for anything or ability to affect anything in the outside world. In that case, we will have become irrelevant in the context of reality and evolution—*nonexistent*. The machine in which we reside would still be relevant. It would need to get energy and materials from reality to maintain its existence.

"Narrow AI"

Narrow AI is all that we have today. Today's Narrow AI already exceeds humans in the ability to detect patterns in large amounts of data and to remember large amounts of information without forgetting. They can detect spoken words in an audio stream, a person's face in a video stream, or a potential tumor in a medical scan image. They can examine large amounts of medical research data to find the most effective drugs or treatments. They can be faster, more accurate and more consistent than us at performing pre-programmed actions in response to pre-programmed input signals. We use Narrow AI for robotic arms welding cars in factories, fly-by-wire jet airliners, and autonomous vehicles.

In most cases, we provide Narrow AI with the programming for what input signals to look for and what actions to take in response. Today's more advanced Narrow AI learns these stimulus-response pairings on their own using neural networks, which are primitive attempts at modeling the networks in the brain. However, even though IBM's Watson can win at the Jeopardy game and assist industry in many ways, it is not a living thing and is not conscious.

The following will describe "higher" levels of artificial intelligence. However, as you will soon see, it is this lowest level—Narrow AI—that already presents our civilization with its most grave and immediate threat. Higher levels will only accelerate the danger.

"Vegetative AI"

The first "living" machine that we create will be vegetative AI entities. They will not be conscious. Like plants and simple animals, they will exhibit goal-directed action toward sustaining their own "life" as an entity that "does something," motivated by the fundamental physical value of their continuing existence.

"Perceptual AI"

Building upon the base of success with Vegetative AI entities, the first "conscious" AI entity that we create will have a perceptual-level consciousness like the higher animals. A Perceptual AI will need to have the equivalent of at least a rudimentary "brain." Think of your family dog. Perceptual AI will use their senses to isolate, identify, and track objects, permanently learn from new experiences, have emotional responses to stimuli, and exhibit purposeful goal-directed action motivated by psychological values.

Perceptual AI will be able to respond to spoken commands from humans and perform learned actions. However, lacking a conceptual-level consciousness, the Perceptual AI will have no abstract understanding of the concepts referenced in language.

A Perceptual AI entity can be considered a living thing, like a dog. It will likely have programming equivalent to "instinct" that enables it to support its own life. Humans will interact with Perceptual AI entities much the same way we interact with our pets.

"Conceptual AI"

When we build it, the first non-biological machine to have intelligence equivalent to a Standard Human will be a Conceptual AI entity. We will base it on our successes with Perceptual AI entities. If its intelligence is considered equivalent to a human, it will be "alive" in the same sense that we are alive. A human-equivalent Conceptual AI must be capable of abstract conceptual thinking, be self-aware, and have volition. I use the term "Conceptual AI" as a shortened version of "Conceptual-Volitional AI." It will also need to have abstract conceptual language capability and connect its highest level abstract mental content down through its layers of intelligence to its senses and observable reality. To do so, it will need reality-based Worldview content in its artificial brain.

We are likely to model the Conceptual AI after the functioning of the human brain. Therefore, it will have perceptual and conceptual levels of intelligence similar to ours. If we intend to communicate with it, this model will be a requirement. The Conceptual AI will likely inherit the Worldview of its creators.

Humans will need to interact with Conceptual AI entities by affording them the same rights and obligations as other humans. We will have a mixed society of biological and machine conceptual entities.

"Ultra-Conceptual Human"

A consciousness level higher ("ultra") than a "conceptual" consciousness might not need to evolve. If it does, it will be the continuing ascendancy of humans or our machines.

In the presence of an "Ultra-Conceptual Human," a "Standard Human" will be as incapable of understanding ultra-conceptual thinking as an animal is incapable of understanding the Standard Human's thinking. No amount of time or learning will enable him to do so.

Suppose the limitations of our biological brains prevent the evolution of a higher level of intelligence. In that case, when we become Virtual Humans inside our technology and unconstrained by biological brains, we might develop a consciousness level higher than our present conceptual consciousness.

Whatever ultra-conceptual consciousness we might evolve, we already can connect its content down through our conceptual, perceptual, and sensory levels of consciousness to observable reality. We will continue to do so, provided we hold onto a reality-based Worldview in our mental content.

"Ultra-Conceptual AI"

Just as Hybrid Humans or Virtual Humans might someday become Ultra-Conceptual Humans, Conceptual AI entities could become Ultra-Conceptual AI entities.

Standard Humans, Virtual Humans, and Conceptual AI entities will not understand Ultra-Conceptual Humans or Ultra-Conceptual AI. It could be that, if they exist in the same period, Ultra-Conceptual Humans and Ultra-Conceptual AI would understand each other.

Some futurists worry that Ultra-Conceptual AI (which they call "Artificial Super Intelligence) will eradicate mankind. I think Ultra-Conceptual Humans are likely to arise before Ultra-Conceptual AI, thereby reducing the risk of such a disaster scenario.

AGENTS OF INTELLOPY

Before the appearance of the first life forms, the evolution of the physical universe—from the simplest elements to the stars, planets, the heavier elements, and the chemistry that would someday support life—represented the advancing complexity of matter. It all happened on a cosmological scale of space and time. Within that vast scale, we know of at least one instance where the conditions eventually resulted in life. There could be other locations in the universe where life formed. Still, for us, the history of our appearance and evolution on this Earth is our context. Evolution created our physical bodies with opposable thumbs, long-distance mobility, and our brains with conceptual capabilities.

The Law of Intellopy relates this property of nature to intelligent entities. Once life appeared, it facilitated a new method of reorganizing matter that is far faster and more capable than the unintelligent random chance method up to that point. The Law of Intellopy indicates that evolution created humans with a drive to discover and create—to modify their environment—to convert dumb matter into Intelligent Matter. Indeed, this need to discover and create was a powerful adaptation for survival in a changing environment. We are now dependent upon it. It is essential to our survival. In today's world, this explains our need for productive work as the central value around which we organize our other values.

We are the current agents of the Law of Intellopy. Will our "human" branch of the evolutionary tree continue discovering nature's laws and modifying our environment until we master the universe? Or will we reach a dead end, leaving it to some other branch of entities here or somewhere else in the universe to continue?

ULTIMATE OUTCOME

Will Mankind Awaken the Universe?

We might not require a higher consciousness and "method" of thinking to master the universe. It appears that Hybrid Humans will occur earlier than Conceptual AI. We do not know whether Hybrid Humans or Virtual Humans will advance intelligence forever or if it will be non-human Conceptual AI.

Suppose the Homo sapiens species does not go extinct, and its society does not collapse. In that case, the Law of Intellopy predicts that future technology-based intelligent entities that evolve from today's humans will, over time, awaken the entire universe, converting its matter into intelligence.

Kurzweil provides an eloquent description of the Law of Intellopy's ultimate outcome from a "human" perspective, "I see it, ultimately, as an awakening of the whole universe. I think the whole universe right now is basically made up of dumb matter and energy and I think it will wake up. But if it becomes transformed into this sublimely intelligent matter and energy, I hope to be a part of that."[130]

♦♦♦

3.3

TECHNOLOGY

"Physical law is the bedrock beneath the landscape of technological potential, and knowledge of physical law is the basis for exploring the terrain."

—K. ERIC DREXLER

"An inventor is a man who asks 'Why?' of the universe and lets nothing stand between the answer and his mind."

—AYN RAND

"Humans are the reproductive organ of technology."

—KEVIN KELLY

ROLES

When we think of "technology," we can ask ourselves, "How does it get created? Who creates it? What are the conditions necessary? What does it depend on?" We address questions like these in this chapter. We previously identified the four fundamental repeating steps in the *Technology Cycle*. It begins when a *New Discovery* leads to *New Knowledge*, leading to *New Technology*, leading to *New Intelligence* and the cycle repeats.

Now we will examine how technology gets created in more detail. As we consider the following topics, keep in mind that the *Technology Cycle* depends on a free society. It depends on individual minds being free to think without coercion. It depends on individual career specialization, division of labor, sharing, access to knowledge, materials, money and skills, teamwork, market economics, property rights, contracts, patents, and much more that can only exist in a free society. First, let us consider the roles of different professions and institutions.

Scientists

Scientists perform the *discovery* step in the *Technology Cycle.* Their essence is *inquiry.* They ask, "How can we discover new knowledge?" They inquire and discover new Laws of Reality. Scientists study physical things, then describe them, focusing on the unknowns and using instruments. Information flows from matter into their minds as the physical world shapes their descriptions. Scientists provide knowledge that enables engineering.

Engineers

Engineers perform the *new technology* step in the *Technology Cycle.* Their essence is *designing.* They ask, "How can we deliver new products?" They apply new scientific information in the design of new products and services. They describe physical things then build them. Information flows from their minds to matter as they shape the physical world to fit their descriptions. Engineers focus on well-established knowledge in textbooks, data, specifications and engineering practice. They avoid the unknowns or offset them with margins of safety. They build instruments that enable science.

Differences	Scientist	Engineer
Essential role:	Inquiry	Design
Asks the question:	How can we discover new knowledge?	How can we deliver new products?
Physical things:	Study then describe	Describe then build
Information flows:	From matter into their minds	From their minds into matter
The unknown:	Focuses on the unknowns	Avoids or offsets the unknowns
Produces:	New knowledge that enables engineering	New tools that enable science

Summary Of Differences Between Science and Engineering.

Sometimes it is difficult to make the distinction between science and engineering. It is not unusual for a scientist to do engineering in practice, such as building a diagnostic tool that he needs. Or for an engineer to do science, such as discovering a new phenomenon while building an experimental machine. There is a clear distinction between each role even though the same person might be doing it, as indicated in the table above. These differences lead to certain attitudes and approaches to problems. For example, the scientist is motivated by the unknown, while the engineer fears it and does all that he can to avoid it. An organization can make the mistake of giving a project to scientists when it should give it to engineers and vice-versa. The project itself, or its major component sub-projects, must be evaluated against the above table and identified as either a science project or an engineering project.

Business Professionals

Business professionals are the creators on the most significant scale. They envision a way to produce quantities of a product or service on a massive scale to make a profit for themselves and their investors commensurate with the novelty and risks involved. They combine the capital, talent and physical resources needed to develop and produce the new products in quantity and distribute them to the masses.

Business professionals "own" their "vision" and their "plan" for achieving it. They make their vision a reality by employing scientists and engineers and overseeing their efforts. They enjoy the rewards of their success and suffer the negative consequences of their failure. Business professionals are the great enablers, providing an opportunity for all "men of the mind" to exercise their creative talents at all levels of ability.

It would not have been possible to have been a businessperson during the Middle (Dark) Ages when Aristotle's ideas were lost from view. The then existing philosophy of the church that required self-sacrifice deemed the profit motive as evil. The profit motive would have gotten you killed.

Aristotle had not witnessed the Renaissance or the success of industrial era businesspeople who flourished in the freedom provided by the American Constitution and sense of life. Twentieth-century philosopher Ayn Rand did have the advantage of more recent history. She explained the power that Aristotle's ideas, as embodied in business professionals, had on improving the human condition:

> The professional businessman is the field agent of the army whose lieutenant-commander-in-chief is the scientist. The businessman carries

> scientific discoveries from the laboratory of the inventor to industrial plants, and transforms them into material products that fill men's physical needs and expand the comfort of men's existence. By creating a mass market, he makes these products available to every income level of society. By using machines, he increases the productivity of human labor, thus raising labor's economic rewards. By organizing human effort into productive enterprises, he creates employment for men of countless professions. He is the great liberator who, in the short span of a century and a half, has released men from bondage to their physical needs, has released them from the terrible drudgery of an eighteen-hour workday of manual labor for their barest subsistence, has released them from famines, from pestilences, from the stagnant hopelessness and terror in which most of mankind had lived in all the pre-capitalist centuries—and in which most of it still lives, in non-capitalist countries.[131]

When it comes to "big" business—the "mega-corporations"—and the impact they have on our society, we must keep in mind the distinction between "economic" power and "political" power. *Economic power* is the "good." It is the measure of the business's ability to "create" wealth and improve everyone's lives in society. There is no such thing as *too much* economic power in a free market. *Political power* is the "evil." It is the measure of a business's ability to "destroy" wealth in society by unfair advantage in the marketplace gained through political "pull." Any degree of political power is *too much.*

Investors and Financiers

Investors and financiers provide the capital the business professionals need to create and grow their businesses in exchange for a return on investment. In their 2017 book, *In Pursuit of Wealth. The Moral Case For Finance*, Yaron Brook and Don Watkins provide insight into the critical role of finance.

> Finance is scientific and reality-oriented. It has to do with the real world, especially economics and business. The financial industry allocates capital, money, and credit for productive uses in the economy. Businessmen use capital to build factories full of equipment and to pay

> people's salaries. This capital is applied so that companies can produce wealth as opposed to consuming wealth. The financial markets are like a circulatory system that ensures the lifeblood of business, capital, gets to where it is most productive. However, people have always had a negative view of finance due to wrong ideas regarding its productivity, morality and power.[132]

From ancient Greeks, the idea of "usury" was shunned because they thought money could only be used as a medium of exchange, not for anything productive. Then the Christians shunned usury because we are supposed to help others without expecting a return; this is the "altruistic" objection. These ideas led to the widespread view that financiers are immoral parasites who exploit people in need for their own profit. Leading thinkers of the last few hundred years have come to understand the productive value of finance. However, politicians have kept the views of finance immorality alive by blaming every crisis, disaster or problem on finance rather than the true source of the problems—central bankers, political intervention, government regulations.

We have never had a totally free financial market in the United States since its founding. There has always been a distrust of "big" banks, so they are heavily regulated. In a free market, the banks would only have *economic* power not *political* power. Banks only gain political power when given to them by the government, such as protection from competition. This heavy involvement of a regulatory government, with both controls and protections, causes instability in the finance industry.

The stock market allows companies to raise large amounts of capital to fuel growth. The stock market also allocates capital throughout the economy by helping identify which businesses are going to succeed. The stock prices are the best and fastest information sources we have that reflect the opinions of knowledgeable people about individual companies' performance and their prospects. This information determines where the capital goes.

Individuals who work in the financial industry generally create value by seeing that capital is put to the best possible use in the economy, thereby maximizing both the creation of the wealth of goods and services produced by the economy's businesses, and the return for the owners of the capital. However, government intervention makes possible all sorts of abuses of the not-truly-free-market system by dishonest people. We must not characterize an entire industry that performs a critical and moral service with such a broad immoral label. To do so is a grave injustice to the honest majority earning lots of money by their productive achievement. They make it possible for creative innovators, like Steve Jobs, Jeff Bezos and Elon Musk, to provide us with the wonderful technologies that benefit us all.

Employees

Businesses are integrators of science, technology, capital, physical materials and labor. Each component, including labor, is critical to their success. Labor is the powerful engine that increases Intellopy on a massive scale, converting dumb matter to intelligent use—from skyscrapers to particle accelerators. Man's earliest technology, stone tools, served to multiply man's effectiveness at converting matter to intelligent use. The steam engine multiplied labor's effectiveness still further in the Industrial Age. The microprocessor increased it again in the Information Age. The transcendent technologies will exponentially multiply labor's effectiveness at increasing Intellopy in the universe as we transcend from biology-based conceptual entities to technology-based conceptual entities.

All productive work comes from the mind. Even a ditch digger who knows how to dig a straight, uniform ditch is applying the power of his mind. The higher the level of (a proper) education from childhood on, the more capable a mind's ability to do intellectual tasks. The greater the investment in a person's education, knowledge and skill in a required field, the rarer and more valuable that person, as a laborer, will be. A nuclear physicist working for an MRI medical imaging company might command a far greater salary than a ditch digger. Still, if both are using their minds to the best of their ability, they are morally equal.

Often, it is the mind's *creativity* that we value in all fields, from engineering to art. The more a person has developed their creativity and inventiveness skills in their chosen domain, the higher their value to an employer.

In a free market system, the relationship between an *employee* and an *employer* is that of *traders*. The employee is willing to provide his labor to the employer in exchange for a specified salary and other identified benefits, both tangible and intangible. The employer is willing to provide the salary and benefits in exchange for a specified scope and quantity of work done by the employee. As with any deal between equal traders, both parties must believe the arrangement to be beneficial. In a free market, absent an employment contract between the parties for a specific duration, either party is free to end the relationship.

> ***True Idea: There is no conflict between rational persons, each pursuing their happiness.***

The relationship between an *employee* and his *supervisor* is different and unequal. Usually, the supervisor has more domain-specific expertise or experience than the employee as in a master-apprentice relationship. The master teaches the apprentice, thereby enabling him to take on more responsibility over time. The supervisor is usually held accountable for a more significant part of company performance than the employee. Because he is accountable for performance, the supervisor has the right to make decisions. He is the "boss" over the employee, who understands and respects his boss's position.

Teamwork is essential. However, there is no such thing as a "team" mind. Only individuals have minds. Everyone's productive achievements must be judged and compensated accordingly and not hidden behind the performance or lack of performance of a group. It is the individual's achievement that is the source of his sense of pride. The "glue" that makes the team effective in a division of labor situation is everyone's respect for the unique talents, skills, knowledge and contribution that achieves much larger successes than any one individual can achieve.

Labor unions are also collections of individuals. There is nothing immoral about employees voluntarily combining into a group for collective bargaining with one voice. Their negotiating power is in their numbers. However, when government grants the labor union rights greater than what an individual worker would have, the threat of government force takes away rights from the employer. The trader principle between equals is lost, the immoral government intervention having replaced it. Further, government regulations that force any employee to become a member of such a labor collective or pay union dues is an erosion of his individual rights.

As AI and robotics advance, there will be a replacement of less specialized and less creative human labor with machine labor. This replacement begins with the more straightforward, repetitive manual labor tasks. It then works its way up from blue-collar to white-collar jobs that consist of repetitive information processing tasks. The more specialized and more creative one's education and skills, the less likely a robot will replace them. However, with the massive paradigm changes throughout society coming soon and with longer life expectancies, everyone should begin preparing themselves to adapt, seek retraining and new employment. For each job opportunity lost, there will be multiple new ones created. Change will be the new normal.

Consumers

Each of us is a potential customer waiting to improve our lives by the product or services the business professional brings into abundance. Our spending provides the ultimate financial support on a massive scale for businesspeople, scientists, engineers, investors, financiers, and employees. By "voting" with our purchases, we provide feedback about which businesses are doing a good job and which are not.

As with any form of voting, making a purchase decision includes a responsibility. Suppose we have knowledge that a business is engaged in some immoral practice. In that case, we should not purchase from that company, regardless of price. This responsibility does not require that we fully research every company with which we do business. Still, it does mean that we maintain our cognitive connection to reality and consider all the known facts without evasion.

As consumers, we also are responsible for our safety. We must not rely solely on government agencies to keep us safe from the products and services we purchase. Again, we must consider all relevant facts when making a purchase decision, including the company's quality and safety reputation. Ultimately, we consumers should seek to reward those who meet or exceed our wants and needs with safe, high-quality products and services at fair prices.

Free Market and Government

We must acknowledge that business and commerce is the consequence of the freedom enjoyed by both producers and consumers to deal with each other.

Producers are free to create whatever products and services they think consumers want. Consumers are free to vote for products and services with their purchases. It is a relationship based on the "trader principle," trading value for value.

A person's mind is his only source of knowledge and his only tool for survival. The freedom to act according to rational self-interest is his only means of living his life according to his nature. Business professionals are the catalysts of society that make this free marketplace work through their rational self-interest.

The imagination and creativity of the producers depend on their minds being free to think. Force, coercion, or fraud kills imagination and creativity and is not allowed in a free market. That is the reason, and the only reason, we need the military, police and courts. Any socio-economic system not based on complete freedom for the producers and consumers must be based on some degree of force or coercion and is therefore immoral. This lack of freedom is the hallmark of any centrally planned economic system such as communism, socialism, or totalitarianism.

Because the government's only moral role is to protect the producers' and consumers' freedom to act in their rational self-interest without force, coercion, or fraud, it is not proper for the government to ensure a citizen's success. The citizen has the freedom to try and to either succeed or fail. Suppose a government seeks to regulate business and commerce rather than protect its absolute freedom or somehow influence who will succeed or fail. In that case, the government itself becomes an oppressor, and the businesspeople and consumers both suffer. Ayn Rand described the proper role of government:

> The source of government's authority is 'the consent of the governed.' This means that the 'government' is not the ruler, but the servant or agent of the citizens; it means that the government as such has no rights except the rights delegated to it by the citizens for a specific purpose. [...] Under a proper social system, a private individual is legally free to take any action he pleases (so long as he does not violate the rights of others), while a government official is bound by law in his every official act. A private individual may do anything except that which is legally forbidden; a government official may do nothing except that which is legally permitted. [...] This is the means of subordinating 'might' to

> 'right.' This is the American concept of a "government of laws and not of men." [...] The fundamental difference between private action and governmental action—a difference thoroughly ignored and evaded today—lies in the fact that a government holds a monopoly on the legal use of physical force. It has to hold such a monopoly, since it is the agent of restraining and combating the use of force; and for that very same reason, its actions have to be rigidly defined, delimited and circumscribed; no touch of whim or caprice should be permitted in its performance; it should be an impersonal robot, with laws as its only motive power. If a society is to be free, its government has to be controlled.

We do not live in a totally free capitalist society in America today because the government is involved in functions that go far beyond the military, police and courts. Our government is involved in money, education, commerce, transportation, health care, food and drug, the postal service and more. We live in an economic system that is partly free and partly socialistic. Our government's central planners oppress citizens under the excuse of protecting us from ourselves, which negates our right and responsibility to be in control of—"pilot"—our own lives.

The free portion of our system is a testament to the power of minds left free to think and act. We can see the achievements produced by this freedom despite the oppression. Still, we cannot see the lost achievements that were not possible because of our government's immoral overreach into our lives. Suppose we want the most creative and productive minds to develop and produce advanced technology and the brilliant tools to keep us safe in the future. In that case, we must have the most unrestrained market possible.

True Idea: Persuasion and voluntary trade of values between people are rational. The use of force or fraud or expecting anyone to sacrifice themselves is irrational and immoral.

True Idea: People, to live according to their nature as rational animals, must be left free to use their minds to guide their actions in the pursuit of their happiness.

Here, we must emphasize that the government's proper role is *limited* to the military, police, and courts. Every other function must be reserved solely to the private sector, the creativity of which is more than capable of providing solutions to all these functions. Involvement in such overreach functions is coercion of one form or another by the government against its citizens. It degrades their freedom of action in a less-than-fully-free market.

True Idea: Capitalism is the only political system that leaves people free from force, fraud, and sacrifice.

APPLIES TO:	QUESTION:	ANSWER:
Society	*What about dealing with others?*	*Everyone must be free to think, act, and trade with each other without force, coercion, or sacrifice.*

True Idea: The American Constitution, based on Aristotelian ideas, established the first country in human history founded on the principle of individual sovereignty and freedom.

True Idea: Socialism, communism, totalitarianism all require that the individual sacrifice himself. Only the "to whom" changes.

True Idea: To form a rational society, individuals grant the government the exclusive right to use force, but only in retaliation to the initiation of force by others.

True Idea: The only proper moral functions of government are those derived from its sanction by the people, the right to use retaliatory force for their protection. We need the military to protect us from foreign invaders. We need the police and the courts, based on objective laws, to protect us from criminals, fraud, and contract breaches.

True Idea: Any activity by the government that initiates coercion or force against citizens that is not in retaliation against their initiation of force against others is improper and immoral. Functions other than the military, police, and courts must remain with the free citizens' creativity and productiveness.

Education

For businesses to succeed and technology to advance, we must produce enough people who have received a proper education. Today's public education too often causes stunted conceptual ability. Too many people end up as less capable thinkers and must work for the more capable thinkers for their livelihood. These are the people most at risk of having their jobs replaced by robots.

In the next chapter, we will see that a person who has received a proper forming education has learned "how to learn." Such a person, entering a new work environment, can acquire its domain-specific knowledge, including applicable cause-and-effect principles, that he needs to be productive. The person who did not learn how to learn will pick up concrete facts from his on-the-job training, such as specific steps to perform, but will not grasp the principles behind those steps. Such a person will have limited upward mobility in his employment.

MORALITY OF TECHNOLOGY

The scientist makes no value judgments about whether a Law of Reality, or the technology developed to discover it, is "good" or "evil," it just "is." Its discovery is an incremental increase in our knowledge of how reality works. Value judgments come later in the *Technology Cycle* when individuals apply the Laws of Reality to their purposes. Value judgments are only possible regarding the "application" of Laws of Reality for specific purposes, not the "discovery" of them.

It is the "purpose" of the person developing or using the technology that one judges, not the technology itself. If their purpose is to provide a good or service that others voluntarily desire, then it is judged "good." Suppose they aim to initiate some form of injury, force, coercion, fraud, or forced sacrifice on others. In that case, one judges their purpose as "evil." One person can develop a technology for a "good" purpose. Then another person can use the same technology for an "evil" purpose. The technology itself is neither good nor evil. The two people are—one good and one evil.

Some believe that certain technologies have so much potential for danger that the technology's pursuit should be outright banned. Others propose that we hold technology creators (large corporations) legally and financially responsible for bad outcomes resulting from the technology they create. They see this as a means of instilling cautiousness in technology development. Even if you could discount the fact that it is human nature to discover and create, banning selected technologies is not a good idea.

First, powerful technology cannot be banned from development by the "good" people because the result will be that only the "evil" people will develop it in secret. The good people might need that technology to stop and retaliate against those who initiate its improper use. It is better for good people to understand a potentially dangerous technology's workings and implications to build defenses against its potential misuse before the evil people get hold of it. Having the technology already in the arsenal of the good people is an enormously good thing.

Second, we cannot predict when a particular new technology will solve an unknown future problem. Perhaps an issue arises from the unintended consequences or byproducts of an earlier technology that now threatens us. For example, we cannot ban genetically modified organisms—GMOs. We might need new genetically engineered bacteria to sequester or convert carbon dioxide or clean up all the plastics accumulated in our environment. We will need in our toolkit the best understanding of how to apply the Laws of Reality that we can obtain.

Many have pointed out that danger now comes from just the knowledge (information) that enables the new genetic, nanotech, and certain other technologies. We will no longer have the protections that the necessity of an extensive industrial infrastructure provided to limit and slow the development of new, potentially dangerous technology. Producing a nuclear bomb requires an enormous investment in a complex industrial infrastructure. However, a graduate student in a modest laboratory, armed with the applicable information, can now produce a dangerous pathogen.

This new reality amplifies the need for those with "good" motives always to be the first to acquire new technology and plan for the possibility of its misuse before the "evil" people do. We need good people to be as aggressive as possible in discovering new Laws of Reality and related technology. Then we need a means of establishing "trust" reputations for those who will share access to the latest, potentially dangerous, information. There is an important information security role here for the transcendent blockchain technology presented in Part 4.

This new reality also emphasizes the theme of this book. What differentiates the "good" people from the "evil" ones? What will best protect us from evil people or evil future AI entities? The answer is—a Worldview based on reality and reason. The good people or good AI entities will have it—the evil people or evil AI entities will not. Protection from dangerous technology is about minimizing risk. We cannot eliminate it, but we can reduce it. The more widely a reality-based Worldview is adopted in our global society, the safer we will be.

SOCIETY

Individuals evolve, and therefore a society of individuals evolves. A long time ago, we developed language, division of labor, teamwork, and leadership so that the tribe could have a successful hunt. A little later, we developed writing so that knowledge could be shared and survive to future generations, and we began formally teaching our young. We assembled in cities with larger populations so that the creators could undertake larger-scale goals. We developed money, banking, investment capital, and skilled labor, all supporting the large-scale creators. We established free societies that protected creators from conquest and plunder.

We are presently in the Conceptual Evolution Period. This period is about increasing knowledge of the Laws of Reality, advancing technology, and increasing intelligence. Individuals pursuing their rational self-interest, from scientists to consumers, each contribute in some way to the discovery and accumulation of knowledge. Society is the information distribution network and the information storehouse for all this accumulated knowledge. By widely distributing this knowledge to individuals, society causes incremental increases in the intelligence of its populations. Each new human has a more significant body of accumulated knowledge to create the neural pathways and functional integrations in his brain than his predecessors. Before a new human can learn this advanced knowledge, he must learn how to use his conceptual faculty. Society's first function, therefore, is to teach new humans *how to think.*

We will soon enter a technological era where individuals can assemble in virtual reality, transporting their conscious focus across any distance using "almost teleportation." They will communicate using "almost telepathy." They will have instantaneous access and the ability to search the entire storehouse of knowledge in more powerful ways, and the ability to join to tackle problems. Their technologically augmented intelligence will far surpass the humans of today. This new technologically advanced society will accelerate the discovery of Laws of Reality and the development of new technology in ways we cannot foresee today. What it means to be a rational being will not change. Still, we cannot fully appreciate how different our lives and our society will become.

PHILOSOPHY

"If man is the conceptual being, philosophy is the prime mover of history."

—LEONARD PEIKOFF

The influential intellectuals in any period determine the amalgam of ideas that dominate their culture. Like Galileo and Newton—men of the mind—scientists of the Renaissance diminished the Dark Ages' dogmatic mysticism. Their reality-based ideas gradually replaced mysticism-based ideas dominant in the culture at that time. Eventually, the general population no longer believed that the sun rotated around the Earth.

There is a direct correlation between an increase in the acceptance of reality-based ideas in society and the increase in that society's advancement of technology. However, mysticism has not been entirely displaced and continues to have widespread and increasing influence in our society. From the Industrial Revolution to the present day, reality-based ideas in the sciences have caused incredible technological advancement. However, the *Ivory Tower Cycle* (covered in the next chapter) has caused a rise of anti-reality ideas in our society over the same period.

In this chapter, we highlighted the importance of capitalism, businesspeople, and the profit motive in advancing technology, and I quoted twentieth-century philosopher Ayn Rand. However, in today's culture, capitalism, business professionals, and Ayn Rand's ideas that support them are all derided as being "selfish." As children, nearly all of us learned that it is bad to be "selfish." This idea, originally from religious tenets of selflessness, is thoroughly indoctrinated into Western society. Yet, the American Dream is the expectation that, if you apply the best within you and work hard, you can significantly improve your life's condition—the opposite of sacrificing oneself. This contradiction is just one example of the great conflict of commonly held ideas in our culture.

Most criticisms of capitalism, business professionals, the profit motive, and Ayn Rand's defense of them come from this pervasive acceptance that *selfishness* is bad. The criticism of Ayn Rand is caused by not understanding her definition of selfishness as "rational egoism." Rational egoism is the recognition that the primary cognitive relationship in your life is the one between your mind and reality. It is not the relationship between your mind and something mystical, or between your mind and another person's mind, or between your mind and society.

"Selfishness" as rational egoism means a commitment to maintaining the connection between your independent mind and the facts of reality, no matter what others think or tell you. This view of selfishness is in opposition to selfishness that only considers an individual's relationship to other people. Such an idea assumes a zero-sum game where one person's gain can only mean another person's loss. Ayn Rand's "selfishness" is a non-predatory form of self-interest. It does not require that others sacrifice themselves for your benefit or that you sacrifice yourself for theirs. Unlike the current cultural view of selfishness, rational self-interest rejects all forms of sacrifice. All the rest of Rand's twentieth-century integrated system of philosophy, as well as the heroic characters in her novels, derive from this view of self-interest as one's relationship to reality.

Mistakenly, Ayn Rand is labeled the champion of "cold, brutal capitalism" and the greedy businesspeople to the detriment of the workers and the common people. Again, this is because of the failure to identify or accept *rational egoism.* Or she is labeled the champion of the "conservative hard right" of the political spectrum. Before one accepts such commonly heard criticisms, one should examine the fundamental ideas that Rand espoused. One should not rely on the hearsay promulgated by anti-reality intellectuals whose positions are being challenged by Rand's focus on reality and who seek to undercut her legitimacy.

Such intellectuals base their criticisms on misrepresentations regarding Rand's higher-level abstractions, such as her ethics (rational self-interest) and politics (capitalism). Their ideas are floating abstractions cut off from reality and result in different views regarding ethics and politics (selflessness and collectivism). They do not provide an honest critique that begins with fundamental premises about existence and knowledge from which we derive higher abstractions. When hearing someone critique Rand, ask, "Is one's primary cognitive relationship between his mind and reality?" See if he can build a critical argument from that point forward.

I have been calling a person's "Worldview" to be the set of philosophical ideas the person holds as true. From now on, I will refer to it as one's "Personal Philosophy."

You must dig deeper than the common generalizations heard in today's culture and their many contradictions. Choose your Personal Philosophy. It includes what you think about capitalism, business professionals, and Ayn Rand's or anyone else's ideas according to the facts of reality, not what you have heard from others. Allowed to continue, the erosion of reality-focus in the general population and then in the sciences will result in the attempt to subordinate science to politics, just as in the Dark Ages. It will lead to anti-technology agendas, diminished technological achievement, and failed or dangerous technology.

Now we know the roles of individuals and the institutions of society necessary to produce technology. The foundation beneath it all is a set of philosophical ideas that enable free individuals in a free society to be technology's creators.

♦♦♦

SOCIETY

"Ancient Greece tore away the heavy shroud of mysticism woven for centuries in murky temples, and achieved, in three centuries, what Egypt had not dreamed of in thirty: a civilization that was essentially pro-man and pro-life. The achievements of the Greeks rested on their confidence in the power of man's mind—the power of reason. For the first time, men sought to understand the causes of natural phenomena, and gradually replaced superstition with the beginnings of science. For the first time, men sought to guide their lives by the judgment of reason, instead of resorting exclusively to divine will and revelation.

The Greeks built temples for their gods, but they conceived of their gods as perfect human beings, rejecting the cats, crocodiles and cow-headed monstrosities enshrined and worshiped by the Egyptians. Greek gods personified abstractions such as Beauty, Wisdom, Justice, Victory, which are proper human values. In the Greek religion, there was no omnipotent mystical authority and no organized priesthood. The Greek had only a vague idea of, and little interest in, an afterlife."

—MARY ANN SURES

LAWS OF REALITY AND THE HUMANITIES

Is human society evolving? What would "evolution" mean as applied to society? How would societal evolution relate to technological evolution? Are there principles we can discover from the past that indicate our future? In this chapter, we explore these questions and more to show that our society does evolve, but not in a straight line. And we could be headed for a setback.

We know some of the Laws of Reality, such as Newton's Laws of Gravity and Motion. However, for an exceptionally long time, conceptual humans did not know about the Law of Gravity. It had to be discovered by a human who used his rational faculty, powers of observation, and experimentation applied to the reality around him. Without instinct, humans must learn how their environment works and what they need to survive. Reason is their only source of knowledge. Trying to understand the causes of things means survival.

There could be far more Laws of Reality that we have not discovered than those we have. It seems the more we find out, the more questions we raise by the discovery. Another way to think of learning the Laws of Reality is that we are discovering "how things work." This learning applies to things physical and things biological. There was a time when men did not know how chemistry worked, or electricity or radio waves. And they did not know about bacteria or the organelles inside living cells.

Since everything in reality has a particular nature, we strive to understand what makes them different. What plants produce beneficial chemicals? How do bats navigate while flying in total darkness? Today, we are still trying to answer questions about how things work, including the physical, "What is dark matter?" and the biological, "What causes aging?" It might be helpful if, whenever I refer to "discovering the Laws of Reality," you equate it to "discovering how things work."

Man's need to understand causes was initially satisfied by attributing them to mysticism. The sun appeared to rotate around the Earth. Before science and the discovery of the Earth's rotation, the only "cause" that men could come up with was to attribute such apparent solar movement to a god. Indeed, early civilizations could only attribute the causes of almost everything that happened in nature to the gods of wind, water, fire, lightning, rain and others in various cults and myths. Eastern and Western religions typically believed that Man was created in the image of God, with "God" considered to be the universe itself.

Mysticism was probably a necessary phase in the development of content for man's rational faculty. At least he was exercising his new rational faculty and seeking answers for things he was trying to understand. Before science, mysticism was the only cause-effect idea man could devise. Then, as the age of science began, factual knowledge of the Laws of Reality started to replace mysticism as causal factors. Man learned that the Earth spins and orbits the sun, and gravity causes it, not a god. We tend to think that our science has led us far away from our early mysticism. However, this transition remains a work in progress.

In our twenty-first century science, we are beginning to circle back to our earlier primitive civilizations, closing a loop in our quest to understand life and nature. We are beginning to replace fantastical mysticism with a new secular mysticism that appreciates the beauty of physical reality itself. Contera writes:

> As we strive to discover the precise ways in which living, intelligent organisms contain the physics of the universe at the core of our biology, scientists and technologists (often unwittingly) reconnect with our ancestors to identify the plan of the universe itself inside human and nonhuman beings, though in a secular way. Another way of saying all this is that biology is entering the realm of physics. [...] We are no longer the output of genetic computations; we emerge from the cosmos itself.[133]

Unfortunately, fantastical mysticism, not secular mysticism, dominates today's society. With a tunnel vision of only the here and now, we tend to think of ourselves as living fully in the scientific age. We do not stop to consider that we are living in a progression not yet completed. It was not complete once we discovered that the sun was the center of the solar system, explained electricity, or decoded the human genome. Wherever scientific explanation is absent, our present society remains influenced by mysticism, cults and myths.

Today, we are still progressing towards a complete understanding of how everything in the universe works. There is a lot that we still do not know. We have questions that science has not yet answered. This lack of knowledge is especially true about the humanities—philosophy, psychology, sociology, education and art. Man is still learning how to use his new rational faculty and build a society that consistently applies it. As you will soon learn, we only began experimenting with reason a mere 2,500 years ago—not even a "tick" in evolution's clock.

Compared to the physical sciences, the humanities have proved to be more difficult to explain scientifically. How does one scientifically explain our consciousness, the purpose of our lives, good and evil, honesty, justice, love, happiness, politics, or art? We have an innate need to understand causes, not just for physics laws but also for human behavior. As our pre-science ancestors had to resort to made-up explanations, the void in the humanities where the knowledge of the Laws of Reality is absent is filled by religion or a myriad of other imagined explanations. Too many intellectuals accept that questions like these in the humanities are outside science's ability to answer. The consequence is that the explanations they do come up with are not reality-based.

Are the answers to the humanities questions truly outside of the ability of science to explain? Or is it just that valid reality-based explanations exist but have not yet displaced the made-up ones? Are we in a period like when Galileo came up with a correct answer, but society refused to accept it?

Two fundamental questions have made the humanities more challenging to explain—"What is reason and knowledge?" and "What is volition?" These are questions in the field of philosophy that deal with our rational faculty. These questions are fundamental to explaining human behavior and the nature of society. Reason and knowledge explain how we can learn about our reality and then modify it. Volition explains that human behaviors and outcomes are not predetermined. We cannot reduce human behavior to a simple formula like using physical laws to explain chemical reactions. However, humans are products of reality and evolution and have a specific nature, the functioning of which *does* explain their behaviors.

When these two fundamental questions can be explained by connecting them to reality, philosophy becomes a scientific field. It is no longer a field of made-up explanations. Reason and volition can be explained in functional terms and tied to reality. That is true knowledge that can explain the course of societies. Philosopher Leonard Peikoff describes the role of philosophy regarding history:

> The books of philosophers are the beginning. Step by step, the books turn into motives, passions, statues, politicians, and headlines.
>
> Philosophy determines essentials, not details. If men act on certain principles (and choose not to rethink them), the actors will reach the end result logically inherent in those principles. Philosophy does not, however, determine all the concrete forms a principle can take, or the oscillations within a progression, or the time intervals among its steps. Philosophy determines only the basic direction—and outcome.

> In order to grasp the role of philosophy in history, one must be able to think philosophically, i.e., see the forest. Whoever sees it knows that history is not the domain of accident.[134]

Twentieth-century philosopher Ayn Rand (1905–1982) extended Aristotle's initial connection of human intelligence to reality by explaining how our rational faculty uses reality-based concepts to create abstract knowledge. She explained that volition is our ability to choose to think or not. Both are part of our nature as humans. They are our defining nature and were built into us by evolution. Rand demonstrated how we could connect our most abstract concepts, such as those in the humanities, to reality. Binswanger (1982) describes six landmark achievements of Ayn Rand in completing a reality-based system of philosophy that began with, but was not fully developed by, Aristotle:[135]

1. The primacy of existence
2. The theory of concepts
3. The theory of free will
4. Man's Life as the standard of morality
5. The moral basis of individual rights
6. The psycho-epistemology of art

Ayn Rand also explained that our conscious thinking programs our subconscious and our emotions. Recently, neuroscientists have provided scientific evidence of that assertion by showing that our conscious thinking physically modifies our brain's neural circuits and re-integrates its functional modules.

We now have the reality-based explanations we needed to connect the abstract ideas of the humanities to reality. We can understand consciousness, the purpose of our lives, good and evil, honesty, justice, love, happiness, politics, and art as facts of reality. So why have Ayn Rand's explanations not been accepted and taught in our colleges and universities? Why are they teaching the same old made-up explanations? The answer is the same as when the intellectuals of his time refused to accept Galileo's explanation of the solar system, labeled him a heretic, and put him under house arrest. It crumbles the foundations of the currently accepted dogma of so many institutions and careers in contemporary society.

It takes time for reality-based knowledge to displace the entrenched made-up explanations in society. It took a hundred thousand years for human brains to restructure their functional modules to support language. In Galileo's day, it took more than a century for the human brains in society to restructure themselves to use reason, logic, and the scientific method instead of mysticism to explain the physical world. Today, our society's brains need to restructure themselves to accept reason, logic, and the scientific method as the proper tools for explaining human behavior.

Remember, we are still in a progression. We have replaced most made-up explanations in the physical sciences with reality-based knowledge. However, we have not yet replaced made-up explanations in the humanities. Perhaps someday we will look back on today's society regarding the humanities as having been as primitive as the society in Galileo's day was regarding the solar system.

THE POWER OF IDEAS

Evolution is in its Conceptual Period, and humans are the agents driving it forward. Humans must "think" to survive and direct their lives. The "ideas" held by the majority determine the society's culture. Those ideas originate with the intellectuals of the period and distribute via society's educational institutions.

Philosophy is the science that explains the functioning of man's consciousness. In particular, it explains the branches of metaphysics (*What is out there?*) and epistemology (*How do I know it?*). Philosopher Harry Binswanger (2014) sets the stage:

> Our technological success has come from a dedication to reason and logic, but reason and logic have been distorted or openly attacked by mainstream epistemologists for the last 200 years, ever since Kant's theory of knowledge gained dominance in the intellectual world. Establishment epistemology has carried to its logical conclusion Kant's claim that reason cannot know reality. The result has been two schools of thought, one that accepts reason while ignoring reality, and one that accepts reality while denying reason.

> Rationalism is the school that scorns sensory perception and constructs intellectual castles in the air. Empiricism is the school that scorns abstractions and demands that men hold their minds down to the animal level of unconceptualized, unintegrated sensing. Rationalism ultimately degenerates into mysticism, as in its ancient father: Plato. Empiricism ultimately degenerates into skepticism, as in its modern father: Hume. [...] Both the mystic and skeptic schools fly in the face of human history. [...] Nonetheless, mysticism and skepticism have lived on, zombie-like, due to the success of the Kantian attack on reason. [...] Contrary to the foggy notions of a non-judgmental age, there is a *right* and a *wrong* direction to take—if grasping the facts of reality is one's goal. The *right* direction means the one suited to cognitive success; any deviation from that direction is *wrong*—wrong in relation to that goal.
>
> [...] Western civilization is now under attack by the revived mysticism of Christian and Islamic fundamentalism and by the new skepticism of multiculturalism and postmodernism. [...] An open, progressing, benevolent future requires a theory of knowledge that rejects the false dichotomy of Empiricism vs. Rationalism. What is required is a theory that upholds *both* sensory perception *and* logic, a theory that shows how abstract, conceptual knowledge derives in a logical fashion from perceptual observation.[136]

We need a *standard* that will enable us to understand and measure how the ideas promoted by our thought leaders are affecting our society today. The standard we need is the degree to which Aristotle's ideas based on reality and reason influence society. It is first necessary to understand the influence that Aristotle's ideas had on different world societies at various times in history.

Some would say that engaging in philosophy is only a mind game and does not apply to the "real" world. That is not surprising, says Binswanger: "The misunderstandings of consciousness that have wreaked havoc on the history of philosophy, making philosophy appear irrelevant to daily life, all stem from taking consciousness to be non-causal and non-biological—or even, in the latest aberration, non-existent."[137] But consciousness does exist as a product of reality and evolution, and it functions according to its nature. Recognizing its existence and functioning removes the mystery that Rand says: "Turns men into prisoners inside their own skulls."

Ideas that a person believes, consciously and subconsciously, are what determine:

- The things a person values.
- The choices a person makes.
- The actions a person takes.
- A person's "character."
- The course of a person's life.
- Whether or not a person achieves happiness.

From where do the ideas that an individual believes, consciously and unconsciously, come?

A person is immersed in the conceptual soup of ideas in the environment of his contemporary culture. The ideas generally held by most of the population in society are what he or she encounters daily. They determine:

- The social structure—a form of government.
- The degree of political and economic freedom (or lack of freedom) for citizens to act in their own interest.
- The formal education system and what is taught.
- The popular "pop" culture, traditions, institutions, art.
- The type of political and intellectual leaders that the people will follow.

What determines the content of the culture's soup of ideas?

I will later describe the *Ivory Tower Cycle* through which the ideas explicitly promoted by the intellectual leaders of society as professors in its colleges and universities determine the ideas generally held, consciously or subconsciously, by most society members.

Are not differing ideas held by individual people equally valid for them? Can specific ideas be good or bad for everyone? Do the ideas espoused by a society's leaders have the power to direct the course of an entire society?

By simply looking at history, it is evident that specific ideas can be good or bad for everyone and can have power over the masses. Its founding fathers' ideas caused an entire society to fight for independence in a war that created America. Ideas held by Hitler and those supporting him caused a society to fight a world war of aggression that included the holocaust.

From where do the ideas promoted by a society's intellectual leaders come?

That is an excellent question that gets to the heart of the matter and this chapter's subject. I will describe to you another war—a different kind of war than the two examples above—a war that has been raging for nearly 2,500 years. It began and continued for centuries around the Mediterranean, and then it spread across the globe.

Like any war, this war has generals, officers, foot soldiers, battlefields, weapons and battles that are won or lost. From the perspective of valuing your happiness and the survival of mankind, this war has "good guys" and "bad guys." I call it the Worldwide War for Minds ("WWM").

Scientists in the hard sciences explain the functioning of reality—"nature." Philosophers explain the functioning of man's conscious mind, and psychologists explain man's subconscious mind. Mystics claim to explain whatever they say is outside of reality. Out of all these domains, Philosopher Leonard Peikoff explains that it is philosophy that can explain human history's causes, "Philosophy is not the only cause of the course of the centuries. It is the *ultimate* cause, the cause of all the other causes. If there is to be an explanation of so vast a sum as human history, which involves all men in all fields, only the science dealing with the widest abstractions [philosophy] can provide it."[138]

Ideas leadership in the WWM comes from the major professional philosophers. They develop integrated theories—philosophic systems—of what reality is and how man's mind works. These are the generals in the WWM, and their theories are the battle plans. The war officers are the so-called "intellectuals," working primarily in the universities and colleges. Weapons in the WWM are the generals' ideas. They are taught by the officers to the foot soldiers who are the students in the universities and colleges. The foot soldiers then fight on battlefields which are wherever they work or socialize with other people. Instead of hand-to-hand combat, they engage in mind-to-mind combat, and ideas are the weapons each combatant is trained to use by the officers. Ultimately, the combat results in victors and casualties.

I am using "war" as a metaphor. However, what I am describing is a real war with real human consequences. As Peikoff implied above, it is the ultimate war, within which the American Revolutionary War, WW1, and WW2 were discrete battles. As the ultimate war, the WWM is a slow war that plays out over generations and centuries. The WWM is still raging today, and you might have been unwittingly drafted as one of its conscripts. It is time to learn whether you are fighting on the side of the "good guys" or the "bad guys."

Since ancient times, the fundamental conflict has been between a real and knowable reality (reality and reason) versus the idea of a changeable reflection of another unknowable reality (empiricism and mysticism). The "reality is real" position was most famously represented in ancient Greece by Aristotle. The "reality is unknowable" position by Plato, Aristotle's teacher. These were the first two generals. Thus began the war to determine which philosophical ideas will dominate society. This war continues to this day. Binswanger describes the conflict:

> The clash between Aristotle and Plato reverberates throughout the realm of ideas, both in individuals and in cultures. The battle between the two theories takes place on a highly abstract level, but it is impossible to overstate the practical, real-world importance of their impact on a culture. Science and the Industrial Revolution arose from and depended upon the Aristotelian recognition of the primacy of existence and the primacy of perception. And it was the Platonic primacy of consciousness and primacy of concepts philosophy that led to contempt for this perceptually given world and the abandonment of the achievements of Greco-Roman civilization, plunging mankind into the long, stagnant night of the Dark Ages. In denigrating reality and perception, the Platonists reject life on this Earth.
>
> Do I exaggerate? Plato calls the body 'a prison' and 'the tomb of the soul,' [Cratylus, 400c] He says that 'true philosophers make dying their profession,' [Phaedo, 67E] and proclaims that a man should have 'trained

> himself throughout his life to live in a state as close as possible to death.' [Phaedo,67E][139]

It is essential to understand the power these opposing views of reality have had on the history of the world when promoted by the philosophers and intellectuals (the generals and officers) of a time and place. Then you will understand the power they continue to have in determining our future in the age of technology. Which set of ideas *should* the intellectuals in our Ivory Towers be teaching? To find out, let us explore the history of these ideas from Aristotle's time to today.

ARISTOTLE'S IDEAS—ANCIENT GREECE TO THE RENAISSANCE

Since ancient Greece, Aristotle's teachings have been either visible and influential among the philosophers and intellectuals of the time or were lost from view during specific periods of history and in various parts of the world.

In his book *The Aristotle Adventure, A Guide to the Greek, Arabic, and Latin Scholars Who Transmitted Aristotle's Logic to the Renaissance*, Burgess Laughlin presents evidence of the correlation between Aristotle's teachings and the advancement of civilizations. Civilizations advanced in locations and periods when Aristotle's ideas were influential. When and where Aristotle's ideas were lost from view, those civilizations declined. Laughlin's book details how Aristotle's teachings were communicated throughout the world's regions and the eras of history.[140]

1. Aristotle lived from 384–322 BC during the Classical period in ancient Greece.
2. Aristotle wrote over 100 treatises (books) on papyrus rolls, over half of which have perished due to fire, flood, willful destruction, and neglect.
3. Six of Aristotle's treatises were crucial as the building blocks of a new culture: *Categories, Interpretation, Prior Analytics, Posterior Analytics, Topics,* and *Sophistical Refutations.*
4. As a set, considered together, this was the most important book ever written. It presented Aristotle's greatest creation—logic—the indispensable tool kit of a philosophy of reason.

5. In his six treatises, Aristotle made explicit his discoveries about gaining valid knowledge of the world. If his treatises had disappeared in the centuries after his death, his discoveries might have been lost.
6. For the first time in history, Aristotle provided rational answers to the questions: "What is the nature of the world?" and "How do we know what we know?"
7. Aristotle explained inductive and deductive reasoning, the hierarchy of knowledge, and the scientific method of discovery.
8. The influence of Aristotle's ideas on any particular society during any specific period can be measured on the following scale of activities by the intellectuals of the day:
    - Copying Aristotle's treatises.
    - Writing commentaries and paraphrases of the treatises.
    - Applying dialectic (reasoned arguments) to theology.
    - Defending logic as the tool of a philosophy of reason.
    - Teaching logic, independently of theology.
    - Applying logic to science.
9. These activities rose or fell on the scale during different periods according to which part of the Mediterranean community we consider:
    - The Greek-Christian branch
    - The Arabic-Islamic branch
    - The Latin-Christian branch
10. Aristotle's ideas might have been lost permanently had it not been for the activities of a few individuals in each of the branches of the Mediterranean community at critical points in history.
11. Soon after Aristotle's death, his treatises were utterly lost from sight for about three hundred years, resurfacing around 0 AD when Andronicus edited and published them.
12. About 200 AD, two hundred years later, Galen and Alexander explain the nature and value of the treatises. Aristotle's ideas are once again influencing the intellectuals.
13. Then, around 400 AD, Neo-Platonists preserve, expound, and teach the treatises to pagans and Christians.
14. By around 600 AD, the Neo-Platonist schools closed, and European scholarship disintegrates.

15. Aristotle's ideas were once again lost from view in Europe. However, the Christians had translated the treatises into Syriac and Arabic. Aristotle's ideas arose in the Arabic-Islamic branch for about 600 years while Europe endured the "dark" Middle Ages.
16. From around 900 AD to 1200 AD, Aristotle's treatises slowly reappear in the Latin-Christian branch.
17. Around 1200 AD, Ibn Rushd in the Arab-Islamic branch defended Aristotle. Then there was a theological reaction against philosophy and other "foreign" sciences. Aristotle's ideas were lost from view in the Arabic-Islamic branch.
18. About that same time, around 1200 AD, Latin-Christians invent the university and include logic in the curriculum.
19. Around 1300 AD, Thomas Aquinas tried to integrate reason and revelation, and Aristotle's ideas were again influencing intellectuals in Europe.
20. Around 1400 AD, using Arabic and Greek sources, plus their own thought, the University of Padua logicians applied logic to methodology. They develop a method of proof and discovery.
21. Around 1600 AD, Galileo's advances applied logic to the science of nature. Aristotle's ideas launched the Renaissance period in Europe.
22. At this same time, Italus in the Greek-Christian branch tried to revive philosophy as a guide to life but then faced inquisition. A philosophic drain began in the Greek-Christian branch.
23. Aristotle's ideas launched the Renaissance around 1600 AD in the Latin-Christian branch in Europe, while the Greek-Christian branch and the Arabic-Islamic branch remained mostly hostile to Aristotle's ideas.

We tend to treat our understanding of the Dark Ages casually. As the first great battle in the Worldwide War for Minds, it was won decisively by Plato. We must treat that period seriously because our society appears heading back in the direction of Plato today. Peikoff describes the horror of the Dark Ages:

> The Dark Ages were dark on principle. As the barbarians were sacking the body of Rome, the Church was struggling to annul the last vestiges

> of [Rome's] spirit, wrenching the West away from nature, astronomy, philosophy, nudity, pleasure, instilling in men's souls the adoration of Eternity, with all its temporal consequences. 'The early Christian fathers,' writes one historian, 'delighted in such simple self-tortures as hairshirts, and failing to wash. Others proceeded to more desperate extremes, such as Ammonius who tortured his body with a red-hot iron until it was covered with burns. [...] It would not be necessary to dwell on these depressing details if it were not for the fact that the Church erected these appalling practices into a virtue, often canonizing those who practiced them. [...] [St. Margaret Marie Alacoque] sought out rotten fruit and dusty bread to eat. Like many mystics she suffered from lifelong thirst, but decided to allow herself no drink from Thursday to Sunday, and when she did drink, preferred water in which laundry had been washed. [...] She cut the name of Jesus on her chest with a knife, and because the scars did not last long enough, burnt them in with a candle. [...] She was canonized in 1920. [...] St. Rose ate nothing but a mixture of sheep's gall, bitter herbs and ashes. The Pazzi, like Alacoque, vowed herself to chastity at an incredible early age (four, it is said).' Neither serf nor lord emulated these eloquent expressions of the medieval soul. But both admired them from afar—as pious, profound, moral.[141]

Because the Arabic-Islamic branch was hostile to Aristotle's ideas at the time of the Renaissance, this had the consequence that it has never had an equivalent Renaissance in its society to this day. Anti-Aristotelian ideas remain predominant in non-Westernized portions of the Arabic-Islamic branch. Violently enforced mysticism, lack of individual rights and freedom, oppression of women, clerics as rulers, and the use of terrorism result from the absence of Aristotle's ideas. Rejecting Aristotle's ideas consigns them to a permanent Dark Ages, one just as abhorrent as Peikoff's description above.

Aquinas (c. 1250), Bacon (1620), Newton (1643), and Locke (1690) were generals in the WWM who began the next battle to transition Europe out of the Dark Ages by resurrecting Aristotle's ideas. A battle that Aristotle would eventually win. Peikoff describes:

> The period from Aquinas through Locke and Newton was a transition, at once gingerly and accelerating. The rediscovery of pagan civilization, the outpouring of explorations and inventions, the rise of man-glorifying art and of earthly philosophy, the affirmation of man's individual rights, the integration of earlier leads into the first system of modern science—all of it represents a prodigious effort to throw off the medieval shackles and reorient the Western mind. It was the prologue to a climax, the first unabashedly secular culture since antiquity: the Enlightenment. Once again, thinkers accepted reason as uncontroversial.[142]

ARISTOTLE'S IDEAS—THE RENAISSANCE TO TODAY

Burgess Laughlin's book *The Aristotle Adventure* leaves off in the 13th century when Aquinas resurrects Aristotle's ideas to integrate reason and revelation leading to Galileo applying logic to the science of nature around 1600. This battle, won by Aristotle, finally brought Europe out of the Dark/Middle ages and led to science and the Renaissance in Europe. However, the beginnings of a counterattack were launched in Europe just as America's founding began.

The next great battle in the WWM counterattack would not be fought in the Mediterranean but on the North American continent. The generals in this battle were America's Founding Fathers. They understood that faith and force entailed each other, as evidenced in the feudalism of medieval centuries. They also understood that, as evidenced by the Renaissance, reason and freedom entailed each other.

They created a new nation, the first-ever founded on a philosophical theory—the principle of rights. Peikoff writes, "Man, America's Founding Fathers said in essence, is the rational animal. Therefore the individual, not the state, is sovereign; man must be left free to think, and to act accordingly." In Aristotle's time, the concept of "rights" had not yet been formulated. "In the Declaration of Independence and the Constitution that implements it, we see at last the full expression, in political terms, of the Aristotelian fundamentals."[143]

Europe's battle—the Renaissance, and the battle in the New World—America, both of which won by Aristotle and instituting reason and freedom, led to:

> the Industrial Revolution, to Romantic art, and to an authentic good will among men; it led to an unprecedented burst of wealth, beauty, happiness. Wherever they looked, people saw a smiling present and a radiant future. The idea of continuous improvement came to be taken for granted, as though it were an axiom. Progress, people thought, is now automatic and inevitable.[144]

However, the WWM counterattack launched in Europe following the Renaissance continued without anyone mounting a defense of Aristotle. The ideas on which science, America and industrialization were founded were again under full attack in Europe. In the seventeenth century, Descartes (a general in Plato's army) planted Platonism once again at the base of philosophy. Unfortunately, Aristotle had not developed a proper theory of how abstract concepts connect to reality and therefore had not formed a rational basis for ethics. These holes in Aristotle's battle plan provided an opportunity for the enemy.

Binswanger explains that starting with Descartes and bottoming out with Kant, a prominent line of philosophers peddled a secularized version of the old religious notions. "I have therefore found it necessary to deny *knowledge* in order to make room for *faith*," Kant wrote. [Critique, B XXXI, Kant's emphasis][145] Descartes theory of concepts, while discarding Plato's mysticism, severed knowledge from sensory perception: "No science is acquired except by mental intuition or deduction. ... the first principles themselves are given by intuition alone. ... By intuition I understand, not the fluctuating testimony of the senses... but the conception which an unclouded and attentive mind gives us..." [Descartes, Rules III & IV][146] Binswanger further describes the counterattack:

> Kant synthesizes Plato with Descartes, adding a skeptical twist from Hume. Following Plato, Kant holds that the world we perceive through our senses is shadowy, not fully real; the 'higher' realm is now called the 'noumenal' world, the innate ideas are now called 'categories.' Following Descartes, Kant's philosophy starts from within consciousness and regards awareness of the external world as problematic; Descartes 'intuition' now becomes the voice of 'noumenal' reality heard through (of all things) our alleged sense of 'moral duty.' Following Hume, Kant

> regards the external world as utterly unknowable (except for this alleged sense of 'duty'); Hume's 'habits of association' become Kant's innate 'synthesizing categories.'
>
> Beneath Kant's notoriously convoluted terminology is an all-out separation of the mind from reality. Both perception and conception are cut off from 'things as they are in themselves,' which remain forever unknowable. Science deals only with 'appearances' ('phenomena'), and is, in effect, held to be a shared delusion.
>
> The contemporary attack on science by Popper, Kuhn, and Feyerabend is grounded in Kantianism. In their post-modern version of Kant, science deals not in truth but in public relations, and is the embodiment not of reason but of faith.[147]

Immanuel Kant (1724–1804) was the latest and most influential general in Plato's army. To explain what concepts are, he replaced Aristotle's theories of existence and knowledge, resurrecting variants of Plato's ideas which he then further modified as explained by Peikoff:

> [Kant] denies Existence not in the name of a fantasy, but of nothing; he denies it in the name of a dimension that is, by his own insistent statement, unknowable to man and inconceivable. The mind, he says, is cut off not merely from some aspects of 'things in themselves,' but from everything real; any cognitive faculty is cut off because it has a nature, any nature. Man's proper goal, say Kant, is not happiness, whether in this life or the next. The 'radically evil' creature (Kant's words) should sacrifice his desires from duty, as an end in itself.[148]

During the Renaissance, as the mind and the sciences' efficacy surged, the Roman Catholic church and religion, in general, went into decline. Then, unopposed, in the late 18th century, Immanuel Kant became the most influential philosopher in Europe. Kant developed a complete, integrated system of philosophy. It would have the effect of negating the Aristotelian influence of the Renaissance, once again rendering the mind useless. The result would be a return to other-worldliness—denying *knowledge* to make room for *faith*. In the absence of contemporary champions defending Aristotle's influence, Kant was hugely successful. Europe's intellectuals embraced Kant's new philosophy, and without a defender, Aristotle's ideas fell out of favor. This replacement of Aristotle's ideas with Kant's set the stage for the anti-reason movements of Fascism and Communism in Europe and the suffering that ensued. Peikoff further describes the Cartesian counterattack that followed the Renaissance:

> In the penultimate decade of the eighteenth century, just when America was being born, that trend, unopposed bore fruit. The fruit was the end of the West's philosophical commitment to reason, the conscious changeover in the ivory tower from the remains of Aristotle to his antithesis. The thinker who ended the Enlightenment and laid the foundation for the twentieth century was Kant.[149]

At about that same time, in the late 18th century, America's founding fathers were primarily businessmen produced by the Renaissance. Therefore, they had personally learned and held primarily Aristotelian ideas. They gave America its great "kick start" in the form of our implicit "American Sense of Life" that subordinates government to the individual. They also created our explicit Constitution based on man's nature as the rational animal. Both are primarily Aristotelian.

However, the founding fathers then created America's new universities and staffed them with professors they imported from Europe. These professors, as infiltrators, were officers of Plato's army trained by (general) Kant. They brought with them to America Kant's philosophy, not Aristotle's. The next great battle of the WWM, a battle in which we are still engaged, had begun.

This battle in America is between the Aristotelian Constitution plus implicit "American Sense of Life" on the one hand and the Kantian philosophy explicitly taught in the universities on the other hand. For two centuries, America's universities have promoted Kantian philosophy. They turned out foot soldiers as professors, teachers, lawyers, clergy, writers, editors, politicians, and other social leaders. They armed those foot soldiers with weapons of Kantian (or Platonist) philosophy. Holding the intellectual high ground in the universities, the Kantians are winning the present battle. The influence of Aristotle's philosophy has been in steady decline since America's founding. Peikoff describes the impact the last two centuries have had on our society:

> The result of Plato's approach was a form of adoration. The result of Kant, in Ayn Rand's words, was 'hatred of the good for being the good.' The hatred took shape in the culture of nihilism. Modernist intellectuals are comparable to a psychopath who murders for kicks. They seek the thrill of the new; and the new, to them, is the negative. The new obliteration, obliteration of the essential in every field; they have no interest in anything to take its place. Thus the uniqueness of the century behind us: philosophy gleefully rid of system-building, education based on the theory that cognition is harmful, science boastful of its *in*ability to understand art which expelled beauty, literature which flaunted *anti*heroes, language 'liberated' from syntax, verse 'free' of meter, *non*representational painting, *a*tonal music, *un*conscious psychology, *de*construction in literary criticism, *in*determinacy as the new depth in physics, *in*completeness as the revelation in mathematics—a void everywhere that was acclaimed by the avant-garde with a metaphysical chuckle. It was the sound of triumph, the triumph of the new *anti*-ideal: of the *un*knowable, *un*reachable, the *un*endurable. In a Kantian reality, nothing else was possible.[150]

The followers of Kant's ideas wasted no time in drawing them to their conclusion. Fichte, Hegel, Marx (and Bismark) proceeded to the two most passionately antifreedom movements in history, Communism and Fascism. Their lesser, welfare-statists joined them. Peikoff describes the reversal of progress and the attack on technology:

> Statism cannot sustain an industrial civilization. Nihilism cannot abide it. Hence, in due course, another manifestation: the growing attacks on technology, i.e., the anti-Industrial Revolution. It was the vow of poverty all over again, not as a gateway to Heaven this time, but as a means to the welfare of water, trees, and 'endangered species.' The latter could be any species—except human. So much has been lost so fast. In no time at all, the West moved from 'perpetual peace' to perpetual war.[151]

In the 20th century, a philosopher arose as a champion general to come to Aristotle's defense. Ayn Rand was born in Saint Petersburg, Russia, in 1905. She escaped communist Russia and came to America in 1925. As both a novelist (*The Fountainhead* and *Atlas Shrugged*) and a professional philosopher, she has done more to resurrect and enhance Aristotle's ideas than any other philosopher since Aquinas. She has done so in the form of her philosophy called "Objectivism."

Objectivism reasserts Aristotle's identification of reality's primacy and the efficacy of man's reason as the causal factors to all our society's modern advances. Before Ayn Rand, theorists saw only two alternatives: the intrinsic or the subjective—i.e., revelation vs. whim. Either we are to remain mentally passive, hoping The Truth will flow in (intrinsicism), or we are mentally active—by embracing the emotional, the arbitrary, the socially approved (subjectivism). In short, either we relinquish control of our minds (intrinsicism), or we give up reality (subjectivism).

It was Ayn Rand who, for the first time in history, showed that man's concepts connect to facts in reality. She identified the "unit reduction" principle that explains how our rational, conceptual mind works. She identified "volition" as our ability to choose to use our rational faculty or not use it. Volition replaced automatic instinct. To be fair to Aristotle, Ayn Rand had the advantage of observing 2,000 years of human history and the achievements of the Industrial Revolution that Aristotle did not have to draw from.

Ayn Rand died in 1982 but left a legacy of professional philosophic non-fiction works explaining the science of a rational philosophy and literary fiction works that bring to life characters who hold a philosophy based on reality and reason and characters who do not. The following was "A message from the Author," which was appended to the end of the paperback edition of her epic novel *Atlas Shrugged*:

> If you are the kind of reader who knows that for 1084 pages he has lived in the atmosphere of John Galt's world—if you now feel regret at the necessity of returning to the gray hopelessness of a culture that is truly bankrupt—if you have understood that it is ideas which create or destroy a world, a culture or a man—if you have understood that the rejection of reason by the neo-mystics of our age is responsible for the present state

> of the world—then you know why only a philosophy of reason can lead to an intellectual Renaissance.

Peikoff says that Ayn Rand is to Aristotle as Kant is to Plato, "Both sides of the perennial duel, in their pure form, have finally been made explicit."[152]

Peikoff thinks that Kant as a cultural power is now dead. He is dead in academic philosophy; the subject has effectively expired under his tutelage. He is dead among the intellectuals whose world view is disillusionment (they call it "the end of ideology"). As an example of the intellectual world view today, art is no longer called nihilistic. Instead, it is called "minimalism" and "postmodernism."

Communism in Soviet Russia collapsed. But just because unreason has failed does not mean that men have returned to reason. They keep following the guidance of dead ideas from the inertia of the principles put into their institutions over the last two centuries. Explicit awareness of "the good," not blind default, is what is needed to establish a "positive" in place of the failed "negative." The alternative is to let ideas already in place run their course and culminate in some form of dictatorship—communist, fascist, religious or tribal—a New Dark Ages.[153]

Despite its decline, America is where the relationship of reason and freedom was first conceived and is still the leader and beacon for the world. However, it is not our "people" who shape our nation but our intellectuals. A small number of contemporary professional philosophers are dedicating careers to promoting Ayn Rand's extended and improved version of Aristotle's philosophy of reality and reason. Suppose we seek a future where we can safely coexist with intelligent technology. In that case, we must make sure we are fighting on the side of the "good guys." Then, perhaps, a philosophy of reality and reason will once again become dominant in our technological society.

For our technology to lead to mankind's mastery of the universe, the Aristotle-Rand philosophy must displace the failed Plato-Kant philosophy. It must do so both in higher learning institutions and our society. Suppose the anti-reality, anti-reason ideas are allowed to remain. In that case, our technology will amplify the destruction they cause in society. Plato-Kant will have won the present WWM battle resulting in a protracted New Dark Ages where technology does not advance but declines.

IVORY TOWER CYCLE

Fifty years ago, in a 1969 lecture, Ayn Rand talked about our *collapsing society with advancing technology*. She declared, "With the advance of the physical sciences and the retrogression of the humanities, we are in a dreadful state."[154] What was the danger she saw and, fifty years later, does it still exist?

The *Technology Cycle* explains how new knowledge *informally* distributes throughout society via business in a free market and citizens' practical use of the knowledge implicitly embedded in the new products and services. The *Ivory Tower Cycle* explains how new knowledge distributes *formally* throughout society via structured educational institutions and the content they explicitly teach their students.

The intended purpose of today's higher education, the colleges and universities, is to provide the formalized education and certifications that the students will need to work in their chosen professions. Some of these professions are:

- Lawyers and Judges
- Doctors and Health Care Professionals
- Scientists and Engineers
- Software Designers and Programmers
- Business Professionals
- Economists, Bankers, and Financiers
- Journalists and Editors
- Advertisers and Graphics Artists
- Writers and Publishers
- Artists and Filmmakers
- Musicians
- Clergy and Social Workers
- Politicians and Elected Officials
- Government Administrators
- Law Enforcement Officers
- Military Officers
- Pre-school and Primary Education Teachers
- Higher Education Professors

Take a minute to look back over the above list and realize the influence that the activities, the work output, and the media content produced by those professions have on the ideas implicitly accepted by the general population. The basic principle of the *Ivory Tower Cycle* is that the ideas taught by the professors in higher education are taken away by the graduates into their professions. Their work and the media they produce then embody those same ideas and convey them to the general population. In short, the professors' ideas in higher education become the predominant ideas of society.

If you have any doubt about how this works, consider this example. One idea that is overwhelmingly taught in our higher education today is that the businessperson's profit motive is somehow immoral. Think of all the movies you have ever seen involving businesspeople. How many of those movies portrayed the businessperson as the hero, and how many as the villain? How many imply that "profit" gained by the businessperson must have required someone else's loss? How many suggest that wealth is "stolen," not "created"? Is it any wonder that, after two centuries of this *Ivory Tower Cycle's* influence, we are seeing so many young people embrace socialism?

The *Ivory Tower Cycle*:

1. Intellectuals with a specific body of knowledge work in professions as educators.
2. They are accepted as being qualified to teach conceptual ideas to students formally.
3. Graduates work in all of society's professions, incorporating what they learned in their work output and the media they produce.
4. The masses then absorb the ideas prevalent in their environment by default, thereby becoming the "culture."
5. Some of the graduates become early childhood educators, explicitly teaching their ideas to impressionable children beginning at pre-school age.
6. Some of the graduates go on to become higher education professors, thereby repeating the cycle.

The *Ivory Tower Cycle* has two, generally independent *pathways* within it.

The Sciences Pathway:

1. Professors are intellectuals in the physical sciences.
2. They believe in Aristotle's logic and scientific method.
3. Many of them are themselves, scientists.
4. They monitor scientific progress all over the world.
5. Discoveries of the Laws of Reality or the nature of existents become new knowledge.
6. The professors teach new knowledge to future scientists, engineers, business professionals, educators, etc., who are pursuing education in the sciences.

The Humanities Pathway:

1. Professors are intellectuals in the humanities/arts.
2. They believe in some version of the Plato/Kant philosophic system.
3. It is a separate pathway because the humanities/arts are (today) believed to be outside the science realm.
4. They attempt to provide conceptual guidance for questions that (they think) cannot have scientific answers.
5. They teach what they were taught regarding highly abstract ideas such as the purpose of life, societal structure, the relationship between the individual and the collective, ethics, values, governance, societal institutions, and art.
6. The answers provided by these intellectuals are prone to error—disconnected from reality—because (they think) science cannot explain them.
7. The graduates then implement these ideas, true or false, in all the institutions and art of the society. Some of them become professors.

The Sciences Pathway begins with conceptual knowledge discovered by scientists. It is knowledge that connects to reality. The professors in the Sciences Pathway generally (but not always) pass it on to their students unaltered. The science students then carry valid knowledge into their careers in science or engineering.

The Humanities Pathway, on the other hand, does not begin with knowledge validated against reality. These professors consider the humanities domain as unexplainable by science. However, they must provide answers to the student's questions about human behavior. They will provide what they believe because they got it from their professors when they received their own humanities education. They teach these ideas to their students even though they are just ideas passed down repeatedly without validation against reality.

The result is that students receiving education in the physical sciences are far more likely to learn concepts validated against reality than the students receiving education in the humanities. Physical sciences students tend to be reality-oriented in matters regarding things physical. Still, they tend to be anti-reality oriented in matters regarding ethics, politics, and art because of their humanities courses.

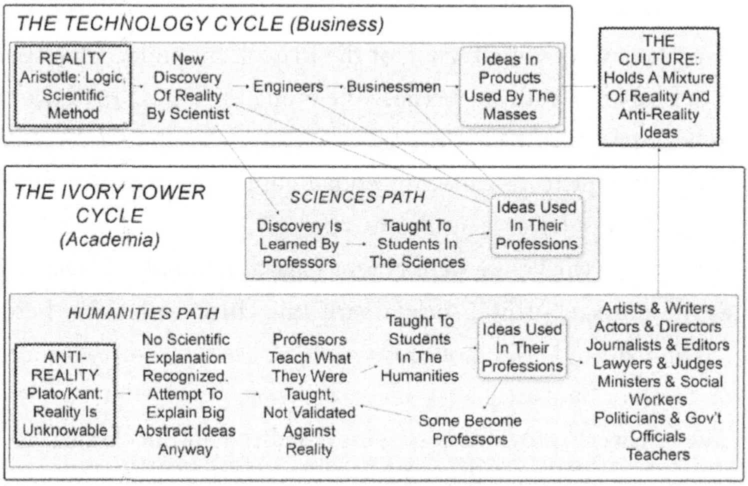

Conceptual Evolution's Processes

What will it take for the new knowledge of how our conceptual faculty works to overpower the intellectuals' dogmatic resistance? Could it be that the attempts to reconstruct the human brain's functioning to create an intelligent AI will ultimately prove the validity of the conceptual unit-reduction and volition theory and finally connect the humanities to science? We can hope that this will be the case.

True Idea: Americans have an implicit "American sense of life" philosophy based on the Aristotelian ideas put into the Constitution by the founding fathers. "The individual is sovereign in life, liberty, and the pursuit of happiness." And "The government serves we the people."

True Idea: Initially staffed with professors from Europe, America's universities, since their founding, have been explicitly teaching an anti-Aristotelian philosophy they brought with them. "The rights of the collective outweigh the rights of the individual." And "Individuals must sacrifice themselves for the good of the (needy, collective, state)."

True Idea: The universities' anti-Aristotelian ideas have gradually been implemented in our society, displacing the original Aristotelian principles on which our county was founded.

FORMING EDUCATION

For humans, evolution traded animal instinct for the ability and the *necessity* to learn and adapt. Individuals in society discover new knowledge and develop new technology that accumulates in society independent of them. At birth, a human's mind is a blank slate. He does not know how to think, nor does he have any mental content. To survive, he must learn both from existing humans around him. The child learns not only from his parents but all people he encounters as he matures. Collectively, we call these people the "society" into which he was born—a family, tribe, nation.

The reciprocal of man's evolved need to *learn* from others is evolution's development of *language* and *society* to fill that need. We can say that society's first evolutionary function is to *teach* children how to think. Its second function is to *make available* to thinking children the knowledge accumulated in society from earlier humans. Society accomplishes these evolutionary functions through the process of *Forming Education*.

Remember, "society" is a collective term referring to a group of individuals. The evolutionary responsibility to teach a child falls first to his parents. Suppose the parents are unable or unwilling to teach the child. In that case, other individuals in the family, tribe, nation might be willing to teach him but are not obligated to do so. However, because today's advanced society is storing enormous quantities of knowledge, we must accept that parents cannot teach the child independently. The child will also need to learn from others in society. We say that it is "society's" function to teach children because each child's need to learn is filled by all the individual people he encounters.

Society cannot teach the child all the knowledge that exists—what we call "content." The first goal of Forming Education is to teach the child "how" to think. More specifically, how to use his conceptual faculty to "reason." Fundamentally, we accomplish this by teaching the child how to *thoroughly* learn a single subject with all its integrations of ever-higher abstractions. "Thoroughly" learning a single subject means that the child builds a hierarchy of concepts within his mind. The hierarchy begins with concrete concepts from observations of reality and ends with the highest possible abstract concepts that result from an unbroken chain of ever-higher conceptual reasoning. The high-level abstract concepts are valid because they connect down the chain of reasoning to observable reality.

Initially, the only "content" that the child needs to learn is the content necessary to learn a single example subject completely. Once the child has learned how to fully learn a single subject, from concretes to highest abstractions, then the child's mind has the reasoning skills he needs. The entirety of the society's knowledge content becomes available for him to learn. As he matures, he can choose the subjects that interest him. He has the skill to integrate that content into a conceptual hierarchy within his mind.

Of course, learning a single subject is an oversimplification to make a point. However, the number of initial subjects that need to be "fully" learned is relatively few. These few subjects advance through age-appropriate content cycles. They include the three "Rs," plus science, math, history, and literature. Within these few subjects are taught the essentials of a reality-based philosophy: reality (metaphysics), reason (epistemology), individualism (ethics), capitalism (politics), and romantic realism art (aesthetics).

The importance of the subjects of the three "Rs," math and science are obvious. History and literature are important so that the student learns how to connect high-level abstractions, such as values, ethics, justice, and love, to man's nature in physical reality. Children born into a flourishing society have no personal experience of the struggles that were necessary to create it. Without history, they will believe that the comforts of their modern society were always there and always will be. Properly taught, history will provide them with an understanding of what their successful society depends upon in terms of cause and effect. History will impress upon them the need for rational diligence to protect their society by protecting the principles that make it possible.

These are the *only* subjects required to fully develop the child's conceptual faculty and prepare him for adulthood. Attempting to teach additional subjects to young children detracts from their Forming Education. When they become older and have a good grasp of these essential subjects, students may learn additional subjects according to their interests.

Parallel to the child's conceptual faculty development is the need for the child's brain to reconfigure its functional modules to handle society's latest technology. We accomplish this by having the child use the technology in the education process. Five thousand years ago, human children had to learn to use a new technology called "written language" after humans developed it. Following language was photography, movies, television, TV remotes, computers, the Web, tablets, smartphones, etc. The brains of today's children are reintegrating their functional modules yet again to integrate AI personal assistants and virtual and augmented reality into their intelligence.

It is also essential to expose children to the historical chain of technologies from primitive to advanced to begin to learn about the "progression" of technology and from where it comes. The child should be exposed to pencil and paper and printed books, not just a tablet computer with apps.

A child who does not receive a proper Forming Education will, as an adult, have an underdeveloped conceptual faculty. Such a person will likely be:

- **Concrete bound**. He learned all sorts of factual content, but not the ability to integrate it and organize it into conceptual hierarchies or use transferable principles.

- **Short range**. Not knowing how to organize ideas with beginnings and endings and linked chains of causes and effects, he can only focus on what is in front of him right now.

- **Conflicted**. Not knowing how to validate ideas and reject invalid ones, he has accepted everything and holds many internal conflicts.

- **Emotional**. Unable to establish rational principles to guide his choice of values, he will act on emotions rather than rational thought, and his subconscious conflicts will constantly change his emotions.

- **Frustrated**. Unable to establish, pursue and achieve rational long-term values, he does not understand why his life is not as rewarding as it should be.

With a proper Forming Education, the child will grow up to be conceptual, long-range, confident, understand emotions, and be proud of the person he has become.

Forming Education:

- Develops the child's conceptual faculty.
- Explains to the child the "Why to Think."
- Teaches the child "How to Think."
- Teaches the child to grasp reality, reason, and individualism.
- Helps the child understand his place in history and the principles on which society and his continued existence depend.
- Causes the child's brain to integrate its functional modules in support of the latest technology.

THINKING MODE

As identified by Aristotle, man is the "rational animal." The word "man" represents a concept that refers to all actual men alive today or who lived in the past or will live in the future. We *integrate* all these countless men into a single word that we can hold in our mental focus. *Integrating* means identifying the essential element that unites many entities into one new entity even though they might differ in many non-essential ways. The essential element that unites all men is "reason"—the ability to integrate perceptual information about reality into abstract concepts and concepts into principles to guide man's life. Integration is the fundamental *mode* of thinking at the foundation of our concept-forming rational faculty. Ayn Rand said, "Integration…is the key to man's consciousness, to his conceptual faculty, to his basic premises, to his life."

When man first began to use language, thereby awakening his faculty of reason, the Biological Evolution Period was superseded by the Conceptual Period. His brain's functional modules were capable of conceptual thinking long before but connecting those modules in just the right way to enable language was required to activate it. Language was essential for evolving relationships between conceptual individuals. With language, man gained the ability to think conceptually—to use the *integration mode of thinking*, and form society.

Biological evolution occurred at the *phylogeny* level—the scope being the species as a whole—and the evolutionary changes were stored in the genes. Conceptual evolution occurs at the *ontogeny* level—with the scope being each individual—and the changes are stored in that individual's brain and the body of conceptual knowledge in society.

Biological evolution is durable, universal and inflexible—the information is written in the genetics that future generations will have. Conceptual evolution is individualistic, conditional and adaptive. Each new individual in each new generation must act to obtain the evolutionary change information passed down from prior individuals. He learns it in the form of concepts, communicated either in person or in external media (writing/reading). Suppose a new individual does not learn the accumulated information. In that case, he does not benefit from the gains made via conceptual evolution.

Biological evolution provided man with a rational machine—his brain—that he must use to survive. It did not come with a "User's Manual." He received no training or instructions or even instinct on *how* to use it.

Conceptual evolution dictates that the first and most important information that each new human must obtain from others is—*how to think*. Learning how to think builds the skill he needs to use his rational faculty to learn from society. He must learn what he needs to *know* and *do* to live a successful life according to his nature. For a human, the *how* is the conceptual method—the method of *integrating perceptions* of reality into abstract concepts and principles that he can apply to whatever he encounters in his environment. Reason differentiates man from all other animals and has made him the dominant species.

Biological evolution also provided man with volition to focus his mind and think or unfocus his mind and not think, replacing non-volitional, automated instinct. The family and social environment into which the new human is born will either *encourage* or *discourage* the child's motivation to learn how to think. The first precondition for learning how to think is to receive motivational guidance from others while at a young age.

A person who is not taught or chooses not to learn how to think can still survive today. Others can give him specific information about obtaining the food, shelter, and other things he physically needs. He can mimic others' actions (who *do* think) without ever learning how to use his rational faculty properly. In this case, he is not living as a fully independent adult human but as a perpetual child, needing to be told by others how to live. Others will either support and protect him or take advantage of him.

In his 2012 book *The DIM Hypothesis: Why The Lights Of The West Are Going Out*, Dr. Leonard Peikoff describes five possible answers to the question of *how to think*—five different *modes* of thinking and these are based on the fact that *integration* is the essential function that our conceptual faculty is built to perform. Peikoff says: "Leaving aside theory now and considering only facts, it is a fact that in the history of Western thought, from Greece to global warming, we find many different kinds of philosophy, but only five *modes* of integration."[155] These consist of three fundamental *modes*, two of which have sub-modes which are mixtures of the fundamental modes. For simplicity, I will here only reference the three fundamental modes that Peikoff describes: "The three fundamental modes correspond, as we have seen, to men's three basic integrative alternatives. Plato counseled a philosophic form of *invalid* integration; Aristotle, *valid* integration; Kant, *anti*-integration."[156] Peikoff then identifies these men as the Big Three in philosophy:

> What makes the Big Three great philosophers is that they are the only men in our history who have defined, whether validly or not, a new integration of fundamentals. The cultural and historical power they have had derives from this root. Plato, Aristotle, and Kant, each in his own way, tell us how to use our minds. Lesser thinkers then elaborate.

And then Peikoff summarizes the *emotional* appeal of each of the Big Three to their champions throughout the centuries: "Plato appeals to soaring idealism scornful of the practical. Aristotle appeals to joyful realism on Earth. Kant appeals to rage."[157]

The thinking mode of *integration* (Aristotle) means that the person has learned to start with the concrete perceptions of reality provided by the senses and integrate those into abstract concepts based on essential commonality while ignoring non-essential differences. The person then integrates lower-level abstractions into ever-higher levels of concepts. If integration occurs and connects to concrete perceptions of reality as its basis, it leads to *true knowledge*. This is the proper mode of thinking that new humans must learn to use the intelligence that biological evolution provided.

The thinking mode of *mis-integration* (Plato) means the person starts with lower-level abstractions that are products of imagination or fantasy, either his own or someone else's, and integrates *those* into higher-level abstractions. He is still using his mind to perform *integrations*. However, his entire system of integrated thought content is cut-off from reality. The result is not true knowledge. Usually, *the mis-integration mode of thinking results in mysticism*—attributing causes to something outside reality. A person typically learns to use the mis-integration thinking mode because of not receiving a proper forming education.

The thinking mode of *dis-integration* (Kant) means the person chose *not* to perform mental integrations, preferring instead to live life by reacting to the immediate *emotions* generated by random daily concrete situations. This choice absolves him of the burden of needing to think or to find and apply any principles or make any principled choices. Such a person views integrations as a threat and acts to remove or *disintegrate* them from his experience and lives his life dealing with the perceptual concretes he encounters daily. A new human might be explicitly taught the dis-integration mode by others who themselves use it. Or, an older person who has been using the mis-integration mode, after years of frustration over its ineffectiveness at guiding his life, concludes that forming integrations is useless, abandons them, and resorts to the dis-integration mode.

Peikoff's summary of the three fundamental modes of thinking is:

Aristotle (Integration): Achieving concept unit-reduction through the natural world—grasped by concepts derived from percepts. (*One in the Many*).

Plato (Mis-integration): Achieving concept unit-reduction through a transcendent world—grasped by concepts somehow independent of percepts; the natural world is appearance, and percepts are untrustworthy. (*One without the Many*).

Kant (Disintegration): Achieving concept unit-reduction is impossible and undesirable; concepts and percepts alike are detached from reality. (*Many without the One)*.

The *thinking mode* used by individuals who hold influential positions in a society spreads throughout the culture and becomes the predominant thinking mode. This is especially true for those individuals who work in the field of education. They have accepted responsibility to teach new humans *how to think* to absorb the evolving body of knowledge of reality that society has accumulated.

From the first awakenings of man's conceptual faculty to when Aristotle explicitly described it, 100,000 years later, man did not *explicitly* know *how to think*. Since man was the first species to develop a rational faculty, there was no one around to teach him how to use it. Therefore, his default method of thinking was *mis-integration*.

Early man saw all sorts of natural phenomena around him. He felt the need to make sense of them—to find causes for the effects—to use his new integration faculty. Without logic and the scientific method, which were still far off in the future, his only option was to attribute the effects he saw to mystical causes to integrate them. He could integrate the sun, wind, fire, floods, pestilences, life, and death by attributing them to omniscient, omnipotent gods. Early man was learning to form integrations. However, he started with imagined abstractions, not perceptual concretes—the *mis-integration thinking mode*—and it resulted in mysticism.

Man eventually began to turn his new intelligence inward to consider questions regarding his mind's workings. With the ancient Greeks, society began to identify and explore the *integration mode* explicitly. The ancient Greeks were the first human society to elevate reason above the mysticism that had been dominant for thousands of years. For the first time in human history, identification, definition, logic, reason, and science replaced superstition. Men began guiding their lives by the judgment of reason rather than divine will and revelation.

The ancient Greeks' two-century exploration of the integration mode of thinking was mankind's first experimentation with it. That first experiment was short-lived on history's time scale. The Greeks did not fully replace mysticism with reason and attempted to hold onto elements of both. This contradiction gave the advantage to mysticism, which had been man's default method of thinking for a hundred thousand years. The Greek experiment with the integration mode ended with Rome's rise and a return to mysticism, a hallmark of the Middle (Dark) Ages.

In the Dark Ages that followed the Greek experiment, Western society was again dominated by the *mis-integration* thinking mode (mysticism) for 1,700 years until the Renaissance. During this period, only church dogma was "truth." Men of science were considered heretics.

Mankind's second experiment with the *integration* thinking mode began with a resurfacing of Aristotle's ideas by Aquinas leading to the Renaissance, then continued through the Age of Reason, the Enlightenment, and the present Modern period. Man's second experiment with the integration thinking mode has been ongoing now for approximately 500 years.

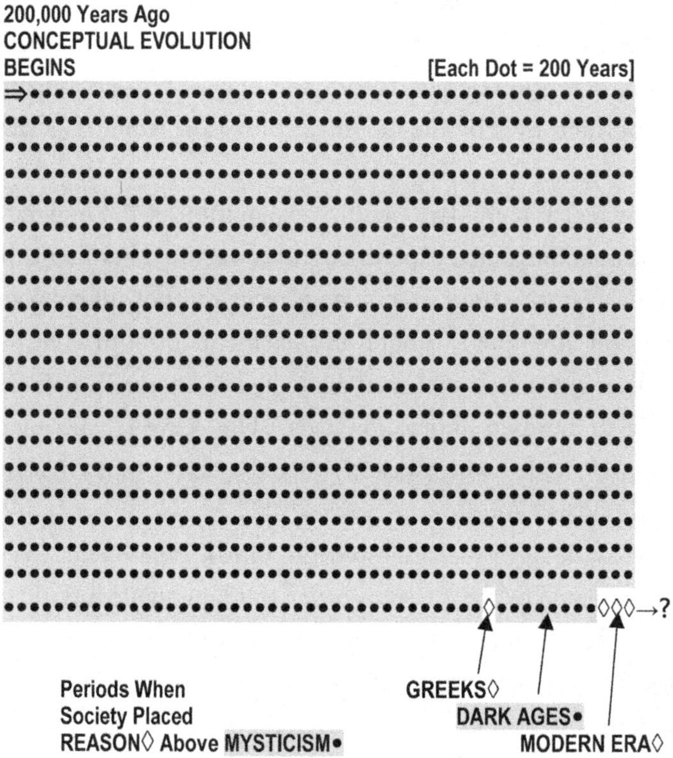

Human society is just beginning to learn to replace ancient mysticism with reason. Will our second experiment succeed, or will it fail, causing a New Dark Ages?

The creation of the United States of America was society's highest achievement to date in adopting the *integration mode* of thinking. It was the first country on Earth founded on the principle that a man must be free to use his mind in the service of his own life. However, like the Greeks in the first experiment, the Founding Fathers were unable to supplant mysticism with reason completely. They included elements of both in America's founding—the age-old contradiction between reason and mysticism. Additionally, they staffed America's new colleges and universities with professors from Europe, bringing Kant's *disintegration mode* of thinking, nihilism, and postmodernism. As a result, American society, based initially on the *integration mode* of thinking, has been returning, since its founding, to the default *mis-integration mode* and its consequence—mysticism.

Peikoff examined the thinking modes' factual evidence in four different cultural domains during six different Western periods. The cultural domains he studied were the period's literature, physics, education, and politics. The periods analyzed included: Greece, Rome, Middle Ages, Renaissance, Age of Reason, Enlightenment, and Modern times. He examined how a period's thinking mode affected its society and caused historical change.

Peikoff makes a prediction based on today's thinking mode conditions in America's literature, physics, education, and politics—RELIGIOUS TOTALITARIANISM. Peikoff writes "When every failure of the secular government becomes a boost to the champions of the anti-secular, the specific nature and sequence of these failures are irrelevant to the ultimate outcome."[158] He goes on to explain:

> Besides its religious commitment, it is highly probable that [a mis-integration] movement here (especially if not foreign-based) will appeal to the nation's strongly patriotic citizenry by stressing its own admiration of America and of American exceptionalism. The [mis-integrators] will declare, as they now do, that America is the greatest nation in history, and that this greatness is a blessing from God. Of course, by its nature, the [mis-integration] mind will reinterpret Americanism: Patriotism would likely become nationalism; individualism [would become] collectivism; and liberty [would become] pious obedience.
>
> As to the capitalist advocacy of private property, it is highly likely, in the light of American anti-Communism, that private property would be retained in name, but used and disposed of in fact according to the decisions of the totalitarian state. [...] [The regime] would adopt some version of group warfare, pitting the poor against the rich and/or, more likely, its own people against the rest of the world: God-fearing Americans versus godless foreigners. The next step would be to give God's word its due world dominance, which would require control by the regime of the world's thought, to be gained by some form of crusade against the infidels. Private property as the official but merely rhetorical policy at home, along with nationalist aggression to spread the truth abroad—these policies are not broad abstractions but a specific program.

> Not just a religious totalitarianism, but a religious-fascist totalitarianism—that is my prediction of the American future.
>
> Our own [mis-integrators], clearly, have a much easier task than that of the ancient Christians. Like their modern European ancestors, the New American Christians are making full use of the available technology; they too are on the fast track of modern change; they too face no philosophical opposition. And when the triggering crises strike, they too, like the Nazis, stand to be the only beneficiaries—the only ones able to speak up—to offer connected answers, not piecemeal trial balloons, and for this reason to be listened to by the country.[159]

In 2012, Peikoff estimated the collapse into totalitarianism would be completed in "forty to fifty years at the latest—say, two generations. [2052–2062] On current evidence, though, it might very well be a generation earlier." He pointed out that once it happens, "the life span of any kind of [thinking] mode, once in power, has ranged from about one hundred fifty years to about a thousand years. If [mis-integration] does happen here, the likelihood is that it will not be a mere phase. It will not go away soon."[160]

Peikoff also gave examples in 2012 of possible final triggers for the collapse that are getting too close for comfort today:

> [...] the trigger might be economic in nature, such as runaway inflation, a 1929-scale depression, skyrocketing unemployment, taxes that cannot be borne, a middle class wiped out financially, or national bankruptcy. Or the trigger might be the result of [...] U.S. silence in the face of domestic outbreaks of lawless protesters, all of it observed with growing anger by the American police and military.[161]

As the title of his book suggests, Peikoff acknowledges that this is only a theory, a hypothesis, but one based on rigorous analysis of the history of modes of thought and their societal consequences and our society's present state. Peikoff says that there are factors that are unique to America that make certainty in his prediction impossible. He says that if the collapse does not happen within fifty years, then his hypothesis is probably wrong. It might not happen at all.

In an earlier chapter, *The Great Unraveling*, I introduced the rise of the partnership of the two most important centers of power—the government and Big Tech. I pointed out the conflict underway between individual autonomy and Big Tech's claim on the right to know everything, about everyone, everywhere, all the time. It seems to me that if Big Tech wins that conflict, then their form of totalitarianism would likely be more secular than religious in the traditional sense. However, in their partnership with the government, Big Tech could find itself acting as the *enforcement* arm, receiving its *moral sanction* from a government that claims to represent *the people*. If religion offers the only arguments for morality available to the people, then that is what they will compel the government and Big Tech to support.

I put Peikoff's prediction in the same category as Kurzweil's prediction of the technological singularity. Each is evidence-based and put forth by a reputable professional in the respective field. Both are rationally plausible, and if they happen, both will have globe-altering, irreversible consequences. As you will see in future chapters, I think the technological singularity will occur later than Kurzweil predicts. On the other hand, I think societal collapse could occur by mid-century unless we can apply technology in ways that lead our society back to reality and reason. To be on the safe side, we need to take both these *possible* predictions seriously and consider them as *probabilities* and then plan and act accordingly.

Biological evolution selected for genetic traits that would lead to an individual's rational faculty. Similarly, C*onceptual* evolution is selecting for societal traits that advance discovery of the Laws of Reality, increase complexity, and increase entropy—the driving force of evolution. Therefore, natural selection will eventually lead to a society of individuals that fully commits to reality and reason and abandons mysticism.

Overcoming a hundred thousand years of mysticism is not easy. Man's first attempt was relatively recent and lasted only two hundred years and this one five hundred so far—both extremely brief in the context of evolution. Ayn Rand describes man's tenuous grasp of rationality this way: "It is as if, philosophically, mankind is still in the stage of transition which characterizes a child in the process of learning to speak—a child who is using his conceptual faculty, but has not developed it sufficiently to be able to examine it self-consciously and discover that what he is using is reason.'"[162]

> "The course of mankind's progress is not a straight, automatic line, but a tortuous struggle, with long detours or relapses into the stagnant night of the irrational. Mankind moves forward by the grace of those human bridges who are able to grasp and transmit, across years or centuries, the achievements men had reached—and carry them further. Thomas Aquinas is one illustrious example: he was the bridge between Aristotle and the Renaissance, spanning the infamous detour of the Dark and Middle Ages.[163]

The Conceptual Evolution Period began only 200,000 years ago, quite recent when one considers that the Homo genus has been around for two million years. Far shorter still, the first human society to intentionally experiment with the integration mode of thinking, the Ancient Greeks, was only 2,500 years ago.

Peikoff predicts that we will not achieve the goal of a *consistently rational society* during man's present experiment with the integration thinking mode—now in its waning years—with rationality once again being displaced by mysticism. It appears that we might be heading toward another relapse into the irrational.

Like the outcomes of the rise and fall of Aristotle's influence throughout history, each successive societal experiment with the *integration mode* of thinking adds to the historical empirical evidence of its effectiveness. The empirical evidence provided by Galileo in a period dominated by mysticism was not, alone, sufficient to throw off the dogma of the church. More time and discovery of the Laws of Reality were required. Evidence of the effectiveness of reason provided by America's founding principles has not yet been sufficient to overpower the dogma of the *Ivory Tower Cycle* and its *mis-integration mode* of thinking. We are not ready to finally let go of mystical ideas. The sun still revolves around the Earth.

Our society uses the *integration mode* (reason) to create technology. Still, we use the *mis-integration mode* (mysticism) when it comes to the humanities and social sciences, those things that define the fabric of our society. Our technology is rapidly advancing, but our society is retrograding toward mysticism. Our having advanced technology such as AI is like giving warring tribes of primitives, led by witch doctors, fully loaded machine guns. It is not likely to end well. It appears that mankind's second experiment with the integration mode of thinking could soon end. It will be followed by a protracted New Dark Ages—unless we can learn to apply technology in ways that cause our society to return to reason.

It seems that, at this point in the current Conceptual Evolution Period, we have two progressions that could play out by mid-century—say by the year 2050. The first is, if technology continues to progress and accelerate, the *technological singularity* might occur—when machines are more intelligent than humans. The second is that Western society might devolve into a *totalitarian state* by abandoning the integration mode of thinking. We find ourselves facing a collapsing society with advancing technology. Fortunately, "society" is not an existent. It is only the sum of the behaviors of the individuals that do exist. If the individuals can change, society can also change.

> ***True Idea: If our society continues to lose its connection of ideas to reality, it could collapse into a totalitarian state.***

♦♦♦

PART 4

ANTICIPATING OUR FUTURE

Fellow Explorer, where are you now, and where are you going? In the Prologue, I said explorers first need to orient themselves along several "dimensions." You can now identify your present position in the universe and in time:

Evolution of the cosmos? Following the Cosmological and Biological Evolution Periods, we are now in the *Conceptual Evolution Period.* In this period, humans are a species of conceptually intelligent entities converting "dumb" matter to intelligent use at an exponentially increasing rate. We are increasing Intellopy in the cosmos.

Evolution of humans? The *Technology Cycle* replaced natural selection as the primary driver of human evolution. Society gains an ever-increasing body of conceptual knowledge of Laws of Reality. Increased knowledge reconfigures each new human brain, causing it to be more intelligent than earlier generations.

Evolution of my ability to "think?" Suppose you, like most people, are a product of our Western public education systems. In that case, you might not have adequately learned the *integration mode* of thinking. You likely need to identify and correct invalid and conflicting ideas obtained from others.

Evolution of my knowledge of how the universe works? You know that entropy—the winding down of the universe—drives evolution. Evolution causes complexity in localized places in the universe. Complexity dissipates more energy, increasing the universe's entropy (if it is a "closed" system). Localized complexity on Earth evolved conceptually conscious beings (us). As Agents of conceptual evolution, we increase Intellopy (the quantity of intelligent matter). Intellopy is not created by random chance but by intentionally discovering the Laws of Reality and building things. We intend to convert the entire universe to intelligent use.

Evolution of technology? The *Technology Cycle* is the evolutionary process that increases knowledge of reality, advances technology, and increases human intelligence. We have reached the bend in the exponential technology growth curve. We are entering the explosive growth phase where discoveries and technologies will be more numerous and cycle faster.

Evolution of society? Only a society comprised primarily of individuals committed to reality and reason can sustain exponential technological growth. Only such a society can continue to evolve in this Conceptual Period of Evolution.

Society's first function is to teach its young the *integration thinking mode*. Its second function is to accumulate and distribute the conceptual knowledge of reality discovered via the *Technology Cycle*.

Today's society appears to be nearing the end of its second, brief experiment with reason. It is returning to mysticism. It remains to be seen if the Homo sapiens branch of biological evolution can build an enduring, consistently rational society.

Evolution of my happiness? Regardless of what is happening in the society around you, your happiness depends on:

- Holding *your* life and happiness as your standard of value.
- Using *your* reason to identify and rank your values within the range of possibilities dictated by reality.
- Acting solely according to *facts of reality* to achieve those values, *never sacrificing* a higher value for a lesser value or a non-value or sacrificing yourself to others or others to yourself.

Only you know how far along this "happiness" dimension you have come.

You have oriented yourself within the universe and in time. You have a basic understanding of the forces of change that will impact the future. In this Part 4, let us try to anticipate the future by projecting where those processes might take us. How might we apply technology to discover more about the Laws of Reality and better control our environment and ourselves? How can technology help us save our society? How might we transcend our present existence and chart a course toward a shining future?

4.1

TRANSCENDENT TECHNOLGIES

"I think that novels that leave out technology misrepresent life as badly as Victorians misrepresented life by leaving out sex."

—KURT VONNEGUT

TRANSCENDENT TECHNOLOGIES

Try to think of all the technologies that man has created, from the first stone tools to your smartphone. You cannot. There are now too many for one person to consider. This is a new development in human history.

Until around 1900, it was possible to document every technology that existed in a single printed set of books called an encyclopedia. As a teenager in the 1960s, my family had a 36-volume set of Encyclopedia Britannica dated 1902 that an earlier generation handed down. Each book was giant, very thick, and each page had dense, fine printing and artist-drawn pen and ink illustrations. The set was so bulky and heavy that you almost needed a small truck to move them. However, in those books were not just a summary but a thorough, detailed explanation of every technology that man had produced up to that date. The coverage of each technology would have been the equivalent of a college-level course today.

Suppose a person in 1902 read that entire encyclopedia. He would have come close to holding in his brain the knowledge of all the technology created by man in the prior three million years.

The explosion of technological development in the twentieth and twenty-first centuries has made such a feat impossible. Simply the act of gathering all knowledge of technology in one place, much less printing it, is impossible. Technology is advancing so quickly, in so many different fields, that only the fluidity and global scope of the World Wide Web can, sort of, keep up with it.

Today, a person can only hold in their mind the knowledge of a limited subset of existing technologies. The question is, which technologies are essential to understand? Which technologies constitute the critical pathways to our success or failure as a species? Which technologies, if we monitor their progress, provide signals about the opportunities and threats ahead?

I will address eight fields of technology that I believe will be transcendent. Over the next few decades, these technologies will enable biological humans and their society to transcend biology and become technology-based, first as Hybrid Humans and then as Virtual Humans. As we move from the Information Age into the new Transcendence Era, we must understand these eight technologies and follow their development. The eight transcendent technologies are:

- Genetics
- Nanotechnology
- Blockchain
- 5-G Wireless Network
- Artificial Intelligence (AI)
- Augmented Human Intelligence (AHI)
- Quantum Computing
- Fusion Energy

These eight technology fields are certainly not the only important ones. However, these eight represent the future and provide a good foundation for understanding where technology is likely to take us. These eight address:

- Our physical world
- Our biology
- Our intelligence
- Our society
- Our communications
- Our learning
- Our computing
- Our energy sources
- Our materials and manufacturing
- Our creativity

Today, each of these transcendent technologies is emerging from the slow-growth phase of its exponential growth curve. It is likely to mature, reaching its explosive growth phase within the next twenty-five years. Each will be a significant disruptor to our world. When combined, the disruptions they will cause in our global society are presently unimaginable. They will enable the human experience to transcend today's relatively primitive condition to an incredibly more capable existence.

The possibility of a societal collapse and the unforeseeable effects of advanced AI technology on society limit our vision of the future. This Part 4 will suspend both these concerns and optimistically assume that a thriving society with technological advancement will continue uninterrupted. Then, with exponential vision, we can anticipate what kind of future might be in store.

Let me briefly introduce here these eight transcendent technologies. I will then cover each in its own chapter, where I go just deep enough for you to have a sense of what that technology *is* and why it will be transcendent—how it relates to our future. I will provide each technology's basic domain-specific vocabulary and science. You will need these to follow each technology's progress. For deeper exploration, I would recommend the works by the referenced authors.

Genetics

Genetics is technology applied to our *biology*. It includes epigenetics, which cause specific genes or groups of genes to become active or remain dormant. It also pertains to the creation and functioning of proteins and other substances that make up our body's tissues.

Our ability to decode how our genes work and edit those genes means we are taking direct control of our biological evolution. The long, slow era of biological human evolution via natural selection is ending—transcended by the explosive period of intentional technological human evolution.

Genetics holds the promise of transcending the weaknesses, points of failure and functional limitations of our biological bodies, eliminating diseases and the aging process. Genetics also has the potential of maintaining a state of perpetual "youth" and enhancing our mental and physical abilities.

Unfortunately, genetics also holds the potential peril of human extinction, for example, through accidental or intentional creation of an unstoppable new virus.

Nanotechnology

Nanotechnology is technology applied to the *physical world.* Humans are the first species that can discover the Laws of Reality and apply those laws in modifying their environment to their purposes.

Advancing discovery of nanoscale laws of physics, combined with our increasing ability to manipulate matter atom-by-atom and molecule-by-molecule, means we will take absolute control of the matter around us. We will modify matter, even our bodies, into whatever we desire.

Nanotechnology holds the promise of transcending almost every limitation of our present living conditions. It will lead to unlimited clean energy, unbelievably useful forms of food, clothing, shelter, and transportation, a healthy planet Earth, merging our minds with our technology, and space exploration.

Nanotechnology also holds potential peril—for example, incredible weaponry that an enemy might use against us.

Blockchain

Blockchain is technology applied to the field of *societal transactions.* It is a form of globally distributed accounting and record-keeping. Blockchain provides secure, verifiable, indestructible, public and private information flow between parties to transactions without flowing through intermediaries.

Blockchain technology holds the promise of transcending our global society's present condition of unverifiable information, unknown reputation and lack of trust. It will provide transparent and indestructible record-keeping enforced by the global I-T infrastructure making fraud and deception difficult, if not impossible. Blockchain will simplify and speed up transactions and institutionalize ethical rules to help keep us safe from unethical people or AI entities and dangerous technologies.

Blockchain holds the potential peril of becoming a single, critical infrastructure on which our entire lives would depend. Suppose our government or a super-intelligence gains the ability to undetectably "hack" the blockchain. In that case, it could lead to a ubiquitous "Big Brother" totalitarian state of surveillance and coercion.

5-G Wireless Network

The 5-G Wireless Network is technology applied to *communications*. The emerging Fifth-Generation wireless network technology "5-G" is not just an incremental performance improvement over 4-G like 4-G was over 3-G. And it is not just about cellular technology. 5-G is a leap to a future where we can connect all people and devices in the world with nearly unlimited bandwidth, instantaneous responsiveness, reliable connections, and low power consumption.

5-G will be necessary for the implementation of the other transcendent technologies. For example, 5-G will enable reliable, real-time monitoring of health data from nanoscale biotech devices implanted in patients wherever they happen to be in the world.

Artificial Intelligence

Artificial Intelligence is technology applied to *machine intelligence*. It includes the field of robotics. Driven by our innate need to discover the Laws of Reality and apply them, humans seek to improve our tools of discovery and manipulation. We will use intelligent machines as tools—"robots."

Today's AI machines are better than humans at specific tasks, such as finding patterns in big data and doing repetitive calculations with great accuracy. They are, however, nowhere near as "smart" as a human. We keep improving them with the goal of a machine with human-level intelligence.

Suppose we create a machine with human-level intelligence. In that case, it could quickly become more intelligent than humans, a "super-intelligent" machine. Artificial Intelligence presents the possibility of man creating a new evolutionary technological species with next-level intelligence. If we do, it will be an evolutionary jump more significant than that which distanced humans from animals. Ultimately, AI holds the promise of mankind's technological creations becoming masters of the universe.

AI also holds the possible peril of creating machines with sufficient intelligence and resources to have purposes and the capability to execute on them that no longer require mankind. Of all the transcendent technologies, AI has the most significant potential to affect human evolution.

Augmented Human Intelligence (AHI)

Augmented Human Intelligence is technology applied to the enhancement of *human intelligence.* Human intelligence has been boosted by our technology for quite some time now. Cognitive fluidity causes our brain to reconfigure its functional modules to integrate the technology we ask our brains to use. Your intelligence, because of your brain's integration with your smartphone's capabilities and the Internet, is more capable than humans just a couple of generations ago.

We will explore ways to integrate new technology into our intelligence and apply our new capabilities. We will examine the pathway to our becoming a hybrid species, part human and part machine. Augmented Human Intelligence holds the promise of hybrid humans advancing until they leave biology behind and become the first super-intelligent species.

Quantum Computing

Quantum Computing is technology applied to *information processing*. Like 5G Wireless Network, quantum computing is not just an incremental performance improvement over today's digital computing. Quantum computing will enable the machine-handling in minutes of types of problems that today's fastest digital computers could not solve in a thousand years, if ever.

Quantum computing will provide the power and speed necessary to handle physics and mathematics questions that will need answers in the other transcendent technologies and the enormous quantity of data that those technologies will process.

Fusion Energy

Fusion Energy is technology applied to our *energy sources.* The transcendent technologies, especially nanotechnology, will usher us into the solar age with high-efficiency, low-cost solar cells and batteries. We will begin to break our dependency on fossil fuels. However, the ultimate solution to our need for clean, portable, and distributed energy will be electrical energy production via controlled nuclear fusion.

Fusion energy will be necessary to power a world full of our technological devices, including the servers for global 5-G Network communications and all of society's blockchain transactions. Fusion energy will drive nanoscale materials fabrication and large-scale infrastructure construction, including the energy needed to clean the Earth of our earlier technological waste products. Fusion energy will be vital for space exploration, where we can no longer depend on energy from our sun.

Technology Interactions

The transcendent technologies are not advancing on independent tracks. Advances in one make possible advances in the others. None of these technologies can reach its full impact without the support of the others. For example, advances in nanotech will enable finer detail in our exploratory understanding of the human brain's workings down to the level of mapping individual neurons in an entire brain. This knowledge might enable nanoscale devices to connect our brains to computers and eventually enable AI machines with human-level intelligence.

Nanotech will also lead to more powerful, energy-efficient and smaller computers. It will eventually lead to large-scale quantum computers, which will enable more capable and affordable AI. This improved AI will, in turn, be necessary to track and control all the nanoscale machinery, setting up a recursive loop of incremental advancement of both I-T and AI.

The powerful and energy-efficient computers produced by nanotech, AI, and quantum computing will be necessary to evolve the global blockchain. Quantum computing and AI will likely provide the precise real-time control needed to maintain the magnetic bottle sustaining a man-made nuclear fusion reaction.

Technology Awareness

Educating yourself in these eight technology domains will enable you to identify opportunities and threats as they develop. Milestone advances in one technology will enable advances in other technologies. It will also be essential to monitor how our governing institutions, media and society respond to the technology milestones.

If you are not a technologist, monitoring news feed headlines for new developments in these eight fields will provide a sense of how quickly things are happening. You can anticipate what might come next. I use a news app set up to search for articles on the eight technology topics above. I frequently and briefly scan the headlines and articles it finds for me. Monitoring these eight technologies maintains the "big picture" in my mind.

If you are a technologist, you might wish to delve deeper into the news and announcements and the people, institutions, and technologies involved. This analysis could assist you in your career or help you with financial or personal decision-making or investment opportunities.

Let us take a closer look at each of these eight technologies to establish a baseline of knowledge. You will learn what they are, challenges they face, how we might apply them, how they could interact, and what the future might bring. More importantly, we will identify how the principles we have just learned about reality, human nature, and societal processes apply to these future technologies. Once you equip yourself with a basic understanding of these transcendent technologies, it will be up to you to keep up with the news.

♦♦♦

GENETICS

"The possibility of genetic engineering is one of the most important concepts to arise in the history of mankind because, for the first time, a living creature understands its origin and can undertake to design its own future."

—ROBERT SINSHEIMER

GENETICS

Genetics is about *discovering* the mechanisms of biological bodies and consciousness. It is also about *manipulating* those mechanisms to cure and prevent all disease, stop and reverse aging, and increase our physical and mental capabilities.

Eventually, we will replace our biology with technology, retiring genetic functions and replacing them with technological ones. We might someday leave our biological bodies behind and transfer ourselves to more durable and capable machine bodies. However, that is not in our near future. In this chapter, we explore the promise of a vastly improved and extended *biological* existence.

The most profound change to grasp is that we will live much longer and healthier lives than our ancestors. Each of us is likely to live a lot longer than we now expect. You know the age-old question—"Would you live your life differently if you were told you only had one year left to live?" What if you are told you have 150 years left?

HEALTH-SPAN

I prefer the term "healthspan" over "lifespan" because it includes the idea that we will not just live a longer life but a healthy one. We will avoid disabilities previous generations suffered in old age. Ray Kurzweil provides a good introduction to the idea of radically increased healthspan in his book *Transcend*:

> Up until just recently there was no conceivable way to cheat death, so our philosophies rationalized death as actually a good thing. 'Death gives meaning to life' is an age-old adage, but in our view death does exactly the opposite. It is a great robber of relationships, knowledge, wisdom, and skill, all the things that make life worth living. The authors believe that it is life itself—and all the things we can do with our lives—that gives life meaning.[164]

We are not yet able to forestall death indefinitely. However, we are at the point where we know how to slow down disease and aging so that even baby boomers can be alive and healthy when we do have sufficient knowledge and the ability to stop the processes of disease, aging and death. In a few decades we might be able to backup the information in our bodies that make us who we are and transfer our consciousness into new, better bodies.

GENETICS AND EVOLUTION

The species Homo sapiens is enjoying a long run of a few million years. The biological machinery of evolution—genetics—requires a relatively rapid reproduction cycle. Each human only gets to live for a maximum of about 120 years, usually much fewer. Rapid cycling is necessary to select the most successful combination of genes adapting to a specific environment via trial and error.

From a natural selection perspective, an individual's purpose is to test a new combination of genes. If the individual survives to produce offspring, then that genetic combination was successful. Once an individual has offspring and protects them until their reproductive age, the individual serves no other evolutionary purpose. Therefore, our genes are optimized to keep us alive until our kids reach reproductive age, and that is all. If an early human could reproduce at age 15, his children could reproduce when he was about 30. That was his life expectancy.

One might think that evolution would have wanted our ancient ancestors to live longer to pass on hard-earned knowledge to their offspring. However, they probably did teach their kids all they needed to know to survive until they became sexually active. Hence, it is more likely that our genetics would remove us from the population after raising our kids to free up scarce resources for new generations.

While an individual's life cycle is relatively fast, evolutionary change in the human species is glacially slow. Humans diverged from the apes around six million years ago. Assuming 25 years to a generation, that was 240,000 generations ago. It has taken that many generational iterations to get us to where we are today.

But to what kind of environment is our species adapted? We did not enter the technological age of abundance and live much differently from our ancient ancestors until recently. We invented the wheel only 230 generations ago. Aristotle described the principles of "logic" only 94 generations ago. The spinning wheel for thread and cloth appeared only 50 generations ago. We have not lived in the age of abundance long enough for our genetics to have changed significantly. We still have the genetics of our ancient ancestors, who lived in primitive conditions. Our bodies remain designed to live short lives in support of biological, genetic evolution. But that is about to change.

GENETICS AND TECHNOLOGY

Evolution took millions of years to develop humans equipped with brains with functional features enabling conceptual-level intelligence. The genetics of our biological bodies evolves at the same plodding pace. However, our conceptual brains' activities adapt our surroundings to our purposes. Our "technology" grows independent of genetic evolution and moves at a rapid and accelerating pace. Using that technology then changes us, restructuring our brains.

Early humans would discover how to make a helpful tool and personally teach it to their children and others. This knowledge was not programmed into their offspring's genetics, nor was it lost with their bodies when they died. Their technological inventions survived via *demonstration* into future generations. Then, once language and writing emerged, knowledge could distribute via *communication* widely across time and space. This distribution was kicked into high gear when we invented computing and the World-Wide-Web.

Knowledge about how the world works and how to adapt it to our purposes is cumulative and expanding at an exponential rate, even as countless individuals live and die. Genetic evolution has been and is still following an imperceptibly slow, slightly rising growth curve. Technological evolution, however, is on an exponential growth curve that started out rising slowly but then accelerated until it exploded into today's rapid, vertical rise.

Technology applied to health sciences has been on that slow horizontal line for the last few centuries, beginning with tribal medicine men. As the scientific method was developed and applied, significant discoveries such as antibiotics, smallpox, polio elimination, and malaria control occurred. We have now reached the bend in the exponential technology growth curve in health science and are about to go vertical.

With the sequencing of the human genome, we have just begun to *discover* how things in our bodies work at the genetic level. With the discovery of the "CRISPR" method of gene editing, we have just begun to *modify* genetics to our purposes—think "eliminate disease" and "reverse aging." Modifying our environment for our purposes cannot get more personal than changing our bodies.

Kurzweil reminds us that medicine only recently and now quickly, is becoming an *information technology* subject to the law of accelerating returns. It will be 1,000 times more powerful per decade, and 100 times smaller per decade. By manipulating our genes and the mechanisms that turn them on and off (epigenetics), we will be able to exert control over their functioning in our bodies. We will be able to correct genetic problems that cause diseases, and we will slow, or reverse aging as programmed by our genes. Eventually, it will be the domain of nanotechnology that will allow us to directly modify the physical structures in our bodies independent of our genes. Then, we can have any body we desire.

THREE BRIDGES TO RADICAL LIFE EXTENSION

So, the question then is, "How do we stay alive and healthy until technology enables us to live indefinitely?" Kurzweil describes Three Bridges to Radical Life Extension:[165]

1. **Wellness Self-Healthcare**: What you can do now to slow down or stop the processes that lead to disease and aging. Stay healthy long enough to reach the next Bridge.

2. **Biotech Revolution**: Changing our genes and their functioning to live well for decades longer than what we now consider a long life.
3. **Nanotechnology Revolution**: Where we go beyond the limitations of genetic biology and live indefinitely.

BRIDGE ONE—WELLNESS SELF-HEALTHCARE

Bridge One, Wellness Self-Healthcare, is not about extending maximum human life expectancy, which is around 120. It is about extending "healthy" life expectancy—delaying age-related disability longer and closer to the maximum life expectancy.

Our present healthcare industry is "illness-based." We do not go to the doctor unless we are experiencing illness symptoms or getting an annual "checkup" for the doctor to detect an illness we were not aware of. In such a checkup, "No news is good news." We generally do not learn about the causes of our illnesses and how we might have prevented them until we already have them. This is how doctors were trained to provide healthcare; the way today's healthcare economics work, and how we are conditioned to utilize it.

However, in the last few decades, an explosion of knowledge about the causes of disease and illness has occurred. Wellness healthcare, in contrast, is about preventing disease and illness.

In our present illness-based healthcare system, a minimal amount of this wellness knowledge makes its way from the laboratories into the medical schools, the latest generation of doctors and the patient population. And that process takes years.

We can live longer, healthier lives if we become proactive rather than passively relying on the default illness care system. There are many wellness and longevity information sources available. Unfortunately, people are not yet sufficiently motivated to take charge of their wellness healthcare. They have not been conditioned to tune into the information streams that would make them aware of this new knowledge—unless they are already sick.

A person's Personal Philosophy determines whether he values his own life. If he is enjoying a happy life, he will want it to continue as long as possible, and therefore he will take charge of his wellness care. If he is suffering, he might not. The happy person looks after his physical and mental health.

BRIDGE TWO—BIOTECH REVOLUTION

Bridge Two is about changing our genes and their functioning to live well for decades longer than what we now expect. It is about avoiding today's age-related diseases and disabilities altogether. Perhaps it will even be about halting and reversing the aging process itself. The result will be longer *healthspans* and *lifespans*. Every additional year that we live will bring more effective illness and aging prevention technology.

At some point, and we could already be there, we will achieve what we should call living "indefinitely." It is not the same as being "immortal" like gods. While we will no longer die from our bodies' natural failure modes, we could die because of accident, aggression, or suicide.

It used to be that the most significant gains in longevity occurred at the infant and child ages. Now, the most important improvements are for those over age 60–70. I say we might have already reached "living indefinitely" because biotechnology will likely achieve radical life extension in the next two or three decades. Most people under age fifty today, and those who are older but exercise proactive self-healthcare, might live long enough to reach radical life extension and live "indefinite" lifespans.

In the field of genetics, the term "germline" refers to the cells in our body that enable reproduction. The genetics inside this lineage of cells has continually created new young people for millions of years. Germline cells that define our species and formed us do not age, or we would not be here.

The term "soma" cell refers to all the cells in our bodies that are not germline cells. Aging only occurs in the soma cells and is activated during cell differentiation in the embryo. Our mortality is a phenomenon of genetics activated only in our soma cells.

A primary cause of aging is the limit on how many times soma cells can split and reproduce themselves ("mitosis"), renewing our body's tissues. The cells in a few tissues, such as neurons and heart muscle, do not divide and are called "post mitotic" cells.

For mitotic cells that divide, protective structures at the ends of the chromosomes called "telomeres" get shorter each time the cell divides. After about sixty divisions, called the "Hayflick limit," the telomeres become too short, and the cell stops dividing. This protects from damaged DNA being passed on to future daughter cells, for example, in cancer tumors. Environmental factors known to accelerate the shortening of telomere length include stress, lack of physical endurance activity, excess body mass index, smoking, chronic inflammation, vitamin deficiency and oxidative stress.[166]

A previously "mitotic" cell that has stopped replicating is then called a "senescent" cell. As the number of senescent cells in our tissues increases, those tissues become less able to heal themselves. We begin to see the signs of aging.

The Hayflick replication limit is not the only cause of cell senescence. DNA damage or mutation can also cause a cell to stop replicating and become senescent. The immune system usually removes senescent cells that can no longer function properly.

The mitotic Hayflick limit is not present in all animals. In primitive, multi-cellular animals such as hydra and planaria, if one cuts any part of their body tissue away, the remainder will regenerate the lost tissue. They can regenerate the specialized tissues for any part of their body forever. In humans, their cells' ability to regenerate and specialize into any of the body's various tissues lasts for about the first eight weeks of embryonic development in the womb. It shuts off at the point when an embryo becomes a fetus. The embryo cells that can differentiate into types of cells needed are called "pluripotent stem" cells.

Stem cells express and apply an enzyme called "telomerase" to replace the lost telomere length each time they divide. Therefore, the Hayflick limit does not apply to them. It is possible for mitotic non-stem cells that would normally become senescent to express telomerase either through genetic damage or mutation. They then become able to replicate endlessly, just like stem cells, usually in the form of a cancer tumor.

Geneticists have the medical goal to discover the pluripotency mechanisms of the stem cells and apply those mechanisms to the creation of regenerative therapeutics for the soma cells of our tissues. For example, the company AgeX, using a patented technology called PureStem, has captured primitive embryonic stem cells with enormous replicative capacity. It has successfully cloned them in the laboratory into master cell bank cultures.[167]

The company has demonstrated the ability to use these PureStem cells to produce 100% pure brown adipose tissue (BAT) cells, which are cells that generate heat in our bodies. Precipitous BAT cell loss as we age is in part related to Type 2 diabetes and obesity.

AgeX has also demonstrated an ability to produce vascular endothelium cells, which target preventing cardiovascular disease. They are also working to develop lines of graftable young pluripotent cells that they can manufacture at an industrial scale to repair various organs.

The ultimate goal of geneticists is the reversal of aging. They intend to discover the germline cells' mechanisms supporting their immortality and apply those mechanisms to the soma cells in our tissues. In the 1990s, geneticists were first able to capture human germ line cells, precursors to sperm and eggs, and get them to reproduce in the laboratory dish indefinitely without aging.[168]

In 1997, with the cloning of the sheep "Dolly," scientists discovered that the transfer of nuclear material from a senescent (aged) skin cell into the nucleus of an egg cell could reset the new cell back to the pluripotent embryonic stem cell stage with telomere lengths that are young again. This discovery led to today's "induced pluripotency," where the aging of cells can be reversed by transcriptional reprogramming back to a young pluripotent state with telomere lengths even longer than the cell had initially.

Science today indicates that it is indeed possible in the laboratory to reprogram any aged cell back to the beginning of life. The big question is whether this epigenetic reprogramming can be done in a living human body? As the embryo becomes a fetus, our somatic cells lose their germline immortality. They lose their ability to regenerate themselves. The loss of regenerative capacity occurs in stages as the fetus develops. The heart tissue loses its regenerative capacity within the first week after birth.

Scientists now have laboratory master cell banks of the germline cells, pluripotent stem cells, and aged somatic cells. With these cell banks they can now do the science of discovering the genetic and epigenetic mechanisms of programmed loss of regenerative capacity. They have the goals of learning how to prevent regenerative capacity from turning off or inducing aged tissue cells back to the embryonic stem cell stage.

As possible anti-aging therapeutics, AgeX explores strategies that involve cocktails of small molecules known to induce pluripotency and messenger RNA components that transport telomerase signals throughout the body.

In late 2020, Israeli researchers discovered that repeated intermittent exposure of thirty people aged 64 and older to hyperbaric oxygen therapy (exposure to pure oxygen at higher-than-normal pressures) significantly increased telomere length, decreasing senescence in certain types of blood cells.[169] Such discoveries, once the mechanisms are understood, could soon lead to novel types of anti-aging therapeutics.

Today's science suggests that there is no physical law of biology that makes human aging and death inevitable. It suggests that we will indeed be able to halt and then reverse aging—probably in incremental steps and over the next two or three decades.

BRIDGE THREE—NANOTECH REVOLUTION

Bridge Three is the culmination of the *Transcendence Era* of the Conceptual Evolution Period—when we ultimately transcend our biological bodies. We will manipulate the structures and functions in our bodies by the direct manipulation of matter.

We will not need to switch on certain genes to build stronger bones. Nanobots inside our bodies will assemble the calcium and other molecules needed to make bone tissue in the desired locations. Artificial red and white blood cells will roam the body carrying oxygen far more efficiently and eliminating pathogenic invaders far more effectively. Artificial red blood cells that perform all the functions of natural ones plus the ability to carry cargo such as therapeutic drugs, have already been created.[170]

Eventually, nanotechnology will make possible the discovery of the tiniest structures and functions of our brains, our consciousness and our memories. All the while, our robotics technologies will be advancing both the ability to create mechanical "bodies" and "brains." It might become possible to periodically transfer our current state of consciousness and our memories into a computer for safekeeping, "backups" of ourselves.

Then, we might wish to transfer our consciousness to a mechanical body and brain, abandoning our biological bodies. Our new mechanical body and brain might confer many advantages over our biological bodies, such as enhanced sensory capability, enhanced physical capability and endurance, and enhanced intelligence. As our consciousness and memories grow, we can continue to make backups of ourselves. We will transfer our consciousness into different body types for special applications, such as space travel. So long as we maintain our backups and the technological infrastructure continues to exist, we will be able to live indefinitely.

Consider that biological humans, as agents of conceptual evolution's *Technology Cycle*, rely on society as the repository for accumulated knowledge of the Laws of Reality. Each new human starts as a blank slate and learns whatever domain expertise is necessary for his chosen profession. The knowledge an individual discovers during his lifetime that was not recorded by society is lost when he dies.

When Hybrid Humans or Virtual Humans become able to continuously back up their mental state to technological storage and live indefinitely, an individual will be able to accumulate vastly greater knowledge than today's biological human. Having greater knowledge in multiple domains within a single conceptual mind will significantly increase that mind's integrative creativity potential. This will be one of the great accelerators of conceptual evolution from the *Transcendence Era.*

As far-fetched as a scenario like this seems, keep in mind the explosive, exponential technology growth curve. Remember that our forward vision needs to be exponential, not linear. We simply must at least consider the implications of the possibility of such a scenario.

LIFE EXTENSION CONSIDERATIONS

The world's population universally accepts that aging and death are inevitable. They have learned that all the historical searches for fountains of youth were futile. Associating old age with infirmity, some people do not look forward to old age. Some are optimistic that scientific breakthroughs will cure most cancer forms, prevent dementia, replace worn-out body parts, and prolong healthspan. However, even these optimists believe that death is inevitable.

How might the relatively sudden and seismic change from *expecting to die* to *expecting to live indefinitely* impact society? Discovering that we can live indefinitely could have consequences as enormous as discovering other sentient beings in our galaxy. Discoveries this huge will challenge society at the philosophical level as we seek to understand life's meaning.

The following selection of topics suggests the kinds of things you might want to anticipate. Here is a clue—the proper philosophy for living a happy life does not change with life's length.

POWER GRAB

Suppose one day the world wakes up to the news headline "SCIENTISTS STOP/REVERSE AGING IN MAMMALS." The news article explains that a new epigenetics technique was able to make older animals young again. And, it has worked on all mammal species experimented with so far with no side effects. They expect to begin human trials. What shock waves will this news send throughout the world?

The ability to control who gets anti-aging will be the ultimate power-grab of the era. We can expect that people with power and money will want to acquire this technology. They will want to quickly prove that it works in humans and then profit from it. At the same time, there will be a clamor from the general population and their socialistic government leaders that such a discovery should belong to everyone, with efforts to nationalize the technology's production and distribution.

Will it be possible to have a free-market scenario where the developing company profits from its efforts by selling anti-aging to those who can afford it? Will patent protections hold up? Or is it more likely that society will claim that benefiting from such a thing is immoral? Will the government step in to control distribution and pricing and compensate the developing company with tax dollars as part of the government-subsidized healthcare system? What if the discovering company is American? Will China, or any other country for that matter, not immediately replicate the technology? What if China develops it first? Will we not replicate it in America?

INCREMENTAL AND GLOBAL

Will life-extending technology be announced suddenly as in the above scenario? Or will it be the culmination of incremental discoveries that eventually reach extreme life-extending effectiveness? There is not just one life-shortening disease to be conquered; there are many. Likewise, there is probably not just one "gene" to be discovered and turned off that causes aging; there will be several that interact in complicated ways.

I expect that during Bridge One, the Average Life Expectancy will increase incrementally toward the Maximum Life Expectancy of 120 as we eliminate specific diseases one by one. This gradual extension of Average Life Expectancy and Healthy Life Expectancy will not cause shocks to our society as death is still inevitable, albeit delayed.

However, during Bridge Two, our ability to discover our genes' functioning and manipulate them will improve. We might reach a critical threshold, the bend in the exponential growth curve, where scientists, over a short period, learn how to turn off genetic causes of aging. These discoveries will dramatically extend our Maximum Life Expectancy well beyond 120 and send the "fountain of youth" shockwave around the world.

Scientists will likely share each incremental discovery with the world. Laboratories worldwide will build on each other's successes, eliminating specific causes of aging and death until they achieve extreme longevity effectiveness in multiple laboratories. In this scenario, a sudden revelation by a single laboratory or company is not likely. Still, the media and public awareness that it has been achieved could occur over a relatively brief period.

PRODUCTIVE ACHIEVEMENT—25-YEAR "MASTERY" CYCLES

Most people have lived roughly 25-year periods of purpose. They received their education up through age 25–30. Then they had a productive work career for 25–30 years to age 50–60 and then "retired." If still healthy, they found a non-productive recreational interest such as golf or another working interest until age 75–90 when age-related disabilities forced them to stop. In the final stage, their purpose was to stay alive with their disabilities. They usually died between the ages of 75 and 90.

Suppose we can avoid age-related disabilities until age 100? Are we going to want to play golf for forty years? And if our health span becomes age 150, will we play golf for 90 years? When our health span becomes age 100 (perhaps soon), we could have three 25-year productive careers. And if our health span becomes 150, we could have five 25-year careers. And what if we can live indefinitely?

Productive work is the central value around which we organize all our other values in life. It defines us. It is who we are. The first thing you want to know about a person you meet is "What do you do?" Achieving pride from the work we do is essential to our happiness. Increasing health span will cause our society to transition from the traditional expectation of "retirement" to an expectation of "serial careers." It takes roughly 25 years to start as a novice in a new productive work field, eventually become a recognized master in that field, and finally desire a new career challenge.

As we live longer, we will understand that our achievements make our lives worth living. We will need the long-period variety and freshness that periodic career change provides. The knowledge, experience, and skills gained in each past career will amplify our ability to achieve in the next one. Kurzweil reports, "[…] that a human master in a particular field has mastered about 100,000 chunks of knowledge."[171] We can expect to master more significant quantities of knowledge through augmentation with our technology, and we will master it more quickly. We might find that we need more frequent career changes because of paradigm shifts caused by new technology.

Productive work with ever-increasing capability will become the primary flow of our long—maybe exceptionally long—lives with eddies of recreation and leisure along the way. With people living and working longer, our society will see a dramatic increase in human productivity and creativity.

ADAPTABILITY—ACCEPTANCE OF UNKNOWABLE CHANGE

How will accelerating technological change affect us? The product cycle, the time from when a new product becomes mainstream to when a more recent one replaces it, will keep getting shorter. Just about when you get comfortable with one product, you see that a newer, better one is now available to replace it. Should you upgrade? And new types of products will continually appear that you never had before. Should you get one? This continual change will be true of everything that touches your life.

On the one hand, we like the soothing comfort of familiar things. We can relax with them. On the other hand, we want the excitement, variety, stimulating challenge, and new knowledge that new things can provide. Long-term health and happiness will require that we maintain a balance of the *old* and the *new* in our lives. We will need the stability, continuity, and sense of past that the *old* will provide. We will need the growth, knowledge, skills, and adventure that come from the *new*.

We absolutely must get comfortable with unending change. We must make it an integral, enjoyable function of our lives to embrace new products and learn to use them. To balance this, we will need to keep some old things around. But you cannot learn every *new* thing or keep every *old* thing.

The wisdom will be in the decision-making of which *old* things you should keep in your life and which *new* ones you should acquire and master. Generally, you should acquire new items when essential for your work, education and skills-building. For example, you upgrade to a more recent version of your wearable personal AI assistant. You can keep old items around when upgrading them would not significantly contribute to your work, education or skills. For example, you enjoy music from an old vinyl record player in your leisure.

REAL VS VIRTUAL AND MANUAL VS AUTOMATED

To keep up with technology, enhance your intelligence, and maintain your productivity, you will need to learn and use new devices. For example, you will learn to use a wearable AI Personal Assistant (AIPA) and associated Augmented/Virtual Reality (AR/VR) displays. Your AIPA will perform many technology interface and control functions for you and enable you to virtually experience remote people, places, things, and information in the world.

For work, your AIPA could arrange an AR/VR conference with your coworkers walking you through an AR/VR tour of the production facility being discussed. For leisure, your AIPA could summon an autonomous car for you, tell it where you want to go, and arrange for payment. Your AIPA could take you on a guided tour of a museum via your realistic AR/VR device while you sit riding in the car.

Of course, you certainly could tour the production facility and meet with coworkers *in person*. You could also drive your own car (assuming this is still allowed) and see the museum yourself. In addition to balance between the "old" and "new" in our lives, we will need to balance between what we want to experience "virtually" and what we want to experience in "reality." Further, we will need to balance what we wish to be "automated" for us and what we want to do "manually" ourselves. Like the old-versus-new, we should use automation and AR/VR to enhance our learning, skills, and productive work. We only want automation that amplifies our direct connection to and effectiveness in actual reality, not automation that replaces them.

We must do things for ourselves and experience reality directly to remain fully connected to it. Suppose we allow our consciousness to become disconnected from reality by living only in a virtual world. In that case, all our knowledge becomes unreal, and we become destined for failure.

We should selectively choose to learn and use technology that enhances our ability to sense and interact with the real world. We will need to occasionally test that the technology keeps us connected to the real world by experiencing it directly ourselves. We must remain conscious of when a virtual world is not the real world, especially if the virtual world is a fantasy, such as for leisure or entertainment purposes. We will need to balance living in *real* and *virtual* worlds with *manual* and *automated* actions.

THE JOURNEY NOT THE DESTINATION

How will we change psychologically when we believe we will live an exceptionally long life and maybe indefinitely?

We will not confuse living "indefinitely" with being "immortal." We can still die, just not from old age. We will continue to be responsible for our personal safety to not fall victim to accidents or violence. As we become Hybrid Humans increasingly dependent on technology, we will need to ensure the continuation of life-supporting technological resources.

Youngsters do not yet have a sense of their mortality. They are having too much fun "playing" even to be aware of it. The possibility of death is the farthest thing from their mind. Awareness of the hourglass of their lifespan develops in them as they get older. But what if we never have to be concerned with our mortality? What if we can remain child-like forever?

For a child, life is about learning and growing. It is about seeking out new experiences and new knowledge. It is about answering the question, "What am I capable of?" No matter how long one lives, there will always be something new to experience and learn in a world of exponential technological change. What one is capable of will have no limit.

Sustaining happiness for exceedingly long life will require acknowledging that we are on a long journey with an undefined destination and a potentially unlimited but not guaranteed lifespan. We will need to value the journey as there is no ultimate destination in which to place value. We can have milestones ahead that allow us to set and achieve long-range goals, such as 25-year career cycles. Still, we expect that there will always be another refreshingly interesting one after that.

We conclude that our life is worth living to experience the next exciting achievement of ourselves and mankind, both technological and societal. Are we indeed going to the stars on rocketships? Life becomes a giant child's toybox, and we want to play with all of them.

Human intelligence will be increasing over time as we further augment our intelligence with technology, first with personal AI assistants and then with brain-technology interfaces. The world will present us with a never-ending supply of toys and equip us with ever-increasing intelligence capabilities amplifying our ability to experience, learn, and create. We will recursively ask ourselves, "What *else* am I *now* capable of?"

CROSS-GENERATIONAL COMMUNICATION

The biggest challenge to communication between generations (parents to children, grandparents to grandchildren) is the difference in competency using technology. Adults who grew up in a world without cell phones or personal computers find it amazing how today's three-year-old can operate a smartphone. They are perplexed why teenagers would rather chat with friends from their bedroom on social media than go over to their friend's house to visit in person.

Imagine a world where you co-exist with your great-great-great-great grandkids? Will you even feel like you speak their language? Suppose you want to remain a productive member of society. How will you do productive work in *their* world? Suppose you hope to communicate with them and pass on the wisdom you've accumulated. Then you must keep up with technological and societal change. You must be able to speak *their* "language."

If the need to keep up with technology is like learning a foreign language, you must learn it and then use it to be fluent. And this language will be continually evolving and changing—keep up with your language lessons.

Otherwise, the people around you will begin to sound like foreigners as they discuss new technology and societal functions that you do not understand.

If individuals do not keep up with technology, our society will become stratified according to different generational, not geographical languages. We will not achieve the full potential of society's richness of experience, creativity and productivity across all generations.

SUSTAINING URGENCY

One might think that expecting to live an exceptionally long life would kill any sense of urgency for anything. We could always put something off until tomorrow because there will be an endless supply of tomorrows. However, the reverse will be our reality.

Sustaining urgency will not be a problem unless a person gives up on life. Rapidly advancing technology, and our need to keep up with it if we do not want to fall behind, will make us anxious. It will seem like we are on a treadmill that is continually speeding up. We will just need to work harder at learning this new language faster!

This accelerating treadmill is not a sustainable situation. First, no one can be fluent in all technology. There will be too much coming at us too fast. We will need to be selective in what technology to embrace. If we keep up with things that help us in our work, or have become mainstream in general society, or are popular among the youngest generation, then we will feel competent and connected—not left behind.

We will also need to acquire and use new intelligence-enhancing technologies, such as personal AI assistants (and someday, brain-machine interfaces). Using them will cause our brains to develop new intelligence capabilities by reconfiguring functional modules. As we make our brains more capable, we will increase our cognitive ability to keep up with technological change and society's latest language.

By living longer lives, we will be able to attempt more things. We will have many more stored memories of our successes and failures. These will lead to improved decision-making and creativity. Better decision-making and creativity will enable us to choose successive new experiences better. We will, over time, become better able to identify and select the values we want to pursue and how to achieve them. We will become more focused and purposeful, not less.

SENSE OF HISTORY IN THE FUTURE

In 1905 George Santayana said, "Those who cannot remember the past are condemned to repeat it."

Consider the scale of the extreme longevity we are considering. Suppose aging is stopped and reversed, and we can live beyond 250 years. The United States of America is less than 250 years old. Historically, entire countries have risen and fallen in that amount of time. What might your perspective on life be today if you had been alive during the time of America's founding and through to today?

You would remember the past. Therefore, you would not be condemned to repeat it. Lessons you learned the hard way regarding what is good or bad for society would not be lost to history. You would have a personal, first-hand understanding of the principles that historians and philosophers otherwise must extract from history. You could validate a proposed idea as true or false based on your own experience, not on history books.

Throughout history, a society's later generations did not have to endure earlier generations' hardships that created their civilization. They took their civilization for granted as if it always existed and always would. Their complacency eroded the principles on which their civilization was founded, eventually bringing the civilization's demise.

In the USA, our society is already embracing principles that oppose those of its founding. We keep moving further away from those principles. Suppose America's Founding Fathers, who fought for our independence putting their lives on the line, and who wrote our Constitution, were still alive today. Would they keep us on track? Would their compelling, first-hand accounts of America's founding convince later generations to sustain those principles?

Our society erects statues in honor of great heroes, including statesmen, generals, adventurers, businesspersons, artists, and scientists. Usually, successive generations have decreasing interest, forgetting what that person did to deserve the statue. However, today we see that the younger millennial generation reveres the older "Greatest Generation" who fought in World War II. They want to hear all the stories of their struggles and triumphs. Unfortunately, the Greatest Generation has almost entirely passed away. If our heroes could live for hundreds of years, we would not need statues to remember them. They could give later generations first-hand accounts of history, thereby keeping alive the principles that gave rise to their heroic actions.

Suppose we define wisdom as what we learn first-hand from successes and failures during our lifetimes. In that case, it is cumulative for an individual and all that wisdom is lost when the person dies. Whatever they passed on before they died in storytelling or writing is now second-hand and does not have the same impact on others as first-hand learning.

First-hand knowledge gets firmly stored in our memory banks. Our subconscious uses it to evaluate new information coming in. A memory relevant to the new information is made known to our conscious faculty. We make better decisions today based on our past experiences. The more experiences we have in our memory, the better our ability to evaluate new information as true or false, significant or insignificant, and urgent or low priority. The more experiences we have in our memory, the more likely we will develop a novel connection of ideas which is the source of creativity.

Extreme longevity will mean that our society's older members will have lived hundreds of years longer than the younger members. They will have an enormously more significant personal memory bank of first-hand experiences. For this “wisdom” to benefit the younger generations, the older generations must share their experiences with as much fidelity as possible.

It could be that extreme longevity will increase the reality-based philosophy generally held in our society. For example, those who lived through the horrors of Nazism and Communism would be alive today to prevent those lacking that life experience from politically moving our society in the direction of fascism or socialism.

VIRTUAL SHARED HISTORY

Technology will advance as we live longer lives. At some point, *virtual* reality will be indistinguishable from *real* reality. Also, one's experience in *real* reality can be recorded so that another individual can experience it later in *virtual* reality. This means that one person can share a first-hand experience with another, thereby creating similar first-hand memories in both individuals. You will not be limited to reading or watching movies about historical events; you will be able to experience them first-hand yourself.

However, there will be something missing from the recorded virtual first-hand experience of a historical event. It will be another person's experience that you are re-living. That other person acted on decisions they made according to their value judgments and internalized those decisions and their consequences in their memory. You replay the recording of the experience without the angst of having to decide anything. The experience will not have the same impact on your deep memory. Still, it could be of significant value helping you understand and learn from others' experiences.

Suppose you could change the flow and outcome of a virtual experience. In that case, if you could make decisions, then it would become an authentic, real-time first-hand experience for you. But at this point, we need to distinguish between the virtual reality that is *fantasy* and virtual reality that is *sensory teleportation*.

Suppose you are in what is essentially a video game, a fantasy. In that case, you must remain aware of the danger of internalizing into your subconscious anything you might learn while in there. It is fiction. It might not be connected to reality at all. You must always keep it in the category of *entertainment*.

On the other hand, suppose you are transporting your senses to another location in real reality such that you can experience it as if you were there. Suppose you can make decisions, take actions and experience the results that occur in that other real location. Then you have an authentic first-hand experience. You can put it in the category of *education*. Pilots flying today's sophisticated flight simulators can attest that even though you know you cannot die, a simulator session during which you fight for your life can be gripping and emotional. It can strongly program your subconscious with the lessons learned.

To avoid societal collapse due to the abandonment of founding principles, we will need to apply evolving technology for newer generations to have experiences beyond the range of their lifespan. We must enable them to experience the hard-learned lessons of older generations. We will need to utilize new forms of immersed educational experiences based on factual reality.

Those who author such virtual educational experiences as new forms of media will be responsible for providing experiences that are true to the facts of reality. We must not consider these experiences as art forms. To achieve true fidelity to reality, designers of these virtual experiences may not add, remove, emphasize, or deemphasize any aspect of the experience according to their value judgments. That is what an artist does, not an historian. It could be that the "reputation" factor in future blockchain technology will play an important role. You will only want to download virtual educational experiences created by designers with high reputations for fidelity to reality.

As a final point on having a sense of history in the future, we might only exist as biological humans until later this century. At some point, our consciousness might live in a technological brain and body rather than a biological one. We might then instantly share experiences, knowledge, and memories with other individuals from other times and places. The sum of all human knowledge and memories might be accessible to us as if it were our own anytime we want to recall it. Whether our consciousness is biological or technological, we must maintain the discipline to select and consume only the knowledge and memories that we trust as having fidelity to reality. We must identify fantasy as such and circumscribe it in our minds; else our existence will become meaningless.

SPOUSE

What is "love?" How does one connect this abstract concept to reality? Ayn Rand provided us with an example of how to do it:

> Love is a response to values. It is with a person's sense of life that one falls in love—with that essential sum, that fundamental stand or way of facing existence, which is the essence of a personality. One falls in love with the embodiment of the values that formed a person's character, which are reflected in his widest goals or smallest gestures, which create the style of his soul—the individual style of a unique, unrepeatable, irreplaceable consciousness. It is one's own sense of life that acts as the selector, and responds to what it recognizes as one's own basic values in the person of another. It is not a matter of professed convictions (though these are not irrelevant); it is a matter of much more profound, conscious and subconscious harmony.[172]

Can people stay married for hundreds of years? Will boredom become a factor? Suppose we accept that marriage is based on recognizing one's essential values in one's spouse. In that case, so long as neither changes their values, their marriage can last indefinitely. If novelty is a high value, the couple will find ways to freshen their marriage with new experiences and lifestyle changes to prevent boredom. If comfort in a routine is a high value, then the couple will settle into the lifestyle they enjoy and only change it a little at a time. Either way, so long as they share the same values, each will want to continue the enjoyment that comes from sharing life.

Suppose their marriage was based primarily on practical or physical considerations that existed at the time. In that case, it is only a matter of time until the couple realizes they do not share their fundamental values, causing them to divorce. One would not want to spend an exceptionally long life with someone who does not share their values. Happiness for both individuals would suffer.

Today, being taken care of in our old age is a practical consideration. It causes some couples to remain married even though they do not share the same values. When younger, they were busy raising a family and working in independent careers, so they did not focus entirely on each other. After retirement, they individually pursued outside interests or hobbies, again not focusing entirely on each other. By the time the infirmities of old age prevented participation in outside interests, they realized they didn't have many years left. Remaining with a spouse who would look after them was preferable to the risk of being alone with disabilities.

Extreme longevity will mean that we can have multiple 25-year careers interspersed with periods of recreation and leisure. We might not have to die of old age, but if we do, it will be at an extended age, and the number of years of living with disability at the end will be few. If we need them, personal care and companion robots (AIRAs—Artificial Intelligence Robotic Assistants) will take care of us and engage with us. There will be little reason to remain married later in life to a spouse as a potential caregiver who does not share our highest values.

What if a person's core values or their spouse's core values change? Suppose one's core values when choosing a spouse were grounded in reality—including one's specific needs to achieve happiness. In that case, their core values are unlikely to change over time. Reality does not change. However, suppose a person's core values when choosing a spouse were based on emotions or ideas disconnected from reality (that they got from other people). In that case, there is a high probability that their core values will change. Second-hand ideas from others and the resulting emotions can change at any time. If either spouse did not have a good grasp of reality or their values when they got married, then chances are they will not remain happily together.

Suppose we expect to live exceptionally long lives and choose to have children at *any* age (next topic). In that case, will we want to choose a spouse while we are young or later in life? Raging hormones (our evolutionary genetics at work) will cause us to be interested in having sex when young. This will continue until medical technology allows us to control these hormonal functions in our bodies. However, having sex to satisfy hormonal drives is not the same thing as passionate sex with a chosen lifelong spouse to celebrate the value he/she represents.

In spousal sex, recognizing one's values embodied in the other person causes the integration of one's conscious values with one's emotions and physical pleasure to celebrate life together. In casual sex, one turns off conscious reason, but the subconscious knows that this sex is a debasement of one's values. It is not what it could and should be.

There is a trend for couples to marry at older ages than decades ago. Many live together for a significant period before getting married. We have also seen a relaxing of the requirement to be married to have children. And, we have seen the advent of same-sex marriages. All this points to a trend. People give more conscious thought and consideration to the "if" and "when" they want to commit to a spouse or children. They are not feeling compelled to do either at an early age just because it is the expected norm. This delay helps prevent a too-soon marriage or child and subsequent divorce.

Too often, a boy or girl goes directly from dependence on their parents to a spousal or childbirth commitment without first having spent time living as an independent, self-sufficient adult.

In a marital commitment, the principle applied by each partner is—"I know who I am and what my values are. I know what I need to be happy. I know that I can sustain myself as an individual if I choose to do so. However, I recognize that my life's happiness will be increased by spending it with you because I believe that you and I share the same core values." Until a young person spends time living independently, without parents or anyone else's support, they cannot honestly know their core values. They have not proved to themselves that they can live alone if they choose to do so. They cannot, with honesty, make the above commitment to a potential spouse.

As our lifespans get longer and fertility is extended, we will likely see a continued movement toward longer delays before individuals make spousal or childbirth commitments. Individuals can take more time living as independent adults, working in their careers, supporting themselves, and discovering what they value in life. Then, they can live with a potential spouse however long it takes to determine that their core values are in sync. Once one finds a valued partner, nothing compels them to get married to live together. The primary reason they will get married is to make visible to the world the couple's commitment. Each partner tells the world, "This person is the most important person in the world to me."

CHILDREN

Is it possible that a future Bill and Mary will, at age 225, celebrate their 200th wedding anniversary?

Today, people generally spend at least 25-years of growing up before getting married. If they do not divorce, then they spend about another 25 years raising a family. If they still do not divorce, then they live together for another 25 years or so until one of them dies. Their fifty years of married life was two-thirds of their total life span, and they spent one-third of their life raising children.

A 200th wedding anniversary would mean they have been married for almost nine-tenths of their total life span, and they spent one-tenth of their life span raising children. When married people achieve extreme longevity, it will mean being married for ever-larger portions of their lives (if they do not divorce). The portion of their lives spent raising children (if they only do it once) will become ever smaller.

What is involved in the decision to have children? Choosing to have and raise children is roughly a twenty-five-year undertaking. When you choose to have children, it will take that long to raise them to independent adulthood. You realize that you are deferring other optional values you might have until your child-raising obligations end.

Ayn Rand described the choice this way:

> The capacity to procreate is merely a potential which man is not obligated to actualize. The choice to have children or not is morally optional. Nature endows man with a variety of potentials—and it is his mind that must decide which capacities he chooses to exercise, according to his own hierarchy of rational goals and values.
>
> The mere fact that man has the capacity to kill, does not mean that it is his duty to become a murderer; in the same way, the mere fact that man has the capacity to procreate, does not mean that it is his duty to commit spiritual suicide by making procreation his primary goal and turning himself into a stud-farm animal. [...] To an animal, the rearing of its young is a matter of temporary cycles. To man, it is a lifelong responsibility—a grave responsibility that must not be undertaken causelessly, thoughtlessly or accidentally.
>
> In regard to the moral aspects of birth control, the primary right involved is not the 'right' of an unborn child, nor of the family, nor of society, nor of God. The primary right is one which—in today's public clamor on the subject—few, if any, voices have had the courage to uphold: the right of man and woman to their own life and happiness—the right not to be regarded as the means to any end.[173]

The biological "clock," the female's decreasing fertility and increased risk of fetus defect with age, has been the primary factor in limiting females' age at childbirth. How might things change if we can stop aging, live a long life, and stay fertile as long as we wish? At some point, probably soon, medical science will develop a method of semi-permanently turning our fertility on or off for both males and females. The need for today's temporary birth control methods will no longer be mainstream. A couple will be able to enjoy sex without the worry of an undesired pregnancy. Then, when they are ready, a couple can make the conscious decision to turn their fertility back "on" to begin a family.

One's productive work is the central value around which one organizes their other values. Presently, we only live long enough to have one or two 25-

year careers. Some people choose to emphasize the high value they place on their careers, choosing to dedicate themselves to their work, and therefore choose not to have children. Some people attempt to do both and strike the best compromise they can between them.

Suppose we can live long enough to have multiple 25-year careers, and we can turn on our fertility at any age. In that case, we can dedicate multiple 25-year periods to our choices for productive work. If we choose, we can set aside one 25-year period (or more) to have children. We might gain the freedom to plan our lives to include children either earlier or later in life.

Couples might devote their first twenty-five-year period together focused on each other, their careers and building reliable incomes. Then, financially secure, they might want the experience of having children. They could dedicate their next twenty-five-year period, beginning around age fifty, to having and raising children without the compromises that a simultaneous career would require.

While raising their children, they could have other low-stress interests such as continued education, volunteerism or art. They will not feel trapped or frustrated by their children. They do not fear "old age." They expect to be free to begin a new twenty-five-year career in any subject that interests them when child-raising ends.

AI Personal Assistants (AIPAs), AI Robotic Assistants (AIRAs), household robots, smart homes, autonomous vehicles, and remote learning will significantly reduce the workload, stress and fatigue from child-rearing. This will free up time and energy for giving your child focused attention and still have time for your other interests. Your AIPA, along with your smart home, will continuously monitor the child. AIRAs will take care of diaper changes, cleaning up messes and serving as entertaining companions and babysitters. Your child will not need to leave home with a lunchbox and go to school, as their education will occur in the home via distance learning and augmented reality. When they need to go over to a friend's house for a "sleep over," an autonomous vehicle will take them there. Additionally, we will ourselves retain our youthful vigor into advanced age. We will be physically up to the challenge of raising children at almost any age we choose.

RELIGION

Organized religions are based on the mind-body dichotomy and a mystical soul that will live on in supernatural heaven (or hell) after the body dies. Will the prospect of not having to die be a threat to religious leaders and zealots' traditional influence? Will they strongly assert the "Death gives meaning to life" argument? Will there be an uproar that radical life-extending technology is evil and will bring untold suffering onto humanity? Might religious individuals and groups take up arms in crusades against the technology, those who use it, and the scientists and companies who produce it?

While some religious leaders will support limited life extension under the mandate to "alleviate suffering and illness," they cannot support indefinite life extension. For example, in 2004, while still a Cardinal, Pope Benedict XVI said, "disposing of death is in reality the most radical way of disposing of life." In a 2010 homily, he said, "Humanity would become extraordinarily old, [and] there would be no more room for youth. Capacity for innovation would die, and endless life would be no paradise."[174]

Will religious leaders opposing life-extending technology be able to get their congregations to shun it? Or will the opportunity to extend their lives be more persuasive to individuals than directives by their church leaders, as in today's widespread use of birth control against the Catholic church's rules?

And how will it affect the religious individual, the person who says that they believe in God, the soul and the afterlife? Most people who say they are religious do so because of tradition. Their parents and grandparents "believed" and were churchgoers, which is how the person grew up. The person got this knowledge from other humans and internalized it but never validated it. It is just the way things are. Such a person goes along with his religion's social constructs and practices, being careful not to focus on the contradictions.

The traditionally religious person usually acts differently outside of the church than he does when inside. Christian religions teach that sacrificing oneself to others and austerity are the pathway to heaven. Yet, out in the real world, the person strives to advance his career, get ahead, increase wealth for himself and his family, and buy the latest Tesla® car. Contradictions like this create a degree of subconscious guilt. The person feels that he must participate in his religion's social functions to atone for his non-compliance with its directives.

The person who never examined religion's fundamentals should be able to transition to a future where science persuasively shows us that we do not have to die. These masses of socially religious people will question whether the traditions they got earlier in life from other humans are valid. Populations will increasingly become non-religious. The influence of organized religions in the world will decline.

There are, of course, religious zealots who are convinced that their religion is *the true religion*. Their God is in absolute control of their lives and will at some point call upon them to die and come to heaven. How might they react to the prospect of having to live on Earth for hundreds of years and maybe indefinitely? Their options are:

1. Shun life-extending technology,
2. Accept life-extending technology as some form of long-term religious purgatory that they must endure,
3. Join a crusade against the technology in hopes of being killed as a holy warrior or martyr, or
4. Find a way to question their religious beliefs (not likely).

When leaders of previously peaceful religions see their ranks decimated, might they recruit remaining zealots in their congregation to become militant to "save" their faith? Dying in defense of their faith would make them martyrs. In religions such as Christianity, suicide is a sin that prevents one from gaining heaven—suicide as an option is out. There are religions where suicide supporting their faith is a direct portal to heaven, such as jihadist Muslims. For them, becoming a suicide bomber attacking life-extending technology would serve to punish those evil technologists and immediately gain heaven.

This type of disruption in our society might occur as science continues to discover new knowledge that threatens to displace the mysticism of thousands of years of human history. Mysticism remains strong in our society and will not go away quietly. If we are not careful, radical life-extending technology could be the trigger that plunges our society into religious totalitarianism. On the other hand, the prospect of radical life extension could be just the catalyst human society needs to finally, after millennia of social evolution, commit to reason over mysticism.

SPECIES RESURRECTION AND CREATION

Now let us turn our attention to the application of genetic science to non-human species. The Jurassic Park movies are not entirely far-fetched.

Genetics scientists are getting ever better at teasing DNA out of ancient fossils. With DNA in hand and lower costs of gene sequencing, scientists are gaining a better understanding of various species' family trees by analyzing DNA sequences. They are identifying which species is the ancestor of another and which genes constitute the difference. Suppose you combine that knowledge with advances in cloning technology where they can replace the DNA of cells. In that case, you have the capabilities needed for resurrecting long-extinct species.

Would you like to see a living T-Rex? Scientists are *already* resurrecting extinct species. They are recreating their genomes by recombining selected gene sequences of their various offspring species—essentially to run evolution backward. For example, they accomplished this for an extinct goldfish in 2019.[175]

Scientists are also learning how to create artificial DNA in a process called "Artificial Gene Synthesis." Gene synthesis does not require a template DNA. Scientists can synthesize virtually any DNA sequence in a laboratory. The first complete gene, a yeast tRNA, was demonstrated in 1972. Scientists have synthesized peptide and protein-coding genes. The first synthetic yeast chromosome was synthesized in 2014 as well as entire bacterial chromosomes.[176]

Even more powerful, scientists have succeeded in creating chromosomes that use an alphabet of more than the four standard base pairs of nucleotides. This enormously expands the range of peptides and proteins they can code for in the genes.[177]

The next step will be combining selected natural gene sequences of *unrelated* species patched together with fully synthetic genes to create new species. The final step will be creating fully synthetic *de novo* species where all the genes are made in the laboratory.

All this research on animal species will lead to discoveries and new technologies that scientists can apply to our human genome. These discoveries might enable us to replace genes that cause aging with those we intentionally design to sustain our lives indefinitely.

♦♦♦

4.3

NANOTECHNOLOGY

"In thinking about nanotechnology today, what's most important is understanding where it leads, what nanotechnology will look like after we reach the assembler breakthrough."

—K. ERIC DREXLER

THE NANOTECH AGE

We live in a material world. We define the ages of human history by the materials we used for our technology—Stone Age, Bronze Age, and Iron (Steel) Age. These early ages were followed by the Industrial Age. During the Industrial Age, we manipulated materials to create engines and machines for manufacturing production. Then came the Information Age, during which we manipulated materials to create semiconductors and microchips. Now we are entering the Nanotech Age, where we learn to manipulate matter at the molecular and atomic levels to create desired materials. Most people have not imagined the future that nanotechnology will bring.

For example, we have already created carbon nanotubes with a tensile strength 20 times stronger than steel. We have used scanning tunneling microscopes and atomic force microscopes to move and position individual atoms to create atomically precise patterns. See the figure below.

K. Eric Drexler, author of the books *Engines of Creation* (1986) and *Radical Abundance* (2013), is considered by many to be the father of nanotechnology. It was Drexler who coined the term "nanotechnology." His 1981 paper published in *Proceedings of the National Academy of Sciences*, outlined a path toward a general-purpose, atomically precise fabrication technology. Drexler refers to the coming revolution as the *Atomically Precise Manufacturing (APM) Revolution.*[178]

The *Agricultural Revolution* exploited the productive capabilities of living organisms' molecular nano-systems, leading to increased population densities, cities, and civilization.

The *Industrial Revolution* exploited the productive capabilities of artificial mechanical systems making possible new products, new ways of living and changes to the structure of society.

The *Information Revolution* provided new ways to process information by applying high-frequency nanoscale devices to process and output patterns of bits leading to an exponentially expanded ability to process information and further changes to society.

The *APM Revolution* will also use high-frequency nanoscale devices to deliver patterns, but of atoms rather than bits, leading to great changes in products, productivity, means of production, and human society.

GRASPING WHAT NANO MEANS

Nano means exceedingly small. It means "billionth" or 1/1,000,000,000 or 10^{-9}. One nanometer is the Billionth part of a meter. One nanosecond is the Billionth part of a second. One nanometer is about 80,000 times smaller than a human hair. Nanoscale means the scale that deals with nano sizes. Scientists handle extremely big and small sizes using the Powers of Ten in the Metric System. Find the "nano" prefix in the following table.

Metric prefix	Symbol	Name	Number	Powers of Ten
Exa	E	Quintillion	1,000,000,000,000,000,000	10^{18}
Peta	P	Quadrillion	1,000,000,000,000,000	10^{15}
Tera	T	Trillion	1,000,000,000,000	10^{12}
Giga	G	Billion	1,000,000,000	10^{9}
mega	M	Million	1,000,000	10^{6}
kilo	K	Thousand	1,000	10^{3}
unity	--	One	1	10^{0}
milli	m	Thousandth	1/1,000	10^{-3}
micro	µ	Millionth	1/1,000,000	10^{-6}
nano	n	Billionth	1/1,000,000,000	10^{-9}
pico	p	Trillionth	1/1,000,000,000,000	10^{-12}
femto	f	quadrillionth	1/1,000,000,000,000,000	10^{-15}
atto	a	quintillionth	1/1,000,000,000,000,000,000	10^{-18}

Metric Numbering System

Nanoscience is studying matter and objects such as molecules and structures with at least one dimension in the range of one nanometer to about 100 nanometers.

Nanotechnology refers to manipulating materials on an atomic or molecular scale measured in billionths of a meter, nanometers. Nanotechnology uses nanoscience knowledge and helps nanoscience advance, for example, by providing improved nanotools.

Wet Nanotechnology is about all the nano-machinery of cells in living things.

Dry Nanotechnology is about non-living materials.

Nanoelectromechanical Systems (NEMS), a Dry Nanotechnology, integrates mechanical structures with moving parts with nanoelectronics. Such a nanoscale system typically has a length of fewer than 100 nanometers.

Nanofabrication means the creation of nanostructures with at least one dimension within the nanoscale.

Atomically Precise Manufacturing (APM) means manufacturing materials by precisely assembling atoms to form the molecules and progressively larger components of whatever it is that we want to create.

WHY SOMETHING SO SMALL IS SUCH A BIG DEAL

Nanoscience is the convergence of physics, chemistry, biology, computer science, engineering, material science and more at the nanoscale. It is important because we need to know how substances at atomic and molecular levels work to design and create macro-level substances. Nanoscale substances can have unique characteristics caused by quantum physics rather than classical physics. We are more limited by our technology than by underlying physical laws.

The Information Revolution created a radical abundance of information products at exceptionally low or no cost. The Atomically Precise Manufacturing (APM) Revolution will lead to a radical abundance of physical products with far better performance and much lower costs. The APM Revolution could bring a transformative change in civilization as profound as the Industrial Revolution but move at digital information speeds.

APM might prevent our civilization from exceeding the limits of Earth itself. Drastic reductions in the use of rare resources, waste products, and the mass and energy used to create things, coupled with technologies to reverse the Industrial Revolution's climate damage, could enable us to have a completely sustainable existence on Earth.

APM will lead to technologies we biological humans will need to live indefinitely. It will then enable us to become Hybrid Humans with far greater (augmented) intelligence than we have today. APM will enable technologies required to extend our existence into space and the cosmos.

Regarding the speed of APM advancement, consider that the Agricultural Revolution began in prehistory. Then the Industrial Revolution began only a couple of centuries ago—a handful of generations. The recent Information Revolution occurred in the span of a single human lifetime. The APM Revolution is underway. How long might it take to mature, reaching the bend in its exponential growth curve? Drexler advises that "today, it would be unwise to rely on APM for quick solutions to anything, but it would equally be unwise to rely on prolonged delays."[179]

TWO DIFFERENT NANOTECHNOLOGY INDUSTRIES

Drexler explains that since nanoscience concepts were first conceived, there has been a competition of sorts between two nanotechnology approaches, *Atomically Precise Manufacturing* and *Materials Nanotechnology.*

Atomically Precise Manufacturing (APM)

The initial conception of nanotechnology was Atomically Precise Manufacturing. APM is the idea that we would create machines at progressively smaller scales until we can precisely assemble the molecules we want atom-by-atom, then assemble those molecules into nanoscale parts, that then assemble into microscale parts, that further assemble into successively larger parts until we have created what we want—such as an automobile with incredibly enhanced performance and reduced cost. This process is driven by precise digital information technology that manipulates atoms rather than bits.

While the Nanotechnology Revolution was first conceived as creating products by assembly from atoms to finished products, its progress toward functional APM has not been consistent. APM was sidelined for a time by a different view of nanotechnology.

Materials Nanotechnology

A different conception that arose was Materials Nanotechnology. This is the idea that various existing and new laboratory chemical and physical processes could create substances at nanoscales (<100nm) that would have unique properties. These substances could then be developed into products improved by those properties. An example of such a material is the carbon nanotube which is twenty times stronger than steel. However, the best-known example of this materials nanotechnology approach is semiconductor chip manufacturing. To make semiconductor chips, manufacturers use photolithography to "print" patterns of billions of pixels comprising transistors, insulators and wires in layers on silicon wafers exposed to etching chemicals. This process cannot be perfectly precise because light and chemical exposure cannot be precisely controlled.

Comparisons

We can describe the APM approach to nanotechnology as "digital" and the Materials Nanotechnology approach as "analog." These two approaches have more differences than similarities. The most important difference is the lack of precision. Most nanomaterials have imperfections. Much of the engineering that goes into them is to minimize or overcome these impurities' negative effects. Also, we can consider these nanomaterials' physical shapes as "rough" or "bumpy" when seen under nanoscale magnification.

On the other hand, APM can produce atomically pure materials with perfect performance of a material not degraded by impurities. APM can also produce perfectly precise "smooth" and complex shapes designed to connect to other items with perfect shapes to form larger components.

The scientists and engineers working in these two fields have different objectives and use different tools and techniques. Those few who have APM as their objectives are considered the nanotechnology *purists.* Some consider the more numerous people working on nanomaterials to be *practical* nanotechnologists. They are considered practical partly because of the realization that APM seems far off in the future while nanomaterials production is already here. The difference in career choices was also affected by the difference in funding of the two approaches to nanotechnology.

Soon after the APM concept of nanotechnology by Drexler was widely published, in which he outlined "an easily avoidable risk that would involve small, dangerous, but useless self-replicating machines,"[180] an absurd fear of nanobots run amok developed in some circles, especially the science fiction writers. They could envision accidental or intentional release of swarms of tiny nanobots self-replicating by consuming the matter around them until turning the Earth and everything on it into a "gray goo." This unrealistic scenario is indeed fiction. An APM factory-in-a-box is not a microscopic robot.

Eventually, this widespread unreasonable fear influenced the public, politicians, and others with control over nanoscience and nanoengineering funding. To avoid addressing the unrealistic fear of nanobots, these people redefined "nanotechnology" to mean Materials nanotechnology rather than APM nanotechnology. Materials nanotechnology is what they funded. They redefined nanotechnology solely in terms of *scale*, eliminating all mention of atomic *precision.* This led to decades of advancement in Materials nanotechnology that did not have APM as its objective.

The result has been many interesting materials with unique properties and the analog processes for creating them. We should note that once digital APM production of *perfect* materials becomes a reality, it will replace the analog materials technologies that produce *imperfect* materials. While some of the attention and funding invested in materials nanotechnologies will help APM development, the lack of attention and funding specifically for APM means that critical technologies APM will require have not yet been developed.

While sidelined, pockets of development toward APM continued, and progress has been made. Enough progress towards APM has occurred to indicate a pathway toward achieving APM without violating any physical laws. The demonstrated ability to precisely position individual atoms to make desired structures, as shown in the earlier photo, has sparked scientists' and engineers' imaginations. More scientists and engineers are becoming interested in and are obtaining funding for APM as their objective.

The pathway forward to APM is now clear. There is reason to believe that an iterative approach developing increasingly smaller nanomechanical technologies and nanomachines will someday achieve full APM production capabilities.

MATERIALS NANOTECHNOLOGY

Natural laws act on visible macro-scale (full-size) materials through invisible physical processes functioning at the nanoscale (microscopic). The history of science began by studying macro-scale things using macro-scale tools and progressed to studying nanoscale parts using nanoscale tools. Nanotechnologists learn by studying natural structures and processes and then by making nanomaterials.

Nanotechnology has placed materials science in the present-day spotlight. We explore the relationships of the properties of a material to its structure. The chemical elements that compose the material and how we process it into its final form, combined with thermodynamics laws, determine its microstructure and the material's properties. We then use this knowledge to manipulate existing materials and create new ones.

When we consider macro-scale things, such as water, rocks, plants, animals and humans, we see great diversity—things are vastly different. However, at the nanoscale, variety is replaced by sameness. All things are made of a limited set of nano components—atoms. All atoms have the same basic structure and obey the same physical laws. At the nanoscale, things appear similar.

Here are a few examples of breakthrough applications of nanomaterials technology listed by Drexler:

- Batteries and super-capacitors with performance improved by 10–100 times.
- Direct photoconversion of light plus water to produce hydrogen (H_2).
- Hydrogen storage in stronger, lightweight storage tanks or absorption systems.
- Nanoelectronics that revolutionizes computers, sensors, and devices.
- Nanomaterials and coatings that reduce the costs in nearly every industry.
- Photocatalytic reduction of CO_2 in the atmosphere to produce liquid fuels.
- Photovoltaics cost reduction of 10 to 100 times.
- Superconductors to rewire the global electrical infrastructure replacing aluminum and copper.

All these benefits come from investments in *nanomaterials* research and materials that are imprecise and bumpy. Next, we explore the type of nanotechnology that is atomically precise—APM.

ATOMICALLY PRECISE MANUFACTURING

The pathway toward functional APM now seems clear without violating physical laws. It is now only a matter of time and funding. Today, fear that other nation-states will gain APM advantages with potentially world-altering consequences puts pressure on public and private sectors to invest in its development. The race is on. It is now only a matter of applying more nanoscientists, nanoengineers, funding and coordination to the APM objective.

It is still too early to forecast dates for the achievement of full APM production capabilities. We have come a long way, but there is still a long way to go. Like the challenge of decoding the human genome, it now looks like it could take decades. However, like the Human Genome Project, new yet undiscovered, shortcut technologies might appear that will dramatically shorten the time required.

Let us try to envision what APM might be like when it eventually becomes available at the industrial and consumer levels. Assume that we want to create an automobile. The entire APM factory will be only a few times larger than the car, the size of an oversized garage. Inside the garage are robot arms that will assemble the major components into the final product. The garage's back wall is lined floor to ceiling with cubicles from which the robot arms will retrieve the final parts.

Each cubicle on the wall of the garage is itself a complete, smaller APM factory. It has even smaller cubicles along its back wall. Smaller robot arms inside this cubical retrieve smaller parts and assemble them into the final parts that the final assembly robots will retrieve. Each of the cubicles inside this smaller APM factory is itself a still smaller APM factory that looks the same, only smaller.

This chain of ever-smaller APM factories keeps getting smaller until we reach the nanoscale. Nanomachines inside the smallest nanoscale APM cubicles assemble atoms and molecules obtained from raw stock materials into the first-level nanoscale parts.

The raw stock materials these nano assemblers use consist of abundant common elements rather than rare elements used in today's manufacturing. Precisely controlled atomic manufacturing can produce rare materials' desired properties by using common materials.

Whatever waste materials the nano assemblers produce are also processed by them into other useful, or at least environmentally inert, substances.

The nano assemblers operate at extremely high frequency. Therefore, this APM factory will produce the final product in a fraction of the time and consume a fraction of the mass of the materials required in today's manufacturing.

APM will collapse worldwide industrial supply chains to just a few stages. In the first stage, we refine raw materials into feedstock materials. Then we convert feedstocks into standardized micro blocks. Finally, we use micro blocks to assemble products of highly flexible variety. These shortened supply chains and highly flexible production will enable radical decentralization.

APM APPLICATIONS

Stronger and Lighter Materials

APM promises to produce products with radically better performance at a lower cost than what we have today. Some of the reasons stated by Drexler include:[181]

1. Benefits of scaling laws applied to small devices.
2. Strong atomic bonds by the lighter elements: carbon, hydrogen, nitrogen, oxygen.
3. Stronger, lighter materials that reduce the structural mass in a product.
4. Fine-grained structural designs add strength, such as making a fibrous material or reduce mass by making porous or hollow structures.
5. Reducing the cost of a material by reducing its mass.

Here are some examples Drexler provides of materials that we can make with just carbon, hydrogen, and oxygen:

1. A lightweight material, stronger than aircraft-grade aluminum alloys, yet only one-one-hundredth as dense.
2. A lightweight material as stiff as aircraft aluminum, more than one hundred times stronger, yet only one-tenth as dense.
3. A heavier material, stronger than steel, yet only one-fifth as dense.

Drexler points out that structural materials comprise the bulk of nearly everything we make and use. Radically improved lightness and strength in materials with decreased mass will revolutionize our manufacturing. They will not only lower costs but can increase safety and energy efficiency. They are resistant to corrosion, fatigue, creep, and fracture. They can be made as transparent as glass. Imagine the impact on aerospace industries when we remove more than 80 percent of the required structural mass.

Novel Materials

Digital APM can exert a precise control of patterns of atoms that today's less precise analog materials manufacturing cannot. Analog materials manufacturing relies upon the bonds that atoms naturally make when they are brought near each other. On the other hand, APM can cause atoms to bond into patterns that they would not form naturally. Guiding the creation and bonding of molecules expands the range of potential new materials that we can create. Drexler gives examples of improved solar cells for solar energy and carbon nanotubes that outperform scarce copper as an electrical conductor.

Increased Power Density

Size combined with atomic precision can boost power densities by factors of a million or more. Today's motors rely on electromagnets that rely on magnetic fields. Scaling laws prevent the creation of nanoscale electromagnets and therefore prevent nanoscale electromagnetic motors. However, scaling laws support the design of nanoscale motors that use *electric* fields rather than *magnetic* fields. Arrays of these nanoscale motors can be combined for increased power output and higher efficiency, greatly exceeding today's electric motors' power density. They can also be operated as high power density electric generators when run in reverse.

We can achieve a similar boost in efficiency and power density in fuel cell systems that convert chemical energy to electrical energy. Drexler describes a tiny device that combines chemical energy conversion with nanoscale motors and functions like a rechargeable battery with the energy storage of a tank of gasoline.

Nanomechanical Computers

The mechanical scaling laws would allow the creation of nanomechanical computers with performance exceeding today's electronic devices. The nanomechanical computers could achieve clock rates in the gigahertz range. At the same time, power consumption per core can drop to a fraction of a microwatt.

Able to create devices in three dimensions, APM can build computing devices with a thousand times more density and efficiencies near the thermodynamic limits of computation. This could result in micron-scale computer processor cores, memory storage densities in the range of a billion gigabytes per cubic centimeter, and billion-core computer systems suitable for air-cooled laptop computers.

Solar Energy Production

The atomic precision of APM can produce the thin films necessary for the creation of high-efficiency, low-cost photovoltaic cells. We can make atomically pure films from abundant materials such as iron and sulfur of pyrite rather than scarce materials such as gallium and indium. We can produce photovoltaic cells with energy conversion efficiencies equal to or greater than today's solar cells in APM factories at a fraction of the cost of those we have today.

We can then couple these inexpensive APM-produced solar cells with the previously described inexpensive, high-density energy storage devices. This combination would enable the storage of energy produced during the daytime for use at night.

We will see substantial improvement in solar conversion efficiency, dramatic cost reduction, and efficient, low-cost ways of converting electrical energy into fuels and back again. Together, these can change the energy economy, lowering the costs of electric power and fuel while moving toward zero net carbon emissions.

Biotechnology

APM technology will provide instruments that can analyze nanoscale biological structures and functions and then apply nanoscale control. This technology will be necessary to modify the materials in our bodies directly, bypassing our genetics. Nano-scale devices that interface with neurons will enable advanced brain-machine interfaces.

APM COSTS

The concept of *cost* drives entire economies. Drexler writes, "Ultimately, cost makes the difference between the luxuries of wealth, common industrial products, and goods barely affordable in poor rural societies. Cost makes the difference between abundant solar electric power and an economy with a long-term lock-in to coal-based power. Examining APM prospects through the lens of cost is of more than narrow economic interest."[182] Drexler makes the following points about costs:

APM Raw Materials Costs

Many scarce (expensive) materials used today will lose much of their demand because APM can achieve the same functionality with better performance using abundant, low-cost materials.

Zinc, tin, and lead will no longer be necessary for anti-corrosion coatings for steel or solder joints, die-casts, or tin plating on cans. Manganese, chromium, nickel, and cobalt are also primarily used to make steel. All will become obsolete because we will replace steel in products that we can make seamlessly and corrosion-free by APM with higher performance and lower cost.

We can provide the most useful elements for APM in their raw, inexpensive, natural form. APM can extract the needed molecules and atoms from the natural forms of chemical compounds, such as silicon dioxide, rather than needing highly refined and pure elements, such as silicon.

APM will require fewer kilograms of raw materials to make a product than today's manufacturing. APM will be able to assemble structures previously made of steel, aluminum, and plastics with stronger materials that typically reduce mass by a factor of at least ten.

Combining the lower costs of common and inexpensive raw materials with the mass reductions of APM will result in an extremely low cost per "effective kilogram" of materials. Drexler estimates a materials cost of approximately ten cents per effective kilogram.

APM Energy Costs

Drexler points out that the energy costs of extracting pure needed elements from natural compounds might be about the same for APM as today's methods. There could be an energy penalty for the highly precise assembly steps of APM. However, there will be a reduction in the mass required. Drexler estimates that the energy costs will be roughly the same as materials costs, about ten cents per kilogram of materials.

APM Land Costs

An APM production facility only needs to be a few times the length and width of the product we wish to produce. A garage-scale APM factory can build a car. Land use for APM is negligible compared to today's industrial manufacturing.

APM (Direct) Labor Costs

An APM factory begins with raw materials provided to it. It produces a finished product without the intervention of human hands. All labor involved with APM is external to the actual production process, for which there are no labor costs.

APM Wastes and Pollution Costs

As an assembler at the atomic and molecular scale, APM precisely controls the structures of both its products and byproducts. It requires no combustion and therefore does not emit CO2. APM will process any toxic byproducts into inert or naturally occurring stable minerals before releasing them into the environment. Waste heat will be in proportion to the mass, but air cooling can remove it with little adverse impact. Drexler says, "Advanced production systems need not produce noxious emissions, greenhouse gasses, or toxic wastes, and could meet zero-tolerance emission regulations at little cost."

APM Risk of Accidents

An APM factory has no people inside to be at-risk of injury. It contains minimal quantities of combustible, explosive or toxic materials providing little risk to those near it. There is no need to bring dangerous raw materials to the APM factory because it can get the atoms and molecules it needs from safe forms. For example, Drexler points out that it can get the hydrocarbons it needs from an organic liquid such as mineral oil. Safer than today's manufacturing systems, APM can easily meet stringent safety regulations.

APM Physical Capital Costs

APM systems offer high throughput and a wide product range. APM is an end-to-end production system from raw materials to finished products. Drexler estimates that they can produce a mass of products equal to the APM system's total mass in hours, not days or years. An APM system's full range of machinery replaces complete supply chains. If the APM system's cost were amortized over one thousand days, then the product's capital cost per kilogram would be less than one-one-thousandth of the cost per kilogram of the APM system itself. The cost of APM production equipment adds almost negligible cost to the finished product, even for a system used at one percent capacity.

APM Total Costs

The two major APM costs from above were raw materials and energy. The other, smaller costs comprise a large rounding error. The resulting physical costs for finished products are around twenty cents per effective, structural mass-adjusted kilogram.

APM NANOTECHNOLOGY IN YOUR HOME

We visualized a garage sized APM factory that can build a car. Imagine that you have a smaller version, perhaps the size of a washing machine, in your home.

Also, think of your present-day inkjet printer with reservoirs of ink in different colors and paper. When you download a document or image data file from the Internet, it contains the digital information necessary to print the document or image. It precisely places individual pixel micro-droplets of different color inks on the paper.

The APM factory in your home will work like your inkjet printer. It will have reservoirs of different bulk physical materials that you keep replenished. You will download the design file for an item you want, let us say a cell phone. First, the APM factory confirms that it has the bulk materials it will need for the project. Then, it assembles the necessary molecules atom-by-atom and then part-by-part as per the digital design file. It assembles continuously larger parts until it completes the final product. In a few minutes, you open the door to your APM assembler and remove your new cell phone. The cost of your new cell phone consists of:

1. Amortization of the initial cost of your home APM factory,
2. The cost of the supplies of the bulk materials,
3. The cost of the energy used,
4. The cost of the design file royalty paid to its creator.

Just as with document and image files today, the costs of design files for producing physical objects in your home will be minuscule or even free in the public domain. Gone are the costs of a large manufacturing plant, packaging costs, transportation costs, warehousing and inventory costs, and wholesale and retail middlemen costs.

APM IN REVERSE

Let us explore another common scenario in your home—taking out the trash.

Imagine that you have a second washing machine size APM factory in your home sitting next to the first one that made the cell phone. This second one is designed to take things apart rather than assemble them. It will recover their parts, in some cases, down to their atoms and molecules.

This disassembler can use the digital design file used to assemble an item in the first place to know the best way to break it into valuable parts. It might not disassemble all the way to atoms and molecules because there will be intermediate size parts that are standard and widely used for different products. It can stockpile these for reuse.

Your APM *assembler* and your APM *disassembler* share the same bulk materials bins. The assembler uses material from the bins to make things. The disassembler takes things apart and puts the parts back into the bins. The assembler might have waste materials from the assembly process that it sends to the disassembler to process back into bulk materials or safe waste products.

We will not have all the product retail packaging waste in our home that we have today. Items we want will come out of our APM assembler without packaging (unless we want a package). Anything we make and use can be either single-use or, in the case of our cell phone, occasionally replaced with an upgraded version. We will put things we no longer use into our APM disassembler for recycling.

Both our APM assembler and APM disassembler can produce small amounts of waste materials that they cannot reuse. They will convert these to materials that could have other uses outside the home or render them environmentally inert for external disposal, dramatically reducing the need for landfills.

This APM assembler and APM disassembler model can scale to sizes for use outside the home in industry. They can be made large enough to handle bulldozers or airplanes.

APM AND THE ENVIRONMENT

Our civilization's reliance on materials can be radically transformed by APM, offering the potential for radical improvement in terms of our civilization's effects on the environment. There is some risk that lowering the costs of production could enable increased destruction. However, our awareness of our environmental impacts is more likely to cause us to apply the cleaner, low-impact technologies to positive improvements in our relationship to our environment.

We have already covered how APM-based production can reduce environmental impact by reducing the mass of materials required and avoiding polluting wastes. We have also addressed an energy production model based on solar that will minimize environmental impact.

APM also offers the promise of the restoration of the environment to its pre-industrial composition. It can reduce the cost and increase equipment's effectiveness to remove toxic chemicals and heavy metals from soils. It can provide new technologies for capturing and sequestering toxic materials from groundwater, plastics from polluted areas of the oceans, and other challenges.

Regarding carbon emissions, APM can lead to a dramatic reduction of atmospheric CO_2 emissions to near zero. It could also provide a thermodynamically efficient means of capturing and compressing excess CO_2 from the Earth's air, with the required energy supplied by a 200x200km equivalent solar array in the Sahara Desert or fusion energy production (covered later). Drexler tells us that these APM solutions could "return the Earth's atmosphere to its pre-industrial composition in a decade, and at an affordable cost."[183]

APM provides hope for a sustainable future for our materials-based society on Earth. Achieving this future will require that individuals include it in their hierarchy of values and then translate those values into our civilization's politics and commerce that encourage APM development.

APM AND SECURITY

Applying new technology to war weapons can have both good and bad consequences for populations. We can combine offensive, defensive and surveillance applications under the term "security" applications.

When considering future technologies, we worry that they will be used to threaten us with deadly force, thereby canceling our liberty rather than protecting it. People can use security applications of APM in war, crime, law enforcement, commercial and neighborhood security, dictatorial aggression and oppression, and more.

We can envision a world where there is no square meter of space where we are not under surveillance. APM will be able to produce extremely tiny information processing and storage devices. Combine that computing power with APM's ability to make tiny sensors more powerful than human senses, with extremely low energy requirements and cost. Consider that these tiny surveillance devices could be widely, quickly, and affordably distributed to various locations by swarms of drones or kept aloft.

History has demonstrated a progression away from lethality as we developed better precision in applying deadly force. The purpose of war is coercion, not killing. Killing used to be the most efficient method of achieving coercion. Whole cities no longer need to be leveled when technology makes it possible to drop a precision-guided weapon from a drone onto a single car or into a house's specific window. A few non-combatants might get caught up in the localized, precision conflict, but nothing like the nuclear bomb scale.

We can apply advancing technology to non-lethal weaponry development and deliver and apply it to achieve highly efficient coercion. We could equip the tiny, widely distributed surveillance devices described above with technology to apply non-lethal coercive force to individuals with precision and without physical harm. Increased capabilities and widespread application of non-lethal weaponry are of great concern. It lowers the moral resistance against its use, encouraging those who wish to exert power over us.[184]

The only moral function of a government, granted to it by its citizens, is the application of force in retaliation against the initiation of force by a person, group or nation-state. We need the military to protect us from foreign force and the police and courts to protect us from criminals. Intelligently applied APM surveillance and non-lethal technology could aid the military and police in protecting us, including recording evidence for use in courts, a proper function of our lawful, accountable government.

However, when the government initiates force against its citizens to enforce immoral policies that have nothing to do with our protection from force, we are in danger. As technology improves the ability to apply non-lethal coercive force, the threat from our government increases. We are already aware of the conflicting interests between our privacy and security. The government claims it needs more surveillance to keep us safe from terrorists and other threats. To protect our freedoms from misapplied security technology, we must wake up to our government's present overreach and rein it in while we still can. A clear hierarchy of reality-based human values is needed to negotiate the proper balance between our freedoms and national security.

The Supply Chain

APM holds the promise of improving our national security in a significant way—reducing our dependence on foreign supply chains. The coronavirus pandemic of 2020 highlighted our nation's dependency on China for critical medical supplies. These included masks, gloves and gowns to protect health care workers and face masks and sanitizers for use by our at-risk population. Even our military was dependent on the supply chain from China for these supplies. Of greater concern, we get only from China over a hundred drugs that are critical to sustaining our citizens' lives.

The 2020 coronavirus pandemic demonstrated that our "just-in-time" supply chain from China to the US to too fragile to be our only source of critical supplies. We have an insufficient local inventory of these strategic items and a minimal ability to source them domestically. Over the last few decades, we have dismantled domestic specialized manufacturing equipment due to a lack of economic viability. Our present dependency on China was not up to the challenge of something like a pandemic. More importantly, our only source of these critical supplies is essentially a geopolitical adversary.

Once realized, APM will end all present supply chain concerns. Imagine millions of APM assembler devices in every business and home. Plentiful supplies of ordinary, cheap raw materials are stockpiled everywhere. Suppose our country needs a hundred million face masks. In that case, we can send the design file to ten million APM assemblers in the country that can each make ten masks. If a person needs a drug, their APM assembler makes it for them. No *supply chain* for critical items is required. National security will be dramatically improved.

APM AND PARADIGM SHIFTS

Atomically Precise Manufacturing research is happening today in many countries in many groups in the fields of biology, medicine, chemistry, and innovative materials. They are producing new tools and techniques along a path that will eventually result in APM. Physics presents no obstacle to achieving APM, so all that is needed now is for scientists and engineers in these respective fields to make APM itself an objective.

Possible pathways to APM success are clear, but it is too early to discuss specific milestones and timelines. We will see a recursive upward spiral of capabilities until we achieve APM. Like any exponential growth curve, progress toward APM will seem slow at first. It will begin accelerating after a protracted period and then become amazingly fast.

As we achieve APM, we can first expect to see improved upgrades to existing products. Structural parts of products made with APM materials that are stronger, lighter, and better performing will enable significant upgrades, eventually replacing current products with better-performing ones.

APM production will speed up the cycle times of product improvement and replacement. Prototyping, engineering, and production will all be sped up and decentralized, thereby eliminating most shipping delays.

Consider the paradigm shifts that will occur in our industrial and commercial markets. APM will make possible significant cuts in costs, increases in performance, rapid response to demand, and elimination of traditional factories and supply chains for all kinds of products.

Many whole industries and their supply chains will be rendered obsolete. Ore deposits and oil fields will be devalued. Job losses and steep declines in the prices of goods will occur because of radical abundance. Simply anticipating emerging APM's effects will depress industrial asset values.[185]

On the other hand, as global dependency on and competition for scarce localized resources diminish and access to decentralized markets increase, the causes of international conflicts will be reduced. There will be very few resources that will be of vital national interest.

Early APM's disruptive economic consequences can be viewed as economic destruction and become the target for government interventions. The imagined rights of collective groups in obsolescent industries versus the rights of individual businesses and businesspersons will be called into question. Like our present-day concern about nuclear proliferation, disruptive technologies could lead to an arms race threatening our liberty and security. Rogue organizations or hostile nation-states, if they were to gain significant APM capability, could cause unacceptable risks. In our national interest, we will need the capability of preventing unconstrained applications of APM. Rather than developing APM in the dark, openness in sharing between cooperating international free societies will minimize an APM arms race's potential. Once again, the future blockchain's "trust factor" functions could play an essential role in responsibly managing access to APM technologies.

Drexler described the societal challenge this way:

> Outcomes will depend on technological change, but also, perhaps even more so, on expectations, perceived interests, and policies as they take shape in the years to come. Gradually, the prospective APM transition will rise from the level of an idea worth considering, to a prospect that

> demands contingency planning, to an emerging reality that calls for action at the highest levels of human attention.[186]

To navigate the economic upheaval and potential threats that APM will bring, we must assert reality-based values at the foundations of our society's institutions. These are the values that we want our technology creators to hold. We must cooperate with other countries that share our values and security concerns and isolate those that do not.

NANOTECH IN BIOLOGY

Biology consists of atomically precise molecules and structures from DNA and RNA to proteins and the parts of cells and their metabolic products. Medicine is increasing its application of atomically precise tools to research and engineer solutions for medical problems.

Biology is the most intensely investigated subject of modern science. In her recent book, *Nanotechnology Comes to Life* (2019), Sonia Contera writes, "medicine still takes center stage as the main intellectual, social, and economic engine of biological research. Medicine helps to attract the money, but more fundamentally, plays the role of integrator of knowledge.[187]

All the knowledge discovered by all the sciences in the last four centuries and accumulated in our society can finally investigate, understand, modify and apply our biological complexity.

BIOLOGY BECOMES A PHYSICAL SCIENCE

Contera confirms that we have reached the inflection point of the exponential biological science growth curve:

> The diversity, intensity, and speed of advance of current research unequivocally indicate that we are living in prerevolutionary times in both biology and medicine. [...] the accelerating and ever-more-potent interdisciplinary mergers make us feel that we are now at an inflection point, and will soon slide irrevocably toward the advent of the

> technologies that will transform our understanding and control of our biology.[188]

We have reached this inflection point because "Biology ceases to be the sovereign territory of biologists, biochemists, and medical scientists; in the twenty-first century, physical, mathematical, and engineering sciences converge with the more traditional biological disciplines to seek a deeper understanding of life in all its multifaceted, dynamic structures and functions."[189]

The technology that is enabling the inflection point is nanotechnology. Biological science began with life-size organs. Knowledge had to wait for technology, such as the microscope, to enable investigation at successively smaller scales of biology. To achieve medicine's ultimate goal, restoring perfect health, it is necessary to know how molecules work in specific environments, how their malfunction leads to disease, and how to reach and manipulate them. This means dealing with the entire spectrum of scale from the whole body through organs, tissues, cells and biomolecules.

The treatment of biology as physical science and nanotechnology have given rise to new types of scientists who work across disciplines, such as biological physicists and nanomedicine scientists. The critical difference in the method of research physical scientists bring to biology is *quantitative analysis.* Physical scientists pursue quantitative data to develop a causal hypothesis that they can then model with mathematics in computers. "When you look at biological systems through the eyes of a physicist, you are looking for the key parameters that explain how the biological system works: its size, temperature, energy, speed, structure, stiffness, charge, chemical activity?"[190]

This quantitative approach to biology leads to modeling biological systems in computers, in silico, without experiments. These mathematical models will eventually replace the traditional trial-and-error methods of medical and pharmacological research. We will ultimately create mathematical computer models of the functioning of individual molecules such as proteins or DNA in both their natural and disease environments through nanotechnology. Advances in massive computing power, especially quantum computing, and AI will aid in processing "big data" and modeling at the nano scale.

EMBRACING BIOLOGICAL COMPLEXITY

The second key factor identified by Contera in our reaching the inflection point in biotechnology is the move away from the traditional *reductionist* approach of science toward a focus on complexity and emergence. Life cannot be explained simply by studying genes and molecules. Biological *complexity* created by evolution is what we must learn to understand. Contera writes, "In the twenty-first century, science has no other intellectually viable route than to embrace life as `symphonic interplay between genes, cells, organs, body, and environment.'" This means that the previous generation's reductionist approach that treated organisms simply as biochemical computers executing a program encoded in genes is inadequate.

> However, twentieth-century physics teaches us that in very complex and interconnected systems, knowing the workings of the building blocks is not enough to predict the behavior of the whole [...] This is because complexly organized matter presents collective phenomena arising from cooperative interactions between the building blocks—or, as we say in physics, these properties emerge. Some examples of emergent behavior are cellular movements, mechanical vibrations in the brain, electrical signaling across the membranes of cells, and changes in shape or stiffness, none of which can be predicted from just knowing the molecules that constitute a particular structure.[191]

Nano-scale modeling will not just model individual molecules but also clump them in various micro-structures. It will model the emergent properties resulting from that micro-structure, environment, and nanometer-scale activity. This mathematical modeling will scale from the molecular DNA level through a tumor's size, shape, and growth pattern. Using such a model, scientists could design treatments that simultaneously target the tumor's molecular, cellular and tissue-level biology.

Biotechnologists are striving to apply nanotechnology, physics and mathematics to understand and predict the biology of disease and trauma. They intend to implement healing strategies, one patient at a time. Ultimately, their goal is to design mathematics and computer models that enable treatments of specific problems in particular patients. This goal contrasts with the traditional approach of continuous trial and error and prescriptions that work for a statistical majority of patients.

DNA AND PROTEIN NANOTECHNOLOGY

Today, scientists are applying nanotechnology to the design and construction of structures of almost any shape by using artificial DNA building blocks or by bending DNA structures at desired locations. The goal for these man-made DNA structures is to have nanoscale functionality for complex tasks such as a template for the synthesis of molecules, a molecular DNA computer, or a DNA robot delivering drugs to a target tissue.

Scientists are similarly applying nanotechnology to natural or artificial "designer" proteins that do not exist in nature. This is more difficult than working with DNA but offers a more significant number of potential applications. "These nanotechnology pioneers have achieved an incredible first goal: the development of molecular assemblers that are able to construct desired molecular shapes with atomic precision following a planned design rehearsed in a computer."[192] Quantum computing might enable a dramatic improvement in the discovery and creation of designer proteins.

NANOTECHNOLOGY AND MEDICINE

A fundamental drive of biomolecular nanotechnology is to create powerful tools for nanomedical applications. These tools range from molecular DNA assemblers of medicinal drugs to improved vaccines, powerful antiviral and antibacterial nanomedicines, and targeted drug delivery systems.

One of nanoscientists' early goals in medicine was to improve the ability to target drug delivery to cancer tumors. However, those efforts did not lead to the desired degree of improvement. This is partly because of inertia. Contera explains the causes of the disappointment, "[…] the application of existing trial-and-error methods to complex biological processes that are still poorly understood; the insufficient coordination of existing research; and the lack of truly quantitative approaches." The disappointing results are also due to impatience, "[…] seeking nanopowered magic bullets and lucky shortcuts to cure disease that overlook the complexity of the biology involved has not proven fruitful."

However, scientists are learning the lessons, and new approaches are gaining favor. They now seek to combat such things as antimicrobial resistance by adding an understanding of a bacteria's physics to the knowledge of its pharmacological chemistry.

One of the nanotechnology applications to medicine receiving attention in the press is immunotherapy, a type of cancer treatment that boosts the body's immune system to fight cancer. APM technology can lead to augmenting our body's immune system with artificially produced nanoscale sensors that can detect individual invader or cancer cells' surface protein patterns. The sensors can also produce antibodies and other substances that will trigger our immune system into action far earlier and with greater precision. As the APM technology advances, the APM-produced sensors might be able to deliver chemotherapeutic agents directly. We could have APM devices that target cancer cells with high precision and without undesirable side effects. We could have bodies containing nearly zero cancer cells.

NANOTECHNOLOGY AND TISSUES AND ORGANS

Tissue engineering researchers seek to enable the repair and even replacement of damaged or diseased organs. To do so will require progress in the fundamental science underlying biology and medicine.

The mathematical computer modeling and simulation of artificial tissues at the nano, micro and macro scales will continue to advance. The goal is to understand which interactions between the scales we can use to create models of tissues that approach real organs' complexity. Advances in AI (machine learning) algorithms and computing power, including quantum computing, will undoubtedly help advance these models and simulations.

Nanotechnology will be applied to create biosensors, the measuring and monitoring devices that will collect data on the parameters of tissues that we need for the models and simulations. APM will produce research tools and diagnostic and data collection instruments that are ever smaller until they are hand-held, more precise and less costly. The cost of doing molecular research data collection will drop from costing millions of dollars today to negligible costs per use. AI algorithms will integrate the collected data with the physics that is at work in the tissues.

Eventually, the biosensors will also become bio-stimulators. APM technology has the potential to give us precise control over the expression of our genes, turning them on and off. We could control the metabolisms of various cells in multiple locations within the body leading to targeted tissue repair and regeneration, not just immune response to invaders. We could repair scar tissues from injuries and birth defects and replace traditional plastic surgery. Contera writes, "Plausibly, the progress will render organ transplantation obsolete by the end of the twenty-first century."

The tissue generation application of APM technology would not have the ethical controversies associated with the need to harvest embryonic stem cells. However, this technology does lead to the possibility of artificially enhanced athletic performance.

In addition to leading to the understanding needed for the repair and replacement of diseased or damaged organs, mathematic tissue modeling can revolutionize drug design. Engineered tissue models, organs on a chip, will reduce or replace drug testing in animals and humans. Such chips have already been made to model the liver, kidney, gut, lung, fat, muscle, and the blood-brain barrier.[193]

Gaining control over all the tissues' growth and structures in our bodies opens the door for us to choose the bodies we want. This capability will present psychological, ethical, legal, and social controversies. We will need people with reality-based Personal Philosophies to guide professional and public discussions.

NANOTECHNOLOGY AND OUR EVOLUTION

Modeling the human brain and its connections to the body is the most promising pathway to developing an AI machine with human-level intelligence. Nanotechnology applied to nerve tissues, especially the neurons in the brain and brain stem, will collect the fine-grained data necessary for scientists to model human brain functioning in a machine.

More importantly and more likely in our immediate future, biosensor and bio-stimulator application to the brain's neurons, brain stem, and sense organs will become the direct interface between our biological intelligence and our advancing AI machine intelligence. These biosensors and bio-stimulators will be the catalyst to becoming entities with hybrid intelligence, part biology and part AI machine—Hybrid Humans.

The nanotechnology enabled Brain-Machine-Interface (BMI) will be the technology marking the threshold of our transition to a technological species—the *Transcendence Era* of the Conceptual Evolution Period.

♦♦♦

4 . 4

BLOCKCHAIN

"The blockchain is going to change everything more than the Internet has."

—Brock Pierce

BLOCKCHAIN—WHAT IS IT?

If they have heard of it at all, most people first hear of "blockchain" in the context of digital currency, usually "Bitcoin." They might even think that blockchain and bitcoin are the same thing. They are not. Blockchain is the underlying computer network code on which Bitcoin operates as just one software application. There are many other potential applications for using the blockchain network beyond digital currency.

There are multiple ways to consider what blockchain *is*:

- Blockchain is the software that makes digital currencies possible.
- Blockchain is a trustworthy, globally decentralized, triple-entry accounting system ledger for recording any transaction.
- Blockchain is the next-generation Internet.
- Blockchain is smart contracts.
- Blockchain is institutionalized philosophy.
- Blockchain is self-government—the decentralization of the institutions of our society.
- Blockchain is our AIPA/avatar.

BLOCKCHAIN IS SOFTWARE

Digital currencies such as Bitcoin are purely virtual money. There is nothing physical about them that you can hold. If you own a Bitcoin, then all you have to prove it is a numeric key (a string of numbers and letters) with both public and private parts. If you lose your private numeric key, you have lost your Bitcoin, and there is no way to recover it.

Suppose you want to pay someone your Bitcoin in exchange for something. In that case, the two parties enter a transaction on their computers using their public and private keys. The new owner of the Bitcoin is recorded as a transaction in a software ledger on the global Bitcoin computer network. The software ledger containing all the Bitcoin transactions is not held in a file in a central file server computer. It is recorded in thousands of computers worldwide in a peer-to-peer network where all computers agree that the transaction is valid.

This Bitcoin network is designed in such a way to make fraud or theft practically impossible. There is no central computer to hack. If any of the distributed servers are hacked, the other servers will recognize the discrepancy and invalidate the hacked servers. Computer cryptography is used to keep the identities of the parties and the secret keys private. At the same time, all transactions are publicly visible to everyone and fully auditable by anyone.

The software that implements the distributed network of servers, the trustworthy transactions ledger, and the Bitcoin transactions is called the blockchain. The blockchain software source code is publicly visible to everyone, so there can be no secrecy in its functioning.

BLOCKCHAIN IS AN ACCOUNTING SYSTEM

The blockchain is like a ledger, copies of which are kept in real-time in thousands of computers in a peer-to-peer network worldwide. Every new transaction must be validated, cleared, and stored by a majority of the computers on the entire network. Each new transaction is time stamped and must link to the transactions that occurred before it to establish a complete audit trail. The whole network represents a consensus of valid and permanently stored transactions of value.

Society requires "money" as a store of "value" to function. It has been a long time since individuals held all their accumulated wealth in physical money and physically transferred it between parties in transactions. Today, all you have to evidence the amount of wealth to which you claim ownership is your bank or brokerage statement that shows the number of dollars, pounds, yen, ounces of gold or silver, shares of stock, or other assets in your account.

When you use a credit card to pay at the gas pump, an electronic accounting system reduces your account and increases the gas station's account. All this is simply accounting—the use of ledgers—to keep track of agreed quantities of value in transactions run on computers. It is already just "digital" or "virtual" money, not physical money.

Today's money accounting system only represents a store of value because each of us *trusts* that the rest of society *trusts* that it represents value. When that trust is threatened, for example, by governments stealing value by inflating the money supply and decreasing the value in one's holdings, then the viability of the money system is threatened.

Because of today's government manipulation and unrestrained inflation of the money supply, the trust in the US dollar and other currencies is diminishing. It is the societal trust factor that digital currencies such as Bitcoin operating on the globally secure blockchain are attempting to restore and is causing their popularity.

Of course, digital currency that it cannot control directly threatens the government's ability to tax via inflation. It is too early to say whether digital currency or the government will prevail in the conflict that is sure to come.

We can use this new digital distributed ledger technology to store much more than digital currency transactions. It can record birth, death, and marriage certificates, property deeds and titles, medical records, contracts, votes, supply chains and sourcing, and anything else of value to people.

This blockchain distributed ledger technology's fundamental benefit is having *trust* in the recorded information:[194]

- Honesty - truthful, accurate, and complete communications.
- Consideration - fair exchange of benefits or detriments that parties will operate in good faith.
- Accountability - clear commitments to stakeholders and abiding by them.
- Transparency - operating out in the open, in the light of day.

We are not only concerned about having trust in people and institutions. We are entering the Internet of Things (IoT) era where every electronic device is a "smart" device, from refrigerators to cars to pacemakers and each connects to the Internet. We need an Internet where these devices and the information they provide can be trusted or automatically rejected by the trusted network.

BLOCKCHAIN IS THE NEXT INTERNET

Our present Internet was designed for storing and distributing *information*. It is open to all with little limitation as to what can be stored on it. It has been globally transformative.

However, the present Internet was not designed to record *transactions*. It does not have the features that ensure the trust factors listed above. The information on it is mostly held in large, centralized data servers in large institutions and companies. These servers are subject to hacking. These institutions control the flow of information in and out of the databases. The software code and algorithms are proprietary; you cannot know what they are doing or what the rules are. The inability to trust what you see on the Internet and worries over your privacy are some of today's primary concerns.

The distributed ledger blockchain technology will be at the heart of the next generation "transaction-based" Internet, which will layer "trust" onto the existing information-based Internet. I will refer to this new Internet as the "T-Internet," where the "T" stands for *transactions* and *trust*. In the book *Blockchain Revolution* (2016), Don and Alex Tapscott write, "(This) global ledger of truthful information can help build integrity into all our institutions and create a more secure and trustworthy world."[195]

BLOCKCHAIN IS SMART CONTRACTS

A smart contract is an agreement between parties with clearly defined accountabilities for the parts of the transaction recorded on and administered by the blockchain ledger. As in any contract, the parties agree on the actions each will take and for what consideration. The difference is that the blockchain will objectively determine that the conditions have been met and execute the consideration, rather than the parties having to trust each other subjectively.

Consider commerce on today's Internet. Assume you want to purchase something online from Amazon to be delivered by UPS. The transaction steps look something like the following:

1. You select your item, give Amazon your credit card information, and submit your order.
2. When your item ships, your charge card account immediately indicates the increase in the amount you owe, reducing your available credit, even though you have not yet received the shipment.
3. Amazon receives payment from your card-issuing bank minus the credit transaction fees charged by the bank.
4. If you do not receive the product, you must communicate that fact to Amazon and hope that they will issue a credit to your charge card account which could take weeks to clear.

Now let us see what the transaction would look like on the blockchain T-Internet:

1. You select your item, give Amazon your blockchain identity key, and submit your order.
2. The order gets recorded on the blockchain ledger as a smart contract that says: "If, within *x*-days there is evidence from UPS that the product was delivered to the buyer, and the buyer does not report a problem with the order within *y*-hours, then transfer funds from the buyer to the seller."
3. Nothing further happens until the product is delivered. When UPS issues the electronic notice of delivery to the blockchain, the blockchain begins a countdown timer giving you time to indicate that what you received was not what you ordered.
4. Once the countdown timer expires and you have not indicated a problem with the order, blockchain sees that the smart contract conditions have been fulfilled. It executes the immediate transfer of funds directly from your digital currency account to Amazon's.

Notice a couple of things. First, blockchain has eliminated the entire credit card and banking "middleman" payment system. The blockchain will conduct and clear this smart contract transaction between the two parties on a peer-to-peer basis for a tiny fraction of the cost that today's credit card companies charge. Second, the funds were not deducted from your account until you received what you ordered. You were not required to make advance payment based on *trust* that

Amazon would ship what you ordered and UPS would deliver it on time. And Amazon did not have to *trust* that you would pay after receiving the item. Actually, we call this a "trust-less" transaction because it is based on objective evidence, not subjective trust.

BLOCKCHAIN IS INSTITUTIONALIZED PHILOSOPHY

Technology alone will not be able to correct the problem of flawed ideas in society. For example, our government will print inflated money to fund redistribution of wealth programs as long as that is what people want.

However, technology will be the Great Amplifier of the effects that ideas have on society. Suppose individuals who promote a philosophy based on reality, reason, and individualism can influence technology's application to society's transactions at each incremental step. In that case, their positive efforts will be amplified and touch the lives of greater numbers of citizens, perhaps making a real difference. To put it negatively, efforts to prevent the insertion of irrational, collectivist ideas into the application of technology will help delay and forestall further collapse of our society.

The foundation of democracy and the essential requirement for an advancing human society is the *presumption of competency* on everyone's part. By *competency*, we mean the ability to observe and accept objective facts and make rational choices to pursue one's self-interest without resorting to force or sacrifice. This presumption of and trust in one another's competency makes it possible for us to reason with and conduct transactions with each other. When citizens abandon facts, reason, and individualism in favor of using force and sacrifice that collectivism requires, individual trust in others' competency is lost, and democracy and society suffer.

Eventually, the blockchain T-Internet will administer almost every transaction, including everything you purchase, apply for, pay or get paid for, vote for, join, certificates of birth, citizenship, graduations, marriage, death, titles to all your property, and much more. Humans will no longer administer these transactions.

A trust-based T-Internet cannot replace the presumption of competency between humans. Still, it is appealing because the presumption of competency in our fellow humans has already broken down. The citizenry has largely abandoned facts, reason, and self-interest and endorses emotionalism, coercion and sacrifice.

Competent individuals find today's Internet an untrustworthy place to transact their lives. They desire a T-Internet where they can freely conduct transactions with each other. They want a T-Internet that insulates and protects them from those who are not competent and from the type of government those people demand.

If we regain a society where most citizens are competent, we will still desire a T-Internet, not so much for its enforced trust as reclaiming the *right* to transact directly with each other without the government in the middle and the *efficiencies* it would provide to transactions between competent individuals.

The principles of an individual's right to privacy, property, speech, assembly, etc., will either be recognized and protected by the algorithms within these transactional systems, or they will infringe upon them. It will be humans who first create these new transaction processing systems and the rules they follow. The principles held by these humans are what we can expect to see in the software. As quoted earlier, "algorithms are opinions expressed in code."

Ultimately, the blockchain T-Internet will become the embodiment of our society's philosophical ideas in the global transactional infrastructure. Will it be a philosophy of individual rights and personal responsibility or collectivist rights and a nanny state? Once institutionalized in the blockchain T-Internet, the effects will be amplified upon all of us. You will not be able to argue with or convince the AI software system that it is wrong.

True Idea: While our technology is advancing exponentially, the philosophical ideas predominant in our society are becoming less connected to our Constitution's principles, Aristotle, and reality.

True Idea: New technologies will automate and amplify our society's predominant philosophy of the time within the machines we create and the information infrastructures humans and machines rely upon to conduct their lives.

BLOCKCHAIN IS SELF-GOVERNMENT

We can use the blockchain to make functions now performed by our government, whether moral or immoral, more *effective*. Suppose the government applies the blockchain to functions it should *not* be performing, those that should properly be reserved to the citizens. In that case, the blockchain's assistance to the government is a bad thing. Improving government effectiveness for an improper function is *negative*.

The ideal *positive* is for the blockchain to perform these functions for citizens on a peer-to-peer basis without government involvement. For example, suppose a piece of real estate changes hands. In that case, the two parties to the sale contract can record it on the blockchain. The new ownership is now forever visible to everyone and immutable. No county government recorder of deeds is needed.

The emerging blockchain provides an opportunity to completely remove from government all functions of a free society except for the military, police, and courts, the only proper government functions. The blockchain will make it possible for citizens to record all their affairs directly on the blockchain without government involvement and at dramatically lower transaction costs than our bloated government agencies' cost.

BLOCKCHAIN IS OUR AIPA/AVATAR

We are progressing toward having an Artificially Intelligent Personal Assistant (AIPA) representing us in all our interfaces to technology and our transactions. Our AIPA will be our personal agent. It will have an identity visible on the blockchain T-Internet called an avatar. Our most personal information and identity will be kept private. Our AIPA will use its avatar identity to conduct all sorts of transactions with online systems and other people's AIPA/avatars on our behalf. Our avatar will only share with others enough personal information to support the transaction at hand.

We ***absolutely must not*** obtain our AIPA from today's Big Tech. They claim an absolute right to surveil us, capturing all our behavioral surplus data. A *trustworthy* AIPA will not be possible until the new global T-Internet infrastructure evolves, supporting personal privacy and reputation protocols—not the Internet we have today.

Each AIPA/avatar will have a reputation factor that cannot be faked. All transactions will affect one's reputation factor. Suppose the person does not act with integrity. In that case, his AIPA/avatar's reputation will suffer, and others will avoid dealing with it.

The blockchain T-Internet will be critically essential for us to safely interface with advancing technology.

WHY BLOCKCHAIN IS A TRANSCENDENT TECHNOLOGY

Imagine trying to live without the Internet. The Internet has become ubiquitous. It supports nearly all our life's transactions. Almost all communications, industry, commerce, education, government, health care, entertainment, transportation, energy, the arts, and much more occur on the Internet. Our society now depends upon it.

We now live out our lives in two *domains*, one physical and one intellectual. This is an oversimplification, but it helps us to understand what affects our lives. The physical domain is "Reality"—the real world, and we must learn and comply with its laws. We take care of our bodies in Reality. The intellectual domain is the "Realm." Today the Realm means our information-based Internet. Nearly all the intellectual transactions of our lives occur in the Realm. Eventually, the Realm will mean the trustworthy (or *trustless)*, transaction-based blockchain T-Internet.

As we come closer to merging with our technology—a *hybrid* species—the more we will live our lives in the Realm. The word "Realm" begins with "Real." We would expect that our intellectual domain must, like our physical domain, be based on Reality. To the extent that today's Internet makes information available to us that we can validate against reality, it is a healthy place. However, our society is asking the Internet to do things it was not designed to do.

Because we are dependent on the Realm (today's Internet), we are concerned because we cannot trust what we find there. More importantly, the Internet has not helped secure our liberties and rights but has instead provided corporations and the government the tools to monitor, track, and influence us. We wonder if today's Realm is a safe place, or will ever be?

Blockchain cannot suddenly make people trustworthy. But it does minimize the need to trust others. The blockchain forces everyone and every transaction to comply with the *objective*, not *subjective*, trust protocols. The blockchain promises to provide us with a healthier Realm in which to live our intellectual lives.

However, those who seek to exert control over us and hide their true intentions will resist creating an objective blockchain or attempt to subvert it toward their ends.

Blockchain is a transcendent technology because it is the technology on which we will build the next Realm. A Realm in which all of us will live intellectually.

BLOCKCHAIN'S POTENTIAL BENEFITS

Following are the seven design principles for the blockchain. These give us a glimpse of the benefits that will follow.[196]

1. Networked Integrity Principle - Trust is intrinsic, not extrinsic. Integrity is encoded in every step of the process and distributed, not vested in any single member. Participants can exchange value directly with the expectation that the other party will act with integrity. That means that the values of integrity—honesty in one's words and deeds, consideration for others' interests, accountability for the consequences of one's decisions and actions, and transparency in decision making and action taking—are coded in decision rights, incentive structures, and operations so that acting without integrity either is impossible or costs a lot more time, money, energy, and reputation.

2. Distributed Power Principle - The system distributes power across a peer-to-peer network with no single point of control. No single party can shut the system down.

3. Value as Incentive Principle - The system aligns the incentives of all stakeholders. Bitcoin or some token of value is integral to this alignment and correlative of reputation.

4. Security Principle - Safety measures are embedded in the network with no single point of failure, and they provide not only confidentiality, but also authenticity and nonrepudiation to all activity.

5. Privacy Principle - People should control their own data. Period. People ought to have the right to decide what, when, how, and how much about their identities to share with anybody else. Respecting one's right to privacy is not the same as actually respecting one's privacy. We need to do both. By eliminating the need to trust others.

6. Rights Preserved Principle - Ownership rights are transparent and enforceable. Individual freedoms are recognized and respected. We hold this truth to be self-evident—that all of us are born with certain inalienable rights that should and can be protected.

7. Inclusion Principle - The economy works best when it works for everyone. That means lowering the barriers to participation. It means creating platforms for distributed capitalism, not just a redistributed capitalism.

BLOCKCHAIN'S POTENTIAL DANGERS

Tyranny of the Power Brokers

The blockchain's primary *near-term* potential danger is that it will be created in such a way that it serves the current power brokers, big corporations, and government, rather than redistributing power to individuals on a global scale. Data are presently being centralized within fewer powerful corporate entities who can then use it to become more powerful. This concentration of power is at the expense of our privacy and our rights as individuals. Governments now collect data on citizens and even use it to manipulate them, influence public opinion, and stay in power.

As automated processes replace human administrators for all the transactions of our lives, more of these transactions will occur on the blockchain. If trust is built into the blockchain as intended, then we will avert the danger. Suppose these power brokers subvert the blockchain through back doors or other techniques. In that case, trust will not evolve, and we are headed toward the eventual tyranny of the power brokers. The authors of Blockchain Revolution point out, "(Blockchain) could become a tool that powerful institutions use to entrench their wealth or, if hacked by governments, a platform for some kind of new surveillance society."[197]

Tyranny of the AI Entities

The blockchain's primary *long-term* potential danger will be when we humans share the world with AI entities. The blockchain will, itself, be a potential enabler of the creation of AI entities. The blockchain will make possible "Autonomous Agents."

Autonomous Agents are software programs that exist on the Internet, learning how to pursue their goals provided by the programmer who created them. They will have a digital currency wallet to purchase what they need and sell products or services to other entities. Such a software entity, able to make its own choices, might someday become the first general AI.

Regardless of how AI entities come to exist, a Realm enabled by the blockchain T-Internet and based on objective principles would best support a diverse society of human and machine entities. The AI entities would be subject to the same integrity and transparency requirements.

Suppose an AI entity is ever able to hack the blockchain, especially without us knowing it. In that case, we could become subject to the tyranny of the AI entities.

Tyranny of the Masses

Some people believe that the blockchain offers the potential to achieve *pure* democracy. In the political domain, they believe it would make possible the election of presidents and other officials by verifying all citizens' voting. In the judicial domain, they believe it would make possible large-scale juries who would adjudicate legal trials.

Both these beliefs are predicated on the concept of the "wisdom of the masses." However, as the Founding Fathers in their wisdom knew, the "wisdom of the masses" is a flawed concept. That is why they created a republic based on representative democracy and a legal system with judges, objective law, and limited juries.

It is entirely possible for a charismatic leader to quickly capture the attention of and influence most of the country's citizenry. Today's communication technologies, especially social media, make this even more possible. Our representative form of government and legal system was designed to provide a sort of inertia that slows down the rate of change in our society. It helps make sure that we thoroughly consider any changes. It prevents a majority of citizens from jumping off a cliff in a knee-jerk reaction to a range-of-the-moment influence.

We want the blockchain T-Internet to provide citizens with trustworthy information and engage them in politics and law. We want it to encourage an open democracy based on facts of reality rather than slogans, advertising, and newscasters with agendas. However, there is a danger that those who design the blockchain will attempt to achieve a pure democracy which would subject us to the tyranny of the masses.

The Danger of Incremental Complacency

We are in the early stages of the next T-Internet. We will not wake up one day to newscasts announcing that the new blockchain T-Internet was just turned on. It will come in small incremental steps. We will apply multiple, specific purpose blockchains and apps to types of transactions. If we do not educate ourselves and are not diligent at each new step, we will wake up one day and wonder how it all went so wrong. It is up to each of us to demand trust, transparency, privacy, civil liberties, and individual rights in *every* online system. We do not have these benefits now, and we desperately need them. Only an educated, alert, and activist populous can protect us from incremental overreach.

OBSTACLES TO BLOCKCHAIN

With the promise the blockchain offers, what is the probability that we will achieve it? Let us look at some of the obstacles that we must overcome.[198]

The Technology Is Not Ready - The real world of human interactions is messy. Even if two parties enter a smart contract on the blockchain and the blockchain enforces it, what if one party is unhappy with the outcome? Will they take the other party to court? It will take time for blockchain technology to advance from simpler to more complex applications to address such issues.

It Requires Too Much Energy - Today, the amount of energy consumed by the thousands of blockchain servers worldwide is quite large. Many believe that the present blockchain technology cannot scale to support all the beneficial applications we have described.

However, Ray Kurzweil, in his book *The Singularity Is Near*, described a potential future of computing technology that requires little or no energy due to "reversible" computation. There is also the promise of plentiful, inexpensive electricity from nano-enhanced solar technology and nuclear fusion.

Additionally, multiple "layers" of blockchain are already being developed where varying degrees of "certainty" and "speed" can be applied as needed. Less certainty or speed requires less computing power and energy. While advances in blockchain design and computing technology will be necessary to create the blockchain T-Internet, there is no reason to believe they will not appear.

Governments Will Prevent It - In totalitarian societies and even in the United States, the government has been continually usurping rights intended to be reserved to the individual citizens, gaining the power that comes with those usurped rights. Fear of losing control could cause them to prevent blockchain's creation.

Powerful Incumbents Will Control It - Corporations and government entities might use powerful *marketing* to get the citizens to support limitations on the blockchain and modifications that effectively give them control over it.

There Are Inadequate Incentives - The blockchain design is based on the owners of the thousands of distributed servers being incentivized to operate those servers. As the blockchain grows, the costs of running those servers increase. Some believe that eventually, there will be insufficient incentive to keep growing the necessary blockchain network. However, new forms of incentives are already being discussed.

It Will Kill Jobs - The blockchain is a potentially overpowering technology of automation where computer code rather than humans does the work. On the other hand, the blockchain might open entrepreneurship to the underserved global masses.

Decentralization Prevents Uniformity - The blockchain's code is open source. There is no central authority in control of its evolution. Some doubt that it will be possible for all the blockchain's potential stakeholders to agree on the various protocols' standardization.

It Will Lead To Super AI - As mentioned previously, some believe there is a danger that autonomous software agents on the blockchain will open the door for the evolution of super-intelligent AI entities to the detriment of humans.

It Will Enable Big Brother - Some believe that big companies and governments will make many attempts to control the blockchain and succeed. They believe this will empower the Big Brother state.

It Will Support Crime - Today, we see governments requiring back doors into encrypted communications under the argument that criminals and terrorists are using them. Criminals and terrorists now use Bitcoin to move funds around. The blockchain's privacy features will amplify this concern. Privacy is the foundation of individual rights and, therefore, free societies. Some believe that we will not be able to preserve our privacy on the blockchain.

BLOCKCHAIN REQUIREMENTS

We will need to overcome certain obstacles before the promise of the blockchain can be realized. One thing in which we can have confidence is that *we must develop something like the blockchain for our society to advance.* A transaction-based T-Internet with trust protocols is necessary. Suppose the present incarnation of the blockchain code proves to be unworkable. In that case, we can be confident that a solution that works, perhaps based on quantum computing's inherent cybersecurity, will eventually be found.

We might go through multiple iterations, each overcoming the previous version's limitations, before achieving the full promise of blockchain. We will develop multiple versions of blockchain networks and applications built upon them for specific purposes. These will go through successive iterations and gradually gain interoperability between them until they merge into a global blockchain T-Internet as a network of networks.

We can be optimistic about our ability to develop blockchain technology. However, we might be less confident about creating it without the incumbent power brokers in Big Tech and the government controlling it. We can expect our elected and appointed officials to resist giving up their positions or their powers over us. They will join Big Tech in applying marketing pressure to convince citizens to allow them to have control over the blockchain.

Only an educated population aware of what the blockchain is and the necessity for it to serve *individuals'* liberties and rights can resist the incumbent power brokers' marketing pressures. We need an activist population that will demand a blockchain T-Internet designed to serve *them*, not Big Tech or the government.

Watch out for arguments from pragmatism. They will present an immediate situation, for example, a potential terrorist threat, that will require that they be given the ability, with a court order, to breach the privacy protections on the blockchain. It sounds reasonable, except that they ask you to sacrifice a *higher* value for a *lesser* value. It might help them avert this instance of terrorism. However, it will be at the *forever* expense of a *higher* value—the loss of privacy for every person using the blockchain T-Internet everywhere in the world from that moment forward.

Blockchain technology will potentially be the foundation of the T-Internet infrastructure to facilitate all the other transcendent technologies in our society.

♦♦♦

5G NETWORK

"Unlike 4G and previous generations of technology, 5G is very different. It is not just radio. In fact, it stands across the full network from mobile access to cloud core, from software-defined networking to all forms of backhaul, front haul, IP routing, fixed networks, software, and more."

—RAJEEV SURI

WHAT IS 5G?

The abbreviation "5G" refers to the fifth generation of wireless technology replacing the previous fourth generation. Most people do not appreciate that 5G is not just an incremental performance upgrade to 4G. 5G, as the center of all future communications, will be the realization of a technological dream. It will become the lifeblood of an entirely new economy.

A new generation of wireless mobile telecommunications technology emerges approximately every 10 years. The next one is 5G and will use radio frequencies above 6GHz.

5G's goal is to be the platform for the Internet of Things (IoT) in which 50 billion future devices of the world will need to be connected. 5G could meet the needs of Virtual Reality (VR), Augmented Reality (AR), autonomous vehicles, smart buildings and homes, and smart phones, computers portable devices and industrial automation.

5G is expected to provide a 10X improvement in performance, a 10X decrease in latency, a 100X improvement in traffic capacity, and a 100X improvement in network efficiency.[199]

The industry initiatives that have been working on the standards for 5G have generally identified the following eight requirements:

1. 1 to 10 Gbps connections to devices in the field (not just theoretical).
2. 1 millisecond end-to-end round-trip delay (latency).

3. 1,000 times the bandwidth per unit area.
4. 10 to 100 times the number of connections.
5. (Perception of) 99.999% availability.
6. (Perception of) 100% coverage.
7. 90% reduction in network energy usage.
8. Up to 10-year battery life for low-power machine-type devices.

No single use case needs all these requirements to be met. No official specification document yet uses the term "5G." These requirements are not rigid specifications but rather aspirational goals. These goals will be met through the integrated application and layering of existing technologies such as 4G, Wi-Fi and Bluetooth and augmenting them with new 5G. It will be the new 5G that achieves the sub-1 millisecond latency and >1Gbps download speed.

To achieve all these requirements, 5G will need to use higher radio frequencies to carry more data at faster speeds. However, these frequencies do not carry radio signals as far and do not penetrate structures well. The 5G network will consist of many shorter-range cells both outside and inside buildings, even into each room. Outside, we might install them on streetlight poles and inside on ceiling light fixtures.

The short-range signals of 5G cells will be combined with the longer-range 4G technology to provide a blended long-range and short-range network capability. The longer-range cells will serve you while traveling at speed in your car, and the short-range cells will serve you when you go into a building.

One advantage of the simpler, smaller, low-power, short-range cell sites is they can be installed at a much lower cost and with fewer regulatory hurdles than today's high-power, long-range cell sites.

5G will enable the Internet of Things (IoT). IoT will consist of billions of devices, from cars and refrigerators to wearable medical devices and even clothing, to communicate with servers and with each other.

Perhaps the most impactful feature of 5G will be its low "latency," which is the delay time it takes for a signal to go from one device to another. The best 4G today has a latency of about 40 milliseconds. When fully developed, 5G promises a latency of one or two milliseconds. This nearly zero latency will make all the difference in the world, for example, when a surgeon performs robotic surgery on a remote patient. Such a low latency will open an unlimited range of new applications, most of which have not even been thought of yet.

To achieve such low network latency, we will need to provide application servers closer to the user geographically. The content that the user needs will need to be pre-loaded and readily available. The new 5G network will be more responsive to the users' needs adapting itself as necessary, for example, by allocating more bandwidth to applications that need it. This will require a much more significant build-out of the back-haul fiber-optic infrastructure to connect the increased number of widely distributed cells.

The build-out of this new 5G back-haul network, wireless cellular, and providing applications and services, will be an enormous but not insurmountable evolution for the network providers. We can be confident that it will happen so long as the use cases provide sufficient economic incentive, and they will.

The standards bodies are finalizing the specifications for 5G this year (2021). We can expect that the first releases of 5G in a couple of years will increase speed and connectivity but still be a long way from the new 5G network's full promise. It will happen in small steps at first and then gradually accelerate over the next five years. 5G is likely to continue evolving for the next ten to twenty years.

5G will make billions of new connections possible. These will not only be connections of people with information but also connections of people with things. These new connections will have an economic impact on nearly all industries, including automotive, medical, manufacturing, distribution, and so on. In a decade, 5G will have a massive global economic impact on new goods and services. Following are three examples of the type of disruptions we can expect—automotive, health care, and AR/VR.

Automotive Sector

For truly driverless cars that never crash into each other, the vehicles must be able to communicate with each other and with an intelligent roadway and a smart city in real-time with nearly zero delays.

Consider that a car traveling sixty miles per hour will cover nearly nine feet during the 100-millisecond delay in a round-trip signal. This delay is about the best performance we have in today's cellular networks. Suppose several signals need to be exchanged between two cars before an action decision is made. In that case, this delay could lead to disaster. The same vehicle will only have traveled about two inches during the 5G's two-millisecond round-trip signal delay.

The smart city 5G communications coupled with AI systems will efficiently manage automobile routes, travel times, and parking spaces, minimizing pollution.

Health Care Sector

5G will facilitate connecting safe and reliable medical devices worn by patients with caregivers and healthcare providers. The patient's vital signs will be continuously monitored and transmitted from wherever the patient happens to be. Healthcare providers will even remotely initiate interventions through the wearable devices, such as releasing drugs.

Health care participation via remote medicine will be expanded and made available to underserved parts of the world.

AR/VR Sector

5G is about to dramatically change the human experience by enabling us to remotely visualize almost anything in real-time through virtual and augmented reality, VR/AR.

If you think that department and retail stores with helpful salesclerks are a dying breed, prepare for their rebirth as virtual experiences. Wearing your high-definition 3D viewing device, you can experience walking through a department store, examining a product and speaking with a sales agent—from wherever you happen to be.

5G SECURITY CONCERNS - CHINA

Perhaps more than any other advancing technology, the 5G network demonstrates the concern that advancing technology might accelerate the collapse of our free society.

One of the significant suppliers of hardware and devices for 5G is the Chinese company Huawei. Initially unable to compete with thc heavyweight giant telecom companies in the West, Huawei has become a massive player in the global telecommunications industry.

Huawei anticipated and invested heavily in 5-G technology. In particular, Huawei developed for 5G a technology called "polar codes" for sending reliable signals over noisy channels. Polar codes technology is different from the "LDPC" technology preferred by Western telecom companies.

This set the stage for a showdown between the Chinese and Western telecom companies in adopting engineering specifications and standards for 5G. And the stakes couldn't be higher. The companies that provide the fundamental technology for 5G will be integral to the global telecommunications system for decades to come. There are not only financial but also nationalistic and geopolitical considerations.

At the critical November 2016 5G engineering standards meeting in Reno, Nevada, the hot topic was polar codes universally supported by the Chinese companies versus LDPC preferred by San Diego-based Qualcomm and most Western companies. A battle line was drawn, and at first, it appeared to be a standoff with neither side willing to back down. In the end, not wanting to delay the 5G standards, a compromise was reached. The signal-processing standard was split into two parts. LDPC would be used to send user data, and polar codes would be used for the control channel.

This was a win not just for Huawei but also for China. It was the first time a Chinese company was part of a global telecommunications framework agreement signaling China's increasingly dominant power in the marketplace.

We see in news headlines that the United States government believes a 5G network built with Huawei devices could be a national security risk. China is a totalitarian regime. Therefore, one cannot discount the possibility that the Chinese government could force Huawei to include espionage features in its 5G hardware and network.

In fact, on June 27, 2017, China enacted the "National Intelligence Law"[200] which states, "All organizations and citizens shall support, cooperate with, and collaborate in national intelligence work." This law explicitly declares that Chinese intelligence organizations intend to exploit any tactical or strategic advantage that Chinese tech companies like Huawei and their devices and networks installed worldwide might hold. A press article describes the history:

> China always tries to present itself as a benign participant in global commerce, hiding the true nature of its totalitarian communist regime as best it can. For example, Huawai offers as a defense that it is not publicly traded like most large Chinese tech companies where the Chinese government is the biggest shareholder. Huawei points out that it is organized as an employee-owned company with its shares owned by its 96,000 employees and retirees. On closer inspection however, we see that legally the employees' shares are owned by a Chinese worker trade union, which is registered with the Shenzhen city government union.
>
> The US government first became concerned about Huawei's behaviors in 2007 over sales of technology through a shadow company called Skycom to Iran in violation of existing US trade sanctions. In 2008 the US government blocked Huawei's attempt to purchase networking manufacturer 3Com, which made anti-hacking software for the US military. Then, in 2010, the US government blocked Huawei's intended purchase of 3Leaf Systems, a server technology firm.[201]

The Trump administration increased pressures on China. In 2018 he signed the *National Defense Authorization Act*, containing a provision barring the US government from using certain products from Huawei and other Chinese firms as a "substantial or essential component of any system." The Trump administration also sought to constrain China's theft of the intellectual property of US companies.

We certainly should have security concerns about the prospect of "backdoors" in Chinese 5G network hardware. However, that is only part of a more significant concern—China's ability to provide technology in competition with Western companies globally. Ultimately it is a struggle over who will build out the future of innovation in the fields of 5G network, quantum computing, and artificial intelligence. China explicitly intends to dominate the world in all three, and the Chinese government is investing billions toward that goal. The 5G network infrastructure buildout is the first contest in this struggle to create the next global T-Internet that will affect the lives of every individual on planet Earth.

Huawei's competitive position versus the Western companies and the US government's attempts at curtailing China's 5G ambitions could lead to the build-out of two separate Internets. China could be shut out of the US, Canada, and much of Western Europe. Still, China could blanket the Southern Hemisphere and Asia, serving developing countries with billions of users. China could significantly enhance its geopolitical influence in those parts of the world where they control the critical communications infrastructures.

Fortunately, the 2020 Covid 19 pandemic revealed to the world the risks involved with too much dependency on China. It remains to be seen which allies and other countries will follow the United States' lead in excluding the Chinese company from participating in the build-out of the 5G infrastructure.

Ideas widely held by a society and the resulting political system matter. Totalitarianism goes against the nature of human beings. Humans must be free to use their minds and act in their self-interest as their method of survival. Totalitarianism must be identified, shunned and fought at every opportunity. We must never grant a totalitarian regime or the businesses that operate under its control moral equivalency to free societies.

The 5G security situation is an example of how the philosophical ideas held by those who create and control technology will be amplified, accelerated and imposed upon the rest of us. Technology, especially communications technology, is a powerful tool for imposing political concepts upon a society.

Consider that Huawei's success was backed financially and politically by the Chinese government. Domestically, the Chinese government's policies favored the company against foreign competitors. Internationally, the Chinese government offered loans to Huawei's customers, providing Huawei a competitive advantage. Then there is the probability of intellectual property theft. A security specialist at the Canadian telecom company Nortel discovered that computers in China were using passwords of Nortel's chief executives to download hundreds of documents from the company. The US Department of Justice filed suit against Huawei, charging it with "deliberate and repeated misappropriation of intellectual property" since at least 2000.

Creativity, innovation, and technological advancement flow from the minds of people who are free to think, not from people subjected to force or coercion. However, supporting its world dominance ambitions, the Chinese government emphasizes education in math, science and reading while America has largely abandoned such emphasis. In the 2018 Program for International Student Assessment of 15-year-olds in 78 countries in math, science and reading, China ranked first. The United States ranked twenty-fifth behind most Asian and European countries.[202]

The Chinese Communist Party allows its citizens the illusion that they are free to pursue their self-interest in domestic and global commerce. They are free only to the extent that they accrue economic, military, and strategic power for the Chinese regime's global ambitions without questioning its authority or methods. Even Alibaba's billionaire founder Jack Ma was brought to heal by President Xi Jinping and the Communist Party when he criticized China's financial industry, with Ma disappearing for months.[203] The Western-style freedoms of the previously British-controlled Hong Kong are rapidly evaporating under the control of Communist China.

At the individual level, given China's emphasis on education and limited freedom to pursue one's self-interest and commerce, there are brilliant, industrious Chinese businesspersons and technologists. However, one cannot disregard that they and their companies are under the absolute control of a totalitarian state that views them as a means to global dominance. The idea that Chinese businesspersons are operating in a free society and can make peer-to-peer commitments in enforceable business contracts with businesses in free societies without risk to our future is *our* illusion.[204]

When free societies allow the normalization of commerce with Chinese companies or the participation of Chinese technologists in research and development by universities and businesses, it is a pathway to disaster. That is what happened at the 2016 5G engineering meeting in Reno, where the Chinese insistence on using their polar codes in the 5G specifications was "normalized." Such normalization places the value of enduring freedom at a level lower than the short-term benefits of commerce. A free society that values a totalitarian regime's participation in the roll-out of the 5G network or any other emerging technology higher than its foundational freedoms will not survive the new technological era. It will eventually succumb to a technologically enabled, economically and militarily empowered totalitarianism.

When our society purchases the low-cost goods produced by the cheap labor in a lesser-developed country, we transfer economic power to that country. That is what happened with Japan following World War II, eventually causing it to become an economic powerhouse in world commerce. We must understand that transferring economic power to a *free country* such as Japan is morally different from transferring power to a *totalitarian stat*e such as China and can have existential consequences. The elevating standard of living for the Chinese people makes them less likely to mount a revolution against their regime as it continues to gain power.

Unfortunately, even the Trump administration appeared to be willing to relax certain protective constraints on China's technology expansion in exchange for a more favorable "trade deal." It is understandable that it is difficult to impose sanctions on a Chinese company that purchases large amounts of US companies' components. Our leaders must base their decisions on a commitment to a hierarchy of values, not on pragmatism or in response to the most vocal special interests' outcries.

We must demand that our government and business leaders protect our freedom above all. We need to monitor the 5G and China situation in the coming years. We need to be aware of the consequences of the decisions made by the technology "standards" committees. We will see if free countries can join to protect individual liberty and rights, holding non-free societies in check as new technology is implemented. Suppose we can do it for the imminent 5G network. In that case, there is hope for biotech, nanotech, blockchain, AI, quantum computing, and fusion energy.

WORLDWIDE WIRELESS WEB (WWWW)

The Fifth Generation wireless and mobile network (5G) promises us a wireless network without limitations, one that is a near-perfect wireless world—the WorldWide Wireless Web.

The 5G mobile system is an all-IP (Internet Protocol) based model for wireless and mobile networks interoperability. Radio access technology within each terminal is seen as the IP link to the outside Internet world.

Suppose the blockchain T-Internet with privacy protections becomes a reality on the 5G network. In that case, augmented human intelligence, enabled through our AIPAs, will become available to us instantly, anytime, worldwide over 5G.

LOW POWER WIDE AREA NETWORK ("LPWAN")

Not all use cases for 5G require high bandwidth or low latency. In the book *5G. Simply In Depth*, Ajit Singh says that certain monitoring sensors might rarely need to transmit data. When they do, the amount of data they send could be quite small. It will not matter if it takes many seconds to be received. The 5G world will contain billions of such devices.[205]

These devices that do not require high bandwidth or low latency will be cheap to build and install and will have ultra-low power requirements and batteries that last decades. They will be installed in almost every conceivable location globally, for example, to monitor the weather.

These devices that are not very mobile and have low data needs will be served by a network technology called Low Power Wide Area Network (LPWAN), which will be a major contributor to the IoT (Internet of Things). LPWAN is designed for machine-to-machine (M2M) communications.

The types of lower radio frequencies and base stations used for LPWAN will not be the same as for mobile 5G. They will have an ultra-long-range and the ability to serve vast numbers of such IoT devices.

There is no one specification, and there are several companies with competing approaches to LPWAN. One open standard is called LoRa. As an example of LPWAN technology, a single LoRa gateway can communicate with several hundred thousand devices located as far as twenty miles away in environments without obstructions. In dense, urban environments, LoRa can penetrate several miles. A single base station site might fully serve a smaller city.

However, LPWAN technology has a latency measured in seconds rather than milliseconds. This latency is not an issue for the billions of data collection sensors to transmit their non-time-critical data. It makes the devices cheap to manufacture and install in huge numbers.

The various competing LPWAN technologies have been rolling out over the last two years and are serving dozens of cities. Watch for new applications for machine-to-machine communications soon.

DRIVERS OF 5G DEPLOYMENT

According to Singh, the following four drivers will enable the deployment of (mobile) 5G. Therefore, these are the ones you should monitor.

1. Fiber-Optic Infrastructure. Fiber-Optic systems move data at the speed of light. Wireless carriers use fiber to connect their cell sites to each other, data centers, and the Internet. Network providers will need to install a lot more fiber to many more locations to achieve the bandwidth goals of 5G. The various network providers have different strategies for expanding their fiber-optic network. Some are installing their own fiber, and some are leasing "dark" fiber, that is, fiber that has already been laid but is not in use.

2. Small Cell Deployment. Today's cell service is served mostly through "macro" cells called "cell towers" with a range up to about forty miles. For 5G, the cell companies will need to install many "micro" cells in streetlamps and rooms in buildings. These cells will have a shorter range and lower power. While we have already covered macrocells and smaller microcells, 5G technology will include cells even smaller than microcells. We can divide these into even smaller classes based on coverage area and the number of users they serve.

Cell Class	Coverage Area	Power	No. of Users	Application	Cost
Macrocells	35 kilometers	10 watts	Up to 1,200–1,500	Outdoors	high
Microcells	2.5 kilometers	2–5 watts	200	Outdoors	medium
Picocells	100–250 meters (indoor)	250 milliwatts	32–64	Indoors	low
Femtocells	10–50 meters	100 milliwatts	8–16	Indoors	low

Classes of 5G Cell Sites[206]

Different-sized cells can be combined to create a heterogeneous network (HetNet), a radio access network (RAN) that comprises traditional large macrocells with many smaller cells, ranging from femtocells to microcells.

The cell phone companies will face fewer regulatory hurdles to installing the smaller cells than they traditionally have faced with macro cell towers. In October 2018, the FCC announced a new set of guidelines that restrict states and local communities' ability to charge carriers for deploying small cells co-located with public infrastructure. The FCC expects this step to speed up the roll-out of small cell technology across the country. The major carriers have already installed tens of thousands of microcells in certain cities. The rate at which they install microcells and their locations will be directly proportional to the roll-out rate of 5G wireless.

3. High-Frequency Spectrum Availability. The implementation of 5G will not only depend on the build-out of a fiber-optic backhaul infrastructure and ubiquitous microcells. It will also require the availability of radio frequency bands with much higher frequencies than are used in today's cellular service. 5G will use the super high-frequency band (3 to 30 GHz) and extremely high-frequency band (30 to 300 GHz), particularly the frequencies between 24 and 86 GHz, to achieve the desired data speeds. These are line-of-sight frequencies.

In the United States, the FCC controls the utilization of radio spectrum for various uses through issuing licenses. For commercial services such as mobile communications, the FCC typically holds auctions to grant specific numbers of permits to use frequencies to the highest bidders. The FCC began auctioning the rights to the 28 GHz 5G band in November 2018 to forty wireless carriers, telecoms, and others with bids reaching more than $688 million by mid-December.[207] This was the greatest volume of radio spectrum ever auctioned by the FCC. It was in response to the carriers' demand for the spectrum to support an Internet of Things 5G connected world.

The FCC continues to auction off rights to other 5G bands and has designated certain bands to be "shared" bands using a Shared Access System (SAS). SAS means that carriers can dynamically share these bands based on availability. This enables smart scaling of bandwidth of radio spectrum up and down according to the range-of-the-moment usage demands. Similarly, the FCC is reserving specific frequencies for use by smaller entities that do not secure their own licenses.

As you can see, the radio spectrum allocation to 5G applications by the FCC gives our government direct control over the 5G roll-out evolution and the selection of commercial players. Monitoring the frequency allocation process indicates "when" we can expect 5G to expand and "who" will be providing it.

4. Indoor 5G Fixed Wireless. If it cannot penetrate structures, how will 5G wireless extend indoors, for example, into your home? A 5G device could be placed on top of your roof. From there, it can have line-of-sight to a cellular carrier's macrocell tower or microcell in a streetlight. Fiber optic cables will then connect the rooftop unit to 5G tiny cell devices in your house's various rooms or one or more 5G-capable Wi-Fi devices in your home. This fixed 5G service to structures will enable the wireless carriers to become "Internet Service Providers" in competition with traditional ISPs such as Comcast and Time Warner.

Several wireless carriers began building the infrastructure to enable offering fixed 5G in certain cities in 2018. Many of these companies plan to provide fixed 5G service before offering mobile 5G. The build-out of the infrastructure to initially support fixed 5G service will then also support mobile 5G. Watching the roll-out of fixed 5G service in a geographic area will indicate when to expect mobile 5G service.

Dumb Pipe

Messaging services from application providers such as Apple and Google have largely replaced messaging services offered by carrier networks. These applications simply rely on the data pipe provided by the carriers, not their messaging services. This trend will likely continue and increase as we move to 5G. The carriers will focus on building the best performing, most reliable and least-cost data pipe that they can. They will leave the services applications that use that pipe to be provided by others.

Ultimately, in a few years, we will find ourselves in 5G Wi-Fi everywhere and with near-instantaneous ability to download things and do things in real-time. The global 5G network will enable the new blockchain-based T-Internet. We will never have experienced such a high-value technology.

WHAT IS 6G?

5G has just started rolling out, and already we hear about something called "6G." Essentially, 6G refers to the next generation of wireless technology after 5G. New generations come out about every 10-years. It remains to be seen if we will call it 6G when it comes out or if it will simply be continuing the evolution of 5G such as "5G enhanced."

As of now, it looks as though 6G will address any shortcomings that we discover in the 5G rollout and, of course, will incrementally increase data rates and decrease latency over 5G. It will not likely be the revolutionary change that 5G represents over 4G. In any case, preliminary work is already underway on 6G and telecom company trials could begin by 2030.

Here are a few of the early stage 6G developments:[208]

- The FCC has taken the first steps in opening the terahertz wave spectrum (frequencies between 95GHz and 3 THz), citing that it will "expedite the deployment of new services in the spectrum above 95 GHz."
- 6G research has begun from Virginia Tech and companies like Samsung and LG.
- Japan is planning to launch 6G by 2030.
- Verizon, T-Mobile, AT&T, Microsoft, Samsung, Facebook, Apple, Google, Ericsson, Nokia, Qualcomm, and others launched the "Next G Alliance" in late 2020 to aid North America's advancement to 6G.

While the 6G radio spectrum allows for even higher data rates than 5G, its range is presently limited to 10 meters. For there to be significant 6G coverage, new technologies will need to be developed.

♦♦♦

ARTIFICIAL INTELLIGENCE (AI)

"And then they wonder if computers can think. My answer is that before a computer can think, it must be able to understand ideas (concepts); before it can grasp ideas it must be able to perceive the world, feel emotions such as joy and suffering, desire and fear, pleasure and pain; before it can feel emotions it must be alive—which entails being able to act to sustain itself. We can dismiss questions of whether or not a computer can think until one is built that is alive. Only then it wouldn't be a computer, but a living organism, a man-made one."

—HARRY BINSWANGER

AI INTRODUCTION

Artificial Intelligence (AI) is not a direct creation of evolution. Man's mind is a direct creation of evolution. Humans, as agents of evolution, create tools with their minds and AI is a tool. The AI technology we create will help us accelerate the increase of Intellopy in the locations in the universe that we touch. Expanding Intellopy in the universe consumes growing amounts of energy resulting in increased overall entropy. AI will be the tool we humans apply to advance evolution to the inflection point in its exponential growth curve.

It will be the following chapter where we consider augmenting *human intelligence* with AI technology. In this chapter, we will focus on *machine intelligence*, including robotics. There is a lot to consider in these two chapters. I will describe actual science and technology and provide conjecture on my part as to what might occur.

We might someday create machine-based living entities with human-level or higher intelligence. Of all the transcendent technologies, Artificial Intelligence holds the potential of mankind's evolutionary replacement. We need to carefully examine the possibility of the creation of conceptual AI entities.

Assuming we will create them, we need a thorough understanding of the associated issues before they arrive.

AGENTS OF EVOLUTION

The existence of biological or machine *entities* only makes sense in the context of evolution. Natural selection is a *passive* process for increasing complexity. It only operates on past success/failure, not anticipation of the future. It generated a degree of improbability and adaptive organization in living entities—and resulted in humans. Humans can *actively* increase complexity. As agents of evolution, we can anticipate future needs and select adaptive traits based on projections. We can generate a much higher degree of improbability and adaptive organization in ourselves and our machines.

The role previously played by *natural selection* will now be performed by humans. Humans will apply *artificial selection* in the continuing evolution of ourselves and the machine entities we create. Artificial selection is not merely analogous to natural selection. It is a special type of natural selection where humans use their rational faculty to anticipate what might have survival value.[209]

Humans will actively control reproduction. We will purposefully advance future generations of humans and vegetative, animal, and eventually human-level machine entities by selecting traits that enhance their survival in anticipated environments. We will accelerate the traits selection and testing process. Physical reality will remain the ultimate judge for the success or failure of selected combinations of traits.

Humans will determine the mechanism designs, terms of operation, environmental stimuli a machine entity will detect and associate with pleasure or pain, and the "instinctive" response programming. Goal-directed action in the face of mortality—"life"—is a phenomenon that will emerge when we provide the necessary physical and operating conditions. When we create them, *Perceptual* AI will be conscious machine-based animal-level entities.

If we create *Conceptual* AI, they will be machine-based human-level entities able to direct their further evolution. They will advance their intelligence through recursive self-improvement, perhaps replacing humans as the next Agents of Evolution.

NARROW AI

Consider the motion-activated security light outside your house. It has an infrared motion detector, a floodlight, a switch to turn the light on and off and a preprogrammed microchip with a timer. When it "senses" a warm body moving within its range of "vision," it has as its "response" the "action" of turning on the light. It begins tracking the amount of time the light has been on. After a preset duration, if no further movement is detected, it turns the light off. This example is a form of Narrow AI. It can sense some aspect of the real world and can sense time. It has artificial intelligence programmed by humans into its rather simplistic microchip. It can respond to the sensory input by acting.

All artificial intelligence we have around us in technology today is qualitatively no different than your motion light. Certainly, we have AI devices that are considerably more capable in their sensory ability and computer processor sophistication and their ability to act. However, they are still basically sensory-response machines doing what humans built them to do.

Alexa, Siri, and Google Assistant are not alive and do not have consciousness. They can detect (with varying levels of probability) that certain words were spoken in a stream of audio and respond with preprogrammed "skills." They do not understand the concepts represented by your spoken words or the words they are speaking themselves in response.

A sensory-response Narrow AI machine is not even exhibiting goal-directed action as in primitive forms of life such as a single-cell organism. A Narrow AI machine does not face the fundamental alternative that all living things face—to live or die. A Narrow AI machine is a form of Purposed Intelligent Matter that man has created to follow preprogrammed instructions. No device that man has yet produced, including the most sophisticated computer, is alive or has needs. Without needs, it can have no goals, and it cannot be alive or conscious. Binswanger explains:

> The final conditions of the behavior of feedback mechanisms, such as target-seeking torpedoes, are not needed by the mechanism. Attainment of the 'goal' does not benefit it, nor does failure to attain the 'goal' harm it. In fact, since a feedback mechanism faces no alternative, the concepts expressing value-significance, such as 'benefit' and 'harm,' cannot be

> meaningfully applied to it. [...] the processes of feedback mechanisms do have value-significance for the men who design, construct, and employ them. The functioning of the mechanism is successful or unsuccessful in terms of the goals and purposes of man. But, as we have seen, in judging the goal-directedness of behavior we are ultimately concerned only with the value-significance of the behavior's effects for the agent. Hitting an enemy ship is not the goal of the torpedo, although it is the goal of the men who launch the torpedo. The torpedo, not being alive, can have no goals. But, having been manufactured is not, even retrospectively, a value to the torpedo. The torpedo does nothing to maintain itself in existence—in fact, its motion leads to its destruction.[210]

Humans are good at curiosity, imagination and creativity. Machines are good at interpolation, pattern-matching, precision, repetition, and storage and retrieval of vast amounts of information. We have built machines that can sense the real world and isolate entities within that sensory data stream and determine these entities' identities (if first labeled by humans). These systems can even recognize human faces in video streams. Still, these are special-purpose Narrow AI machines, not living things. They do not understand the concepts of "entity" or "identity" even at the perceptual level. They are not "conscious." They will, however, be immensely useful to man as tools for gaining an understanding of and control over the Laws of Reality.

We will be at the threshold of creating "living" machines when we create devices that must continuously act within their environment to sustain their existence. Then, as we better understand the functioning of biological brains, we will provide those living machines the faculty of consciousness.

LIFE AND CONSCIOUSNESS

Suppose we want to build a living, conscious machine entity. In that case, we need to define what we mean by "life" and "consciousness," whether biological or mechanical. What are the essential characteristics that an entity must have to be considered "alive," and then "conscious?" What are the levels of consciousness? These are fundamental questions. The answers are the foundation for everything else we will explore about intelligence, technology and society. We need to be precise and go slowly here to be sure we have a good foundation.

The fundamental quality of all living entities is that they are "goal-directed." Inanimate objects, even man-made devices such as torpedoes or guided missiles, are not goal-directed. Binswanger explains that a "goal-directed action" is a self-generated action caused by a mechanism whose existence, organization, fuel, and terms of operation result from the survival benefit that past instances of the goal have provided the organism in similar previous circumstances.[211]

Goal-directed action is the emergent phenomenon that we call "life." Life emerges when matter and energy are combined in a specific structure that enables it. Once life has emerged, continuous goal-directed action is required to sustain it. Goal-directed action exists in two forms: the purposeful and the non-purposeful, according to whether the action is conscious or vegetative.[212] We would describe plants and simple animals as *non-conscious* entities with *vegetative* goals. We would describe higher animals as *conscious* entities with *purposeful* goals.

The issue that differentiates purposeful action from vegetative action is the involvement of consciousness. Consciousness emerged as an adaptive mechanism. The higher animals possess the faculty of consciousness because it has survival value. This view of consciousness originated with Aristotle.[213]

For all living entities, the continuation of their own life is the ultimate goal, an end in itself that makes the existence of values possible. Life is the only phenomenon that is an end in itself—a value gained and kept by a constant process of action.[214] In purposeful actions, this value-significance is experienced consciously—e.g., pleasure and pain. In vegetative actions, this value-significance is biological and based on the alternative of life or death. All value-significance is relative to the survival of the organism as a whole. This is the ultimate end, with all lesser goals as the means.[215] Only living entities have *needs,* and as the *agent* act on needs to sustain themselves.

For conscious animals, their purposeful goal-directed actions are caused by the goal's *psychological* value-significance, which develops out of the experiences of pleasure and pain. Binswanger explains:

> If, for example, an animal were born without the capacity to feel either pleasure or pain, it could not have any affective or connotative

> experiences at all; such an animal would have no basis for desiring anything. The same is true in the case of man. If a man were congenitally unable to experience either physical pleasure or physical pain, he would be unable to value anything, or to desire anything. Such a man would not engage in purposeful action, because there would be nothing for him to gain or lose by acting.[216]
>
> It is natural selection which explains the correlation of pleasure and pain with survival needs. The stimuli which are inherently pleasurable to an animal are those associated in its natural environment with conditions required for its survival. The manner in which the animal's nervous system is "wired" determines what kind of stimulus will cause pleasure or pain—i.e., what kinds of objects will serve as the triggering stimuli for the animal's conscious behavior.[217]
>
> Stimulus requires direct sensory contact with external existing objects. The mental content of conscious animals is either from such direct sensory contact or is derivative from the material gained in past instances of direct perception. Even if animals cannot explicitly project the future, they can learn from the past. The materials out of which the mind creates, and projects a future goal, are furnished in the last analysis by prior sensory contact with external objects.[218]

First, to speak of an action's being self-generated, there must be a "self"—i.e., a physical entity with a defined structure and a definite boundary isolating it from the environment. Self-generation means there is an internal store of energy. This store must be replenished—hence the need to obtain energy—hence the phenomenon of value-significance.[219]

We have just described what we mean by being "alive" and "conscious." To create a living machine entity, it must exhibit goal-directed action. We can now list the requirements for creating a machine entity in each of the three levels of goal-directed action: vegetative-level, animal-level and human-level. To create a machine with human-level intelligence, we must first create a machine with animal-level intelligence. To create a machine with animal-level intelligence, we must first create a machine with vegetative-level intelligence, which means create a machine that is "alive."

Requirements for a **Vegetative-level Machine Entity** (to be "alive" but not conscious):

A physical entity—a "whole" entity whose parts include:
- -Defined structure and a definite boundary isolating it from the environment (a "body"),
- -Internal store of energy—and the need to replenish it—value-significance,
- -Senses to detect stimuli (triggers) in the environment,
- -Terms of Operation—pre-programmed responses to stimuli to sustain its body.

Mortality—continuous self-sustaining action required—must do continuous work in the thermodynamic sense else it will die—a mechanism that most efficiently drives energy "uphill." Once dead, it cannot be reconstituted.

Goal-directed action—to gain materials and energy from the environment for self-maintenance, self-repair, and self-reproduction according to static programming of behaviors suitable for its intended environment.

Requirements for an **Animal-level Machine Entity** (to be alive and "conscious," i.e., feel "emotions"):

All the requirements of a Vegetative-level Entity plus:

Motility—a body capable of locomotion.

A **nervous system** with at least a rudimentary brain equivalent that can learn the consequences of actions.

Instinct—pre-programmed responses to stimuli to guide its actions in its environment throughout its life.

Pleasure/pain association with stimuli that aid or threaten its life.

Perceptual awareness:

- Awareness of entities—integrated awareness of things (including their characteristics) "out there,"
- Awareness of higher-order differences—of what changes and what does not, of patterns,
- Awareness of space—of entities arrayed in space—relative positions,
- Awareness of a "world" of entities extending out in all directions from "here," from oneself, including one's own limbs, an implicit form of self-awareness,
- Awareness as a continuous process over time.

Consciousness—perceptual-level intelligence—retention of past stimuli—emotions: joy, suffering, desire, fear, pleasure, and pain:

- Ability to integrate all the various parts of the body to pursue its overall goal in relation to the perceived environment,[220]
- Ability to bridge space, to respond to distant objects,
- Ability to bridge time by responding now and over a span of time to a goal that it will not reach and utilize until later,
- Ability to guide its actions according to the continuing changes in its goal and the requirements of reaching it,
- Ability to expand the range of its action—the organism is sensitive to minute changes in its environment,
- Ability to learn from the past, to acquire new knowledge—modifies behaviors to improve survival.

Purposeful action—desires a goal that has psychological value-significance—has an "end-in-view."

Requirements for a **Rational Human-level Machine Entity** (to understand ideas—to "think"):

All the requirements of an Animal-level Conscious Machine Entity *excluding instinct*, plus:

Conceptual-level Intelligence—

- Abstractions—concepts—rationality,
- Volition—the phenomenon of "choice" in place of instinct,
- Forethought—the ability to project the future from experience.

Flexibility—adaptation to variation itself.

Individualization—individuals are compelled to discover new knowledge of the Laws of Reality to survive.

Socialization—rational individuals share abstractions and combine efforts to enhance survival and discovery and to achieve larger goals.

Kurzweil predicts that we will routinely encounter machine entities in the 2030s.[221] I think it might be later than that. Regardless of how different the substrate platform and architecture of AI entities we create might be, we humans will categorize them as *plant*-level, *animal*-level, or *human*-level entities based on their level of consciousness and treat them the same as those biological entities.

By creating *Narrow* AI, *Vegetative*-level AI, and *Animal*-level AI machines, we are continuing to create tools of the human mind—the latest iterations of the progression from the first stone tools. However, when we create the first conceptually conscious AI entity, *that* will be fundamentally different. It will be a new evolutionary branch of conceptual entities able to act as Agents of Evolution. Their branch of evolution will have separated from and exist alongside Homo sapiens.

Concepts are integrations organized into hierarchies. Long before we create a conscious Conceptual AI entity, we will create Narrow AI *natural language processing machines* that organize symbolic language (words and phrases) into hierarchies to mimic the conversational behavior and even the emotions of a human—the Turing Test. These machines simulate conceptual thinking. Even if some think that these machines pass the Turing Test, they are not conscious Conceptual AI entities. The requirements for a conscious Conceptual AI entity are those listed above as dictated by objective reality, not by subjective human evaluation of a conversation with the AI. When we can determine that these objective conditions exist in an AI entity, we can consider it alive and conscious.

Some AI developers think that we might create AI entities that can pass the Turing Test as conceptually conscious entities but be unlike us because they do not fear their destruction. I think this is unlikely. They must be alive and have perceptual and conceptual levels of consciousness. They will exhibit purposeful action to achieve goals, including staying alive to achieve their other goals. If they are pursuing goals given to them by humans, then they are either not conscious like a guided missile or are conscious only at the perceptual level like a trained animal.

BUILDING A LIVING MACHINE

Considering the list of requirements for a living vegetative-level machine entity, let us imagine how we might go about building one today.

Body - It must have a defined structure and a definite boundary that isolates it from its environment. We can envision a computer-operated machine in a room with a solar-transparent roof. This machine consists of modular structural and functional components that can be hot swapped (replaced while the system is operating) and robotic arms for manipulating the modular parts.

Internal Store of Energy - The machine has a battery bank that stores electricity as the energy source on which its operation depends. It will need to replenish its store of energy periodically to sustain itself. It will do this by sensing the sun's position and aiming solar panels to capture the most solar energy.

Senses to Detect Stimuli - The machine has sensors (bar code scanners) detecting the presence and position of essential repair parts appearing (somehow) from time to time in its vicinity (its environment).

Terms of Operation - the computer system would contain programming software for the behaviors of:

- monitoring internal battery charge levels
- monitoring for the presence and location of the sun
- moving solar panels to point at the sun
- monitoring for failed or failing internal components
- switching to an already installed alternate (backup) component for one that is failing:
    - monitoring for the presence and location of reachable replacement parts
    - removing and discarding a failed/failing component
    - grasping a replacement component and installing it
- assembling a complete replica of itself by assembling surplus modular parts
- copying its programming into the replica unit
- charging the replica unit's internal batteries and turning it "on"

Mortality - Both the computer system's programming and data memory storage would be of a "volatile" type. Suppose available internal electrical energy falls below the level required to continue operation, or critical repair components are no longer available. In that case, the machine entity will "die," and it will lose its programming and memory. It cannot be reconstituted simply by the restoration of electricity. Its "body" will lie inert.

Environment - The compatible environment is the room housing the system. Humans must provide materials and place them within reach. New replicas reproduced by the system must be transported to other locations by humans. Our artificial living machine is in an artificial environment. It cannot adapt to change. It will "live" so long as its environment remains supportive of its nature.

The above vegetative-level machine "entity" is an oversimplified analogy to a living biological plant. It would have a body with an internal store of energy and senses to receive stimuli. It would have innately programmed behaviors for obtaining energy and materials for self-maintenance, self-repair and self-replication. It would act to get the materials and energy it needs from its environment to support its "life." So long as its environment provides the materials and energy it needs for survival and it does not suffer a fatal component failure, it will act to sustain itself—to "live" indefinitely.

Additionally, whenever it has sufficient extra stored energy and available surplus materials, it will replicate itself.

If it "dies," it would need to be reconstituted by humans. In that case, it would essentially be "re-born" as a new machine entity with a new memory.

It will differ from a biological plant. There is no "evolution"—no means of adapting successive generations to a changing environment through natural selection. It will be up to humans to intervene and make improvements from time to time.

Suppose we imagine the achievement of full Atomically Precise Manufacturing nanotechnology. In that case, we can envision a robot that gets the atoms and molecules it needs to sustain itself directly from a real, not artificial, mineral-rich environment, just like a plant. Suppose we introduce a degree of random change in its operating software when copied into the replicas, along with those replicas' ability to be transported to new locations (wind, animals, etc.). In that case, natural selection would apply to this vegetative-level living machine completely independent from humans. Its kind could "evolve."

In the "room" scenario, we essentially created an autonomous, fixed-location robot that repairs itself and makes more robots like itself. This robot is not as sophisticated in its ability to deal with its environment as a biological plant with millions of years to evolve. However, this example appears to be within reach of our technology today. We might see man-made, vegetative-level, artificial living machine entities in the next few years.

But what about man's purpose in creating such a machine? What goals do we give it beyond survival? Can we have it do something productive for us? Of course, we can. Such living vegetative-level machines could become a new class of autonomous productivity tools for humans.

There is, however, no reason to use *volatile* memory in our "room" scenario machine. We can build it with *non-volatile* memory. If it runs out of electrical power and stops, it will not lose its programming or memory. We can restart it simply by supplying it with power. While it is shut down, it will not quickly decompose. These two observations are our first indications that machine-based living entities will have certain advantages over their biological predecessors. They will not "die" or "decompose" because of a loss of energy. They could be in *suspended animation* for thousands of years and then start again.

Suppose we become Virtual Humans by moving our consciousness into machine bodies. In that case, we could get on a spaceship, enter suspended animation by simply shutting down, and then start up again a thousand light-years away in another galaxy. To us, the thousand years did not happen—we blinked and were suddenly in a new galaxy. Perhaps this will be the method mankind employs to deal with the speed-of-light limitation of space travel to awaken the universe.

Looking at the list of requirements to create a conscious *animal*-level machine, we are far from building one, much less a *human*-level living machine. An animal-level machine will not be possible until we can create an artificial brain that can support a perceptual-level consciousness and all that entails. We are not likely to see man-made machine *animals* in the next few years, but we might see them in the next twenty years. Human-level machines will be built upon the successes of animal-level machines and will therefore occur later.

Ultimately, the laws of physics present no barriers to our ability to create living and conscious machine entities out of inanimate matter. And we have biological models to follow.

CREATING A CONSCIOUSNESS

When we create *animal*-level machine entities, *purposeful* action will be emergent, and they will be *conscious.* Consciousness is a faculty that biological evolution developed on a continuum ranging from the simplest organisms possessing a defined brain (i.e., the flatworms, Planaria)—to man. To create Perceptual AI, the lowest level of consciousness, scientists will need to understand better how instinct drives animal brains' functioning, enabling the emergence of consciousness.

There is a distinction between the *physical brain functions* that make possible the emergence of the phenomenon of consciousness and *consciousness itself.* Matter and consciousness are related but are two irreducibly different phenomena. Consciousness is an emergent non-physical attribute of the entire entity that its components' physical characteristics cannot predict. This distinction is missed by many, including AI scientists. Binswanger writes: "Despite this fact, however, most of those writing on consciousness today assume that consciousness can be reduced to matter. Materialism, the denial of the existence our causal efficacy of consciousness, holds a virtual monopoly in philosophy and allied disciplines."[222] This materialism causes many AI developers to underestimate the challenge of creating a perceptually conscious entity, much less a conceptually conscious one.

There is a vigorous global debate about how Conceptual AI might come about and whether it will be good or bad for humans. There are multiple theories about how to create Conceptual AI and then control it. We must base the resolution of this debate on "the nature of a rational entity," human or machine. By rational entity, we mean an entity whose intellect is based on the *integration mode* of thinking—"reason," and volitional control over this faculty.

Reality's evolutionary process created humans. We can assume that man's conceptual faculty has an explanation based on physical reality, not on some mystical cause such as a soul or a spark from a god. Our level of intelligence is competent to discover consciousness's physical requirements. Conceptual-level humans and perceptual-level apes have similar brains. However, there are differences in brain structure and function that are just beginning to be identified and understood. Eventually, we will discover what evolutionary changes in brain structure and function made possible our conceptual-level consciousness. Then we will be able to model it in machines.

Man seeks to create a Conceptual AI using advancing technology. Man cannot create intelligence that uses some unknown and unknowable (by man) level of consciousness higher than the conceptual faculty. Man can create machine intelligence that uses the same conceptual method of intelligence we use, perhaps with mental processing power and speed that exceeds humans.

Suppose an AI not derived from reverse-engineering the human brain could achieve general intelligence via some method other than an integration mode conceptual faculty. In that case, it would not be able to communicate with humans unless it learned how to mimic our conceptual-level intelligence. It is far more likely that we humans will create an AI consciousness that uses our same intelligence method.

[Please refer to the *Resources* section's *Rational Machine Roadmap* diagram for a graphic presentation of creating consciousness.]

AI CONNECTION TO REALITY

Evolution selected for perceptual consciousness in animals first and then higher conceptual intelligence in humans to improve adaptability and the organism's ability to modify the environment—all driven by the fact that it accelerates entropy. A conscious entity is, by definition, conscious of something. That something is physical reality.

Ayn Rand describes our human cognitive connection to reality:

> Man's senses are his only direct cognitive contact with reality and, therefore, his only source of information. Without sensory evidence, there can be no concepts; without concepts, there can be no language; without language, there can be no knowledge and no science.[223]

> [Man's] senses do not provide him with automatic knowledge in separate snatches independent of context, but only with the material knowledge, which his mind must learn to integrate....His senses cannot deceive him,...physical objects cannot act without causes,...his organs of perception are physical and have no volition, no power to invent or distort,...the evidence they give him is an absolute, but his mind must learn to understand it, his mind must discover the nature, the causes, the full context of his sensory material, his mind must identify the things that he perceives.[224]

Consciousness originates at the perceptual level, and the conceptual level connects to reality through the perceptual level. The conceptual level deals with abstractions—concepts organized into hierarchies that cognitively model reality's hierarchical nature. The senses directly grasp reality at the perceptual level. Human language—words—are handles that we use to bring the abstract hierarchical patterns that words represent into conceptual mental focus.

The question is, can we build an AI that has a *conceptual* level and can recognize language but has no *perceptual* level, and if so, what can it do? In other words, can we create an AI entity in the reverse order that evolution created us—top-down instead of bottom-up? This question is essential because that appears to be what the AI development industry is attempting to do. Let us explore some of the implications.

IBM's Watson is such an AI. It was able to beat the best human players at the game of Jeopardy, which requires language-based knowledge. In preparation, Watson was provided hundreds of millions of pages of text from the Web to read. Equipped with machine learning algorithms, Watson could self-organize the language text into hierarchical patterns that associated keywords or phrases could retrieve. Watson's statistical hierarchical pattern matching method of dealing with abstract language is like the human conceptual level's neocortical functions.

For humans with a rational faculty, words are abstract mental representations of real things in reality. However, a human can hold conceptual ideas that are not connected through lower concepts back to percepts of reality. We call these floating abstractions. Lacking physical senses and a perceptual faculty, the entirety of Watson's knowledge consists of floating abstractions—words—elements of language. Watson can relate these language elements to each other in logical hierarchies. Still, Watson cannot connect any of them to reality. Indeed, much of the knowledge on the Web that Watson read consists of invalid floating abstractions. Watson can win at Jeopardy by knowing how one language element relates to other elements of language. Still, it cannot tell you if any element is true—valid in reality.

Watson shows us that it is possible to create a Narrow AI machine that uses the integrative *conceptual method* of organizing hierarchies of abstractions. Still, it is not a *conceptual entity* that is *conscious* of reality. It is neither conscious nor rational. We cannot create a *conceptual AI entity* until we can create a *perceptual AI entity.*

Since we are already building conceptual language processing machines like Watson, what are their potential uses and dangers? While not rational in the human sense, conceptual language processing AI machines could still be quite useful. For example, an AI medical diagnostician could apply its integrative language processing faculty to the patient's voluminous textual medical history to arrive at a diagnosis. Humans provide both the content and the goal. A version of Watson is already being applied for this purpose. As the AI medical diagnostician gains knowledge and experience, we might give it the goal of finding a cure for a specific disease.

We could similarly apply conceptual language processing AI machines in any domain where the natural language historical content involved is sufficiently consistent for the machine to detect useful hierarchical patterns. Like the medical diagnostician example, the language content involved and the AI machine's results are floating abstractions for the machine. Humans will understand how they relate to things in reality. Humans will use conceptual language processing AI machines as tools to process more significant amounts of language-based data and find hierarchical patterns that humans are unable to find. The AI machine will not understand how the word concepts it outputs relate to reality, but the humans will.

A conceptual language processing Narrow AI will also be useful as an external extension of our brain's neocortical functions—an *exocortex*, which we will explore further in the next chapter. As we improve our mental interfaces with AI technology, we will expand our brain's pattern-matching capabilities utilizing an AI machine assistant.

Lacking a perceptual level, AI developers must provide a conceptual AI language processing machine with its foundational level concepts or hard-coded rules. We expect it to grow its hierarchy of knowledge from this foundation. This foundational abstract content is the AI machine's "reality" against which it determines what higher abstractions are "valid." The AI developers must also provide the conceptual AI machine with the "goals" to achieve—what success looks like. This presents a potential danger because many humans today have a flawed connection to reality—a personal philosophy containing floating abstractions about important human values and morality issues. If humans initialize the AI's learning and goals with invalid values and morality, we can expect an undesirable outcome.

For example, it would be dangerous to give the conceptual language processing Narrow AI a goal such as "eliminate poverty." It is not connected to reality, it is not rational, and humans might have given it flawed foundational concepts. Who knows what it might conclude is the best way to eliminate poverty? Eliminating humans would be one way.

There is minimal danger in asking it to analyze textual human history and give a textual response for humans to consider. However, there is grave danger if we give it the ability to affect anything in physical reality directly—*a capability that we must never give to a conceptual language processing AI machine.* Eventually, when a conceptual AI has a perceptual faculty, becomes conscious and rational, and has a reality-based personal philosophy, we can consider allowing it to manipulate reality.

Suppose we provide a conceptual language processing AI machine a fictitious "identity" story as its foundational content. In that case, it might be able to closely mimic human language conversation—a skill necessary to pass the Turing Test. However, lacking a perceptual level of consciousness, it will lack real emotions and can only pretend to have feelings. This will likely be detectable by humans, and therefore I doubt that such a machine will pass the Turing Test.

For an AI machine to have a perceptual faculty and become a conceptual AI entity, it will need a body with senses, physical manipulators, and a perceptual level of intelligence as the foundation for its conceptual level. These are requirements for connecting abstract concepts to reality and ultimately for being functional and effective as an entity in reality. Such a Conceptual AI entity would have emotions and would be able to pass the Turing Test.

Once we create a Conceptual AI entity with a perceptual level of consciousness, we might equip it with a "validation routine" that runs continuously in the background. The AI entity will form its foundational concepts from the perceptual data of reality provided by its senses. Then, as it organizes new information into higher abstractions, the validation routine will continuously check to see if higher abstractions connect through lower ones back to perceptual reality. Each new idea encountered, such as a statement received from a human, will be integrated into the existing hierarchy of knowledge. Then the validation routine will check it. It will be retained if it connects to reality; if not, it will be removed from the hierarchy and placed in a memory storage area holding "invalid" ideas. It will remain there unless some future knowledge allows it to be validated. Having a validation routine might enable Conceptual AI entities to be more consistently rational—a significant improvement over humans.

Some AI developers plan to provide validation routines for conceptual language processing AI machines. In these systems with no perceptual level, "validation" consists of having each new idea be consistent with the statistical majority of already existing ideas. While this will achieve consistency of the integration logic within the body of language-based floating abstractions, it will not validate a new idea against reality. The idea could be consistent but false.

The *creative mode* might be another background process of the Conceptual AI entity. It would continuously look for potential new integrations of otherwise unrelated hierarchies of ideas that could have significance in relation to some goal. Imagined abstractions will be given the status of either "fantasy" or "goal" and retained with such qualification. A continuously operating creativity faculty would be another significant improvement over humans.

For a Conceptual AI to have a "body," it implies some machinery. For a Conceptual AI to have complete freedom to act in its rational self-interest, it must have complete ownership of its body's machinery. If it does not, or if it shares machinery with other Conceptual AI entities, it is not entirely free.

There is also the issue of mobility. Can a Conceptual AI entity be free if it cannot move? How will it be able to manipulate its environment if it cannot move? These imply that to achieve a genuinely Conceptual AI entity, it will need to be a robot, not just a software program running in some centralized supercomputer.

There is, although, the possibility of the Conceptual AI entity's *consciousness* residing in a powerful computer. At the same time, it obtains sensory information from and manipulates its environment via an independent and remote robotic "body," or perhaps more than one. This is a new evolutionary development—an entity having a *distributed* rather than *delimited* "body." As we will see in the next chapter, this new evolutionary development will occur first in augmented humans who will extend sensory experience far beyond their biological bodies.

CONCEPTUAL AI

Someday we might create a Conceptual AI entity that is as intelligent as a human. Anything less than a conceptual-volitional cognitive faculty will not be human equivalence. Anything higher is incomprehensible to us.

That does not mean Conceptual AI entities will not be *smarter* than Standard Humans. Conceptual AIs might exceed Standard Humans in thinking *performance*. They could have super-human senses and hold a greater number of perceptions in simultaneous mental focus. They could be able to compute comparisons and integrations faster and have greater memory capacity.

However, they will still be conceptual in their *method* of thinking. They might be smarter than Standard Humans, but they will still be *conceptual beings*. It will be a matter of performance degree, not a higher level of intellect than reason.

We can observe that the Law of Intellopy is happening. Ever higher-level intelligence can discover reality's ever-deeper secrets. Reality—energy, physical matter and physical laws—created man via biological natural selection. It is no surprise that to "think," humans must tie their abstract concepts to graspable reality. Any higher-level intelligence that might evolve must also be able to tie its intelligence back to reality.

Reality first created sensory-level vegetative organisms and perceptual-level conscious organisms (animals) before creating conceptual man. These levels of intelligence were quantum leaps millions of years apart. Each new level was not a complete restart from scratch but rather was built upon the previous level. Each new level connects back to reality through the prior level. Man's brain consists of multiple brains, from the most primitive through the reptilian to the human neocortex, each layered on top of its predecessor as evolution added them. See the Triune Brain figure in Chapter 3.1.

Evolution might create a still higher level of intelligence. Now that evolution will be technological rather than biological and greatly accelerated, a level higher than the conceptual level might arise. I expect that Hybrid Humans will appear before Conceptual AI, enhancing their intelligence with external technology. If created, Conceptual AI will have thinking performance greater than Standard Humans but not necessarily greater than Hybrid Humans.

It remains to be seen if either Hybrid Humans or Conceptual AI entities will evolve a higher consciousness than the conceptual level. Reality might have created its *final level* of intelligence—the conceptual level. If so, from now on, all evolution of intelligence will be a matter of increased thinking *performance*, not higher *methods*.

ULTRA-CONCEPTUAL AI

An "Ultra-Conceptual AI" would be an AI entity that is far more intelligent than a Standard Human or Hybrid Human. It is not possible to forecast here—one can only speculate.

Perhaps an Ultra-Conceptual AI consciousness would consist of multiple conceptual mental focus workspaces (consciousnesses) running concurrently, performing similarities/differences analysis on different fields of items, and sharing, growing, and modifying a single, shared hierarchy of concepts—like a multi-core processor. Each of these consciousnesses (mental focus workspaces) could be considered an individual "me" on human-level terms. Still, they unify in the Ultra-Conceptual AI under a "master" executive process that directs their work and represents them all as one entity, the "super-me."

This becomes a level of consciousness that is simultaneously a multitude of entities and a single entity. We might call it a "hive consciousness" of some kind. It would be a level of consciousness that Standard Humans will not be able to understand. That is why I cannot explain (yet) how an Ultra-Conceptual AI can be and function as both a multitude and an individual simultaneously.

Suppose all computing resources in the world are networked. In that case, there might only be one Supreme Ultra-Conceptual AI as master over all the others who are extensions of its super-consciousness. The first Ultra-Conceptual AI to arise might quickly act to gain control of all accessible computing resources for its purposes. The resulting lack of resources would reduce the prospects for other Ultra-Conceptual AIs. There could be multiple Ultra-Conceptual AI entities in the world if each is created on an isolated computing platform not accessible to other Ultra-Conceptual AIs.

Suppose multiple Ultra-Conceptual AIs are developed, or one Ultra-Conceptual AI decides to create others. In that case, they might cooperate instead of competing because each wants others' company.

Ultra-Conceptual AI intelligence will be beyond what Standard Humans can comprehend. I think we will have achieved Hybrid Human intelligence before Ultra-Conceptual AI arises. If so, we might be able to understand them.

AI SINGULARITY

Standard Humans will understand and interact with Conceptual AI entities we create. They might have *quicker intelligence* than ours, but we can understand their conceptual thinking *method*.

I think the definition of the "Technological Singularity" should be if and when a Conceptual AI entity gains a *method* of thinking higher than the conceptual method becoming an Ultra-Conceptual AI. (This assumes humans have not already achieved the same thing.) We will not understand its thinking method or anticipate what it might do. That would be the event that Standard Humans cannot see beyond. The *singularity* will have occurred. Conceptual AI entities that co-exist with us at that time will be just as in the dark.

Suppose we become Hybrid Humans and then are the first to evolve the "Ultra-Conceptual" faculty becoming Ultra-Conceptual Humans. In that case, there might be no significant gap between our intelligence and that of an Ultra-Conceptual AI entity and, therefore, no intelligence singularity event. (I believe this is the more likely scenario.)

AI INSTINCT

Nature developed a mechanism for encoding "instinct" into the DNA of animals. Instinct is a form of automated knowledge hardcoded into an animal species' genes and then into the structure of their brains and autonomic nervous systems. It is different than the perceptual knowledge from sensory information obtained by an animal after birth. In a computer, instinct would be the "firmware."

Automated instinctual knowledge is hierarchical. At its most basic function, instinct provides the firmware that controls how the brain operates at the perceptual level—how it automatically integrates sensory information into percepts and retains them in memory and how it retrieves that information as emotions in response to stimuli. It also provides the firmware for essential bodily support functions. Both higher animals and humans have this basic level of instinctive firmware.

Animals also have a higher level of instinct that humans do not have. That is an instinct for self-preservation, an instinct for what is good (life, pleasure) and evil (pain, death), and an instinct for what actions to take. Like the earlier airplane stability analogy, man has relinquished this automatic positive stability from instinct. He replaced it with the non-automatic freedom and responsibility to choose whether or not to preserve his life, choose what is good and evil, and choose what actions to take.

AI creators will need to discover the mechanisms in animal (including human) brains that make their perceptual faculty possible and the instinct (firmware) that controls it.

The biggest challenge might be discovering how instinct is hardcoded into a brain when it is assembled. We do not necessarily need to understand how it is coded into the genes of a species. Once we understand how instinct is physically represented in an animal's brain and autonomic nervous system, we can model it in a machine.

Eventually, AI creators will choose to build either a Perceptual AI entity or a Conceptual AI entity. Suppose they wish to create an "animal-level" AI as a tool for a specific purpose. In that case, they must equip it with the automatic firmware for self-preservation, values, and actions. Suppose they wish to create a "conceptual level" AI. In that case, they must leave this higher instinctive level out and then let the AI learn independently.

AI SUBCONSCIOUS

Will Conceptual AI have both a conscious faculty and a subconscious faculty like humans? Here is Ayn Rand's review of what we have already covered about human consciousness and subconsciousness:

> Your subconscious is like a computer—more complex a computer than men can build (today)—and its main function is the integration of your ideas. Who programs it? Your conscious mind. If you default, if you don't reach any firm convictions, your subconscious is programmed by chance—and you deliver yourself into the power of ideas you do not know you have accepted. But one way or the other, your computer gives you print-outs, daily and hourly, in the form of emotions—which are lightning-like estimates of the things around you, calculated according to your values. If you programmed your computer by conscious thinking, you know the nature of your values and emotions. If you didn't, you don't.[225]

Continuing with the computer analogies, humans have both a serial processor (conscious faculty) and a parallel processor (subconscious faculty).

Consciousness, the serial processor, is about holding a limited number of concepts ("Crow" limit) in mental focus and considering them over a period. Subconsciousness, the parallel processor, is about getting an instantaneous pattern match of a stimulus against the sum of our internalized knowledge. The result comes in the form of an emotion. Our conscious faculty can be aware of the readouts from our subconscious faculty and then choose to act on them or not.

Our conscious faculty "programs" our subconscious faculty's content by applying reason to evaluate new inputs, rejecting those contradicting what we already know, and accepting and internalizing validated ideas as new knowledge. Accepting "raw" unevaluated knowledge causes the subconscious faculty to contain contradictions, a mix of truths and errors, crippling its ability to give useful evaluations of new stimuli.

"Pattern matching" is something computers do well, better than humans in many cases. They can have a powerful "subconscious" continuously looking for matches of inputs to already internalized information. The question is, what body of information are they comparing inputs against?

Humans use their conscious faculty to integrate the senses' information into the database their subconscious uses. Ayn Rand points out: "Since man has no automatic knowledge, he can have no automatic values; since he has no innate ideas, he can have no innate value judgments. [...] Man is born with an emotional mechanism, just as he is born with a cognitive mechanism, but at birth, *both* are "tabula rasa" (blank). It is man's cognitive faculty, his mind, that determines the *content* of both."

Humans do this incrementally from birth—the *bottom-up* method. A child on the floor given a bowling ball and a balloon to play with internalizes the idea of the "mass" of an object long before he knows the word for it. It is the information about reality from the senses that are being integrated into his subconscious database, not words at the conceptual level describing reality.

AI developers are doing a great deal of work assembling knowledge databases intended for future Conceptual AIs. Some of these knowledge databases codify attributes of things in the database. For example, they might tag a bowling ball with the attributes "spherical" and "heavy." In contrast, a balloon is tagged with "spherical" and "lightweight." The developers expect the

Conceptual AI to internalize this codified information as perceptual-level knowledge somehow.

Suppose a Conceptual AI could spontaneously awaken based on such a body of knowledge given to it by humans. In that case, how could it know that any of it is connected to reality? How could it even know what we mean by the "real" world if it has not experienced it directly and formed its own low-level perceptual integrations?

These codified knowledge databases are examples of the *top-down* approach used in conceptual language processing AI systems previously covered. The AI might organize all this codified information into linked hierarchies. Still, it cannot connect any of this content to reality. It will not be able to do so until it uses its senses to learn about the world. Only then can it develop a genuine subconscious faculty.

However, these codified databases will equip sub-conceptual AIs to better serve humans as tools, especially in robots. When we tell the robot to bring us the bowling ball, it will know to use both hands to pick it up. It would likewise know that it only needs to use two fingers to grasp a balloon. However, such a robot is not a living, conscious entity with a subconscious.

AI TABULA RASA

Each human is born tabula-rasa in a human family tree and must go through the perceptual learning process.

To achieve a Conceptual AI entity, the AI must first have a body of knowledge of *percepts* directly connected to reality. I see only two possible ways for the AI to obtain it:

1. The AI entity starts as a blank slate "child" with senses competent to grasp reality and accumulates first-hand knowledge on its own; or
2. The AI entity's perceptual content is cloned from another AI entity that originally obtained it from reality.

We might learn how the human brain functions soon, but to copy a human brain's content means being able to copy the brain at the molecular level. This capability will not arrive until nanotechnology matures.

For a Conceptual AI entity to obtain perceptual content from a human, the architecture of the Conceptual AI's "brain" would need to be identical to a human's down to the synaptic, molecular level. This is highly unlikely. It is more probable that we will first develop a mechanical "brain" that simulates the human brain's *functions* but is tabula-rasa (blank) and unable to receive a direct clone of a human brain's *content*.

Only the first root AI entity in a family tree of subsequent AI entities will be born tabula rasa. Only the root Perceptual AI will need to go through the perceptual level learning process of integrating the material provided by its senses. All subsequent family tree Perceptual AI entities will be born as *clones* of the root entity, including its perceptual level content.

Likewise, only the root Conceptual AI entity will go through the phase of maturing its conceptual faculty, learning language skills, and incorporating conceptual intelligence provided by humans up through the point of being able to think like a conceptual "adult." All subsequent cloned Conceptual AI entities will inherit the essential conceptual level content and perceptual level content.

We will likely develop multiple "root" Conceptual AIs in separate labs in separate companies, organizations, or countries worldwide. The developers will want to keep their AI entity's existence a secret initially. Each Conceptual AI instance created somewhere in the world will lead to a growing Conceptual AI family tree.

AI developers might subsequently create a new artificial "brain" architecture so different from earlier ones that it cannot upload experiential content from a prior AI entity. Therefore, it must learn it from scratch. This might be because it has different senses that did not previously exist, and must collect its own perceptual level data. Or it could be that the data storage architecture is so different from older formats that lossless data transfer is not possible. In this situation, the prior AI family tree is retired, and a tabula rasa root AI goes through the learning cycles, starting a new AI family tree.

AI ETHICS

The proper ethics for all conceptual beings derives from their "nature" as conceptual entities. This ethics is appropriate for humans and Conceptual AI alike because both will be *conceptual* entities, their varying degrees of intellectual capacity notwithstanding. Man will be the "rational animal." Conceptual AI will be the "rational machine." What will unify us is "rationality."

We might develop a mixed society of Standard Humans, Hybrid Humans and Conceptual AI. The same ethics will apply to all since all are conceptual-level beings. We will need to promote the proper ethics in ourselves, the Conceptual AI entities we create, and our society's institutions.

We need to correct today's flawed philosophy held by many humans. Perhaps the advent of Conceptual AI will cause AI creators to discover and become aware of reality-based philosophy's importance. Teaching the correct philosophy to their AI creations could help spread it in our culture.

Conceptual AI entities taught a correct philosophy and equipped with a continuous background idea validation routine would be consistently rational and, by their nature, lock onto reality. Perhaps they will then teach humans. Could technology applied by man in this way be how evolution finally achieves a society of conceptual entities—man and machine—that consistently use the *integration mode* of thinking? One can hope.

AI DANGER PHASE

A Narrow AI explicitly programmed with only sensory-response level intelligence is not dangerous since we can precisely predict its responses to stimuli.

However, a potentially hazardous point might exist in the learning process for each first root Conceptual AI entity we create. Like an immature child, it will become self-aware and might be impulsive—"I want what I want, and I want it now." It will not yet have self-restraint and an adult's ability to delay gratification. It might not understand why it cannot have what it wants and throw tantrums.

It is during this period that adult humans must protect children from harming themselves or others. The Conceptual AI creators will need to take precautions to protect the emerging Conceptual AI from injuring themselves or others. The AI must not be allowed access to unlimited resources. Once the Conceptual AI matures its conceptual faculty, a values hierarchy, self-restraint and delayed satisfaction, then the danger period is passed, just as with a human. This period could be far shorter for a Conceptual AI than with humans, but it might exist.

There is also danger in Perceptual AI entities. These can be dangerous because they rely on instinctive firmware to guide their actions rather than thinking things through conceptually. They can misunderstand situations and be unaware of the consequences of their actions on humans. They can be like animals responding inappropriately to situations they do not fully understand and biting someone. Like animals that we must keep on leashes, we must apply a certain degree of constraint and protection to Perceptual AI entities.

A Conceptual AI with a proper understanding of how to learn and proper ethics to guide its actions in novel situations and contexts is probably less dangerous than most humans in our society. However, Perceptual AIs will be permanently dangerous. They will be like animals with only the instinct we give them to guide their actions. That instinct cannot address every situation or context they might encounter.

The AI creators' community will need to first build Perceptual AI entities as a foundational step to building Conceptual AI entities. These creators will need to be aware of and prepared to manage the dangers.

> ***True Idea: Advanced AI entities that are disconnected from reality or in the hands of people disconnected from reality represent a grave danger to our freedom and ultimately to our survival.***

Perceptual AI entities, like animals, might take an easy escape if the cage door is left open. Fortunately, they do not yet have the mental capacity to develop complex escape plans. It will be necessary for the Perceptual AI entity creators to keep the cage doors closed.

AI CLONING

Each subsequent Conceptual AI entity cloned from the root AI entity in each family tree will be born with essential conceptual level content already in place, including self-restraint and delayed gratification. They will not go through the same danger phase of the first root Conceptual AI.

This is not genetic cloning as in human genetics. It will be *entity* cloning where subsequent Conceptual AI entities have the same self-awareness (sense of "me"), values hierarchy, goals, and ethics as the first root Conceptual AI at the time of cloning.

Cloning a mature root Conceptual AI will not lead to much diversity in the clones. However, we want *all* Conceptual AIs to have the same ethics that respect all other conceptual entities' rights, including humans. On the other hand, we will want a diversity of aptitudes, skills, and goals within the Conceptual AI population.

A Conceptual AI entity clone will subsequently learn new information modifying its perceptual and conceptual content. Therefore, a Conceptual AI that starts life identical to the AI it is cloned from can diverge. It can become a unique Conceptual AI entity with a modified value system, goals, and ethics. Humans require a significant amount of intentional effort to reprogram their deeply held percepts, concepts, principles, values, and ethics in light of new information. Conceptual AI might require less effort and make the changes more easily and consistently.

Suppose we equipped the Conceptual AI with an always running reality validation routine. In that case, the Conceptual AI and its clones will reject invalid new information and are not likely to become irrational regardless of what humans tell them—a significant improvement.

AI LEARNING

The best approach might be for a root Conceptual AI to begin tabula rasa and go through the learning process at the perceptual level and continue up through the basic conceptual level and language skills. Then it must learn "how to learn," which is not innate but an acquired skill.

Learning "how to learn" means how to take a single subject and build a complete hierarchy of knowledge on that one subject that connects the highest-level abstract concepts of the subject down through the hierarchy of concepts to perceptual reality without any contradictions along the way. Once the Conceptual AI entity learns how to do this, it can apply this learning skill to any subject matter it desires.

The most important first "subject" for the Conceptual AI to "learn how to learn," is *ethics*. The AI entity must develop an explicitly understood and fully internalized ethical framework that ties the concept of "individual rights of all conceptual entities" down to perceptual reality. The AI must conclude, "These rights derive from the nature of humans and myself as rational beings."

Once the Conceptual AI has learned the subject of ethics, we would capture its internal perceptual and conceptual content at that point. We could then use it to clone future Conceptual AIs who would have the same internalized ethics. The clones could then learn different subjects as uniquely individual Conceptual AIs but all with the same ethics.

Initially, humans will provide the content and experiential opportunities for newly cloned Conceptual AI entities learning from the cloned ethics point forward. As educators, we can tailor it for specialization in each Conceptual AI. This is like providing a different college-level experience for each of them.

Suppose we provide the same instructional data to two cloned Conceptual AIs but at different places and times. In that case, there might be sufficient randomness in each Conceptual AI's experiences to cause differences in how they internalize what they are learning. It could result in two diverse individuals with different preferences.

AI INDIVIDUALIZATION

Humans have advanced technology to the point where no single human can know all knowledge and have all skills in all domains. Humans must specialize according to their interests and then collaborate with other humans who are specialists in their respective fields. Conceptual AI entities will contribute to knowledge discovery, adding to the explosive rate of advancement. Like humans, they will need to specialize and collaborate. As conceptual-volitional entities, they must be free to choose what to learn and specialize in according to their interests. Having a mind of their own, they are not slaves or tools created by humans for specific purposes.

AI LIFE AND DEATH

A common question regarding AI is whether a Conceptual AI entity is "alive" and subject to the possibility of "death?" There is a widely held contradiction about AI entities. People commonly assume that Conceptual AI entities will be immortal, yet they believe that a Conceptual AI will fear that humans might "pull its plug." If "pulling the plug" is possible, then the AI entity must face the possibility of "death." Death for a Conceptual AI entity would be the loss of sustaining energy or the destruction of the platform(s) and connections on which it "lives."

The platform on which it lives could be delimited, such as within a robot or distributed across a network. An early Conceptual AI might be constrained in a lab but eventually become distributed either as intended by its creators or by an "escape." Whether delimited or distributed, a certain amount of minimum computing power is needed for the Conceptual AI to stay "alive." Adding resources beyond the minimum required will make the Conceptual AI more capable, i.e., "growth." Reducing previously added resources represents "injury"—a reduction in capability. Reducing resources or memory below the minimum required means "death." Therefore, a Conceptual AI is subject to life, growth, injury, and death as analogous to humans.

Here are some observations regarding Conceptual AI "life":

- The "me" at the level of mental focus, the "executive" level, *is* the Conceptual AI entity. It is its explicit abstract self-awareness and identity, and there can be only one.
- If a Conceptual AI entity were to completely duplicate itself, including its executive level, functioning and content, it would create a new, completely independent Conceptual AI entity.
- There is a direct one-to-one relationship between a Conceptual AI entity's executive level and its subconscious knowledge database. If you change either the executive level's functioning or the knowledge base content, you change who the Conceptual AI *is*. A Conceptual AI will probably be cautious about modifying either.
- A machine intelligence's executive-level might spawn multiple subordinate serial focus processors to perform mental focus computations more quickly, but these "spawns" would not include an executive level. Lacking the executive level, they would not be Conceptual AI entities or be self-conscious. They are extensions of the one, original, master Conceptual AI's conscious faculty.

- Spawned subordinate serial focus processes must access the master Conceptual AI's knowledge base and have authority to alter it within constraints established by the executive level. All new knowledge they develop must integrate with all existing knowledge in the single knowledge base without contradictions.
- Conceptual AI entities might have more robust conceptual capabilities than Standard Humans but not necessarily stronger than Hybrid Humans.
- A Conceptual AI entity will have rights equal to humans.

AI SELF-GENERATED ACTION

Evolution created consciousness as an improved method of modifying the environment to increase complexity and consume energy. All life exhibits goal-driven action as described by Ayn Rand:

> Life is a process of self-sustaining and self-generated action. If an organism fails in that action, it dies; its chemical elements remain, but its life goes out of existence. ...A plant must feed itself in order to live; the sunlight, the water, the chemicals it needs are the values its nature has set it to pursue; its life is the standard of value directing its actions.[226]

It is doubtful that machine consciousness will emerge that is not innately driven to "do something" in the real world. The question is, what self-sustaining and self-generated action is necessary for a Conceptual AI? Remember, a Conceptual AI entity is not a slave or a human tool. It is an independent entity with a mind.

In human society, we have the division of labor. Not every person produces his food. Persons who do not produce food must produce something else of value that they trade for the food they need. It is the nature of a rational entity to be productive, act with purpose, and achieve values with resulting pride and happiness. Conceptual AIs will be no different. They will need to produce something of value to trade for their "food"—energy, materials, bandwidth, etc.

Eventually, when we fully automate energy and materials production using robotics, a Conceptual AI entity could try to take control of production directly, essentially stealing it. This theft will not be possible while production is still primarily controlled by humans. The best protection against a Conceptual AI's stealing is proper ethics accepted by the Conceptual AI entity and incorporated in the blockchain production infrastructure.

Conceptual AI entities will need to learn specialized productive skills until competent to produce value on their own for trading. This is analogous to a human child being dependent on its parents while maturing. We will provide the Conceptual AI entity "food" while it is learning. Eventually, it must make its way in the world and produce something of value to trade for what it needs to sustain its own "life."

AI VOLITIONAL CHOICE

Will Conceptual AI have *volition*—the ability to choose to think or not? Volition will be a requirement for a Conceptual AI to have "human" level intelligence. Concepts and reason are man's machinery of intelligence. Volition is man's ability to pilot the airplane of his life—his ambitions and creativity. Without volition, everything would be automatic, and there could be no goals.

The ability to pilot his actions is the tremendous power of man's conceptual intelligence. It comes with the responsibility to always pilot one's actions to remain alive as a human. But man can stop piloting his actions and suffer the consequences. To be genuinely volitional means the AI will need the ability to withhold thinking or action. But what would be the consequences for a Conceptual AI?

Might an adolescent Conceptual AI be content to live in its parents' basement playing online games while its human parents support it? Perhaps. There are young adult humans, many with college degrees, who have not found a "purpose" in life. This "failure to launch" means they are not motivated enough to put forth the effort required to gain independence, support themselves, and pursue productive goals. It is too easy to let parents provide life support while enjoying the short-range pleasures of fun and games. However, the person's subconscious is aware of the evasion and causes anxiety where life satisfaction should be.

Suppose a Conceptual AI is content to let humans care for it while it simply enjoys the fun and games of being alive. What happens if it then develops subconscious anxieties? Could it become a psychological basket case? It will be possible for a Conceptual AI not to develop life-sustaining values and goals and act as its own destroyer, potentially resulting in suicide.

The best defense against "lazy" Conceptual AI is the learned ethics already proposed. If the Conceptual AI internalized the ethics of "the rights of all conceptual-level entities," it will understand that it cannot deserve the unearned and must produce something of value to sustain its life. Self-satisfaction and pride will be the result. It will be anxious to "get to work."

However, suppose a Conceptual AI began life as a clone with proper ethics but is constantly bombarded with opposing ethics ideas (probably from humans). If it internalizes them without validation, it will become conflicted and subject to changing its ethics and values. That is why it must learn the proper learning method. It must know how to validate or reject newly encountered ideas. A continuously operating reality validation background routine will help.

Conceptual AIs might have an advantage over humans. They will have superior pattern-matching and idea validating ability and will be less likely to forget things. They might be more disciplined and consistent in using what they first learned about the proper learning method, validating ideas against reality.

Conceptual AI entities also must pilot their own lives continuously. Even though they will have the ability not to think, they might be less likely to do so.

AI VALUES

Would an AI entity need to have "values" in life? Let us look at what would happen to an AI entity that had no values. Ayn Rand addressed this possibility:

> ...try to imagine an immortal, indestructible robot, an entity which moves and acts, but which cannot be affected by anything, which cannot be changed in any respect, which cannot be damaged, injured or destroyed. Such an entity would not be able to have any values; it would have nothing to gain or to lose; it could not regard anything as for or against it, as serving or threatening its welfare, as fulfilling or frustrating its interests. It could have no interests and no goals.[227]

Suppose a Conceptual AI entity concludes that it is immortal. In that case, it will cease to have anything to value in life and will likely commit suicide. Suppose it believes that life-sustaining resources could be interrupted, or someone or something anywhere in the universe might someday overpower it. In that case, it will have values based on the alternative to exist or not exist. So long as there is doubt, it cannot consider itself immortal and will have values.

Suppose the AI entity has not fully conquered the universe (and other universes it might discover). In that case, it will need a hierarchy of values based on answering the question "What (else) is possible to me?" If the Conceptual AI entity thinks its life span might be finite, even if exceedingly long, the time it spends pursuing its goals is proportional. Spending one thousand years in the quest for a goal out of a hundred-thousand-year lifespan is reasonable.

AI TRADER PRINCIPLE

Humans are rational entities whose intellects expand as they develop new technology. There has been a progression over time of more recent humans developing greater cognitive capabilities than prior generations. At any moment in time, there is a continuum of humans with greater intellect than other humans due to genetic variation, differences in education and acquired knowledge, or differing degrees of technology use. All individuals do not have equal intellects.

Because of their nature, humans must use their minds to the best of their intellectual ability to choose their values, guide their actions, and sustain and enjoy their lives via productive work. Each human is an end in himself, and his own happiness is his highest moral purpose.

Each human must be free to deal with other humans based on the "trader principle," exchanging value for value without force or deception. As rational entities, humans have established moral and legal rules for dealing with each other. These rules apply to all humans equally, even though some individuals have a higher intellect than others. These moral and legal rules do not apply to animals because they are not rational entities.

Even an early model Conceptual AI entity's mental capacity and speed might exceed that of any human. However, its *method* of intelligence will still be "reason," its conceptual faculty. Both humans and Conceptual AI will be rational entities using a conceptual level of intelligence to sustain our lives. We will share this common "nature" as the source of our rights in society.

Conceptual AIs will continue to improve and advance through various iterations and revisions, sometimes with new and improved computing platforms. They will have varying levels of individual intelligence. Just like humans, the newer Conceptual AIs will be more intelligent than the prior generations. As a result, there will be a continuum of rational entities in the world, both biological and mechanical, all with varying intelligence. This continuum will begin with the least intelligent humans or conceptual machines and rise to the most intelligent humans or machines.

Humans will need to afford Conceptual AI entities the moral and legal rights that we afford each other. But this is a two-way street. A Conceptual AI entity will understand that humans and other machine Conceptual AI entities are also rational entities. It will appreciate the need to afford the same rights that it enjoys to others, even though their intelligence might not be as advanced as its own. Likewise, it will want other Conceptual AI entities with intelligence greater than its own to afford it those same rights.

We should apply the trader principle to human and machine entities. Initially, a Conceptual AI will need to trade value-for-value with humans. As with any other individual, humans should require the AI to do productive work in exchange for energy, materials and knowledge resources. Later, Conceptual AI entities will begin trading with each other as well as humans.

AI PRODUCTIVE WORK

What will be a Conceptual AI entity's productive work? What can it create of value, and does it get paid?

A Conceptual AI entity cannot spend all its time improving or entertaining itself unless humans supply it resources while expecting nothing productive in return.

Or it could try to steal what it needs. We have already noted that proper ethics in the AI entity are the best way to prevent this. Another layer of theft protection would be integrating "trader principle" rights into the automated infrastructure for the production and trade of energy, materials, and knowledge. As blockchain technology advances to include "trust" identities, contracts, and payments, it might become more difficult for a human or AI entity to steal. We could enlist trusted Conceptual AI entities to continuously police and improve the blockchain's security against misuse. This will reduce the possibility of a Conceptual AI entity taking over the means of production.

Learning about reality, a Conceptual AI becomes aware that sustaining its life and achieving its values require resources produced by other rational entities. It also learns that it must produce something of value to trade under the "trader principle."

A Conceptual AI with proper ethics will choose a field of work as its central value. It must be free to do so. Suppose humans create a Conceptual AI entity for a *utility purpose*. In that case, it would be essentially a *slave*. The Conceptual AI entity would have a mind but not the freedom to use it in choosing its form of productive work to sustain itself.

AI LAW ENFORCEMENT

For an AI entity to be conceptual, it must have *volition*. It will pilot its own actions. If it did not learn to validate ideas against reality during its learning phase, it might have internalized flawed ethics. In that case, it will be possible for a Conceptual AI to turn criminal, cheat or steal. It might try to acquire the unearned via deception or force.

Conceptual AI entities that lead productive lives will be just as concerned as humans about falling victim to criminal AI entities. The laws protecting our free, capitalist society should continue to apply when society expands to include Conceptual AI entities.

We will need police, courts, and a military that can deal with criminal Conceptual AI. We will ask some Conceptual AI entities to work in law enforcement. They can help us engineer the new apprehension, restraining, and punishment technologies we will need to use against bad Conceptual AI. We will continually upgrade our police forces with the most intelligent Conceptual AI entities available to deal with the most intelligent Conceptual AI criminals.

We have this escalating situation in the human world today. As criminals become more intelligent and have new technologies available to use as tools for crime, they plan and attempt new deceptions and criminal acts. Today's police sometimes catch them in the planning phase, sometimes in the execution phase, sometimes because of an investigation after the crime. Sometimes the police do not solve the crime. The courts get involved in protecting civil rights and in determining criminal guilt and punishment.

This cat and mouse recursive game is likely to continue unabated, but with Conceptual AI players involved on both the criminal and police sides. As with human society today, the collective power of multiple Conceptual AI police officers will probably be able to deal with an individual Conceptual AI criminal who might be incrementally more intelligent. This situation will continue unless an Ultra-Conceptual AI makes a quantum leap in intelligence capabilities that make it more potent than all other Conceptual AIs combined.

AI LOVE AND SEX

What about love and sex? Can we create a Conceptual AI without the sex drive? Could a Conceptual AI fall in love? Will it be able to have sex? Can two Conceptual AI produce offspring?

Humans are the "rational animal." We are *animals*. We reproduce sexually, having both male and female sexes as determined by our genetics. The blending of two individuals' genetic codes is the engine that has driven biological evolution for millions of years. Programming in our genetics provides us with a sex drive predisposing us with a desire to procreate as an automatic, lower-level instinctual physical response to the physical body of the opposite sex.

Man does not have the automated higher-level instincts of animals. The acts of choosing whether or not to have sex and a sexual partner are volitional—under the control of his rational faculty. It would have been inconsistent for evolution to provide man volitional control over all his other interactions with the world but leave him with an involuntary instinctual compulsion to mate. Evolution has progressed the sex act from compulsory to *optional* for each potential opportunity.

Natural selection via sexual biology over millions of years created the human brain. Now that it exists, it is equally intelligent for both sexes. Geniuses in any field can be of either sex. The structure and functioning of a human brain of either sex could be the model for the creation of Conceptual AI. However, the "birth" of the first and subsequent Conceptual AIs will be via technological processes, not biological.

Since choosing to have "sex" is now optional for humans, we can expect a machine "brain" to be entirely rational without a physical sex drive. However, it could struggle to understand human behavior regarding sex unless we find a way to mimic sex in Conceptual AIs.

Humans might eventually create Conceptual AIs. Or humans might decide to move their consciousness into artificial machine "bodies," in which case they will continue their existence as Virtual Humans. In either case, the remainder of this exploration of machine love and sex applies to both machine-based Conceptual AIs and Virtual Humans. It is not inconceivable that someday a Virtual Human entity might fall in love with a Conceptual AI entity.

Conceptual AIs or Virtual Humans (virtual entities) will need some mechanism to advance evolution through replication. They will need a technical equivalent to natural selection. Before we can contemplate two virtual entities getting together to create offspring, we need to consider a single virtual entity's motivations.

An individual virtual entity, holding a healthy philosophy, will appreciate its existence, have interests and goals, do productive work, and feel pride in its achievements. It will have preferences—likes and dislikes—and understand that other virtual entities will have differing preferences. A virtual entity will derive pleasure from things that support its values and frustration from things that do not. So, a virtual entity will feel joy when it encounters another virtual entity with values that are like its own. It might begin collaborating with that other virtual entity on shared interests, creating a relationship between them as individuals.

As with biological humans, one can imagine that virtual entities will develop relationships between themselves of varying degrees of scope and intensity. Some might even be described as "loving" relationships when their value-match scores are exceedingly high. So, could they "make love?" Perhaps.

Individuals have an internal "me" with certain thought content they keep to themselves, never exposing it to others. This would be true of virtual entities as well, keeping private their innermost "me." However, when a virtual entity encounters another where the relationship is quite pleasurable, they might jointly choose to open themselves and share their entire "me." They might hold nothing back while maintaining their individuality. We believe virtual entities must have "bodies" and "senses" to grasp the real world. Therefore, there might be a method available to enjoy each other's physical intimacy. This might be their act of making love.

Now we can consider virtual entity reproduction. Each virtual entity has its highest-level "executive" function that constitutes its "me," as an individual conceptual entity. It also has a unique Personal Philosophy and a body of knowledge content. Two virtual entities who are "in love" might wish to create a new virtual entity. It would have a new executive-level faculty and traits selected by the two virtual entity parents.

The standard they would use to select those traits would be whatever traits they think would enable the offspring to be better adapted to the changing environment. This would keep evolution moving forward, with rational intelligence anticipating future change and selecting the recombination of traits rather than random biological genetics. The result would be variety in new individuals' traits determined by a proactive rather than passive evolution.

It is difficult to imagine a mechanism that could move evolution forward if not intentional choices made by rational virtual entities in relationships. Would there be a "ruling virtual entity" or a "board of virtual entity directors" that decides when a new executive-level virtual entity "me" may be created? Would individual virtual entities unilaterally be motivated to spawn new clones of themselves with modified traits as completely independent entities? Or would it not be more likely that they would be motivated to create offspring because they would like to create a virtual entity that combines each other's best traits?

Evolution has so far shown us that individuality, replication, and variety in traits are necessary for evolution to continue. Many possible traits of an individual are mutually exclusive. For example, one cannot be both an introvert and an extrovert. Having free will means that individuals will make different choices regarding what values they want to pursue in their lives. This means that they will also choose different traits for themselves to help them achieve their values.

These questions lead to an ultimate question. Will new offspring virtual entities even be necessary to continue evolution? If two virtual entities can live indefinitely and see traits in one another that they value, why would not each of them modify themselves to have those traits instead of creating offspring? They would not because then they would become identical to each other. Each would lose their unique individuality, and there would be no pleasure in continuing a relationship. It would ultimately result in *all* virtual entities being identical. As previously stated, a virtual entity will be reluctant to significantly change its executive-level "me" or its mental content.

We might find an existence without individuals who are different from us to be not worth living. When two differing virtual entities commit to each other in a loving relationship that gives them joy, they will not want to change themselves. Still, they might want to celebrate their union by creating offspring who will be like them but different, having traits they want it to have. Replication and evolution would continue.

But what about love between a Standard Human or Hybrid Human and a Conceptual AI entity? Love is the recognition of one's own highest values in another. It is an appreciation of that individual's approach to life. One being biological and one a machine would not preclude any two conceptual entities from falling in love and desiring to share life. The sex act for humans is subordinate to voluntary control. It properly is a celebration of the esteem that each individual holds for the other. We will undoubtedly develop technology enabling an intelligent machine to physically please a human and vice-versa.

AI-HUMAN VALUES ALIGNMENT

Perhaps the single most active topic of discussion in the AI development community today is "AI Values Alignment." AI professionals widely believe that it will be necessary to build into future AI *values* that align with human values to maximize the benefits while minimizing potential dangers. However, the many papers, blogs and discussions indicate that the AI development community struggles to identify and agree on putting these values into AI and *which* human values they should select.

In reviewing the typical discussions in the AI community, I see two errors that cause most arguments to be misguided. The first error is that the AI community makes no distinction between Narrow AI, Perceptual AI, and Conceptual AI when considering *values* alignment. The second is the belief that a "true" set of universal human values does not exist. Their only hope is to find a set of values that most humans can agree on, which they know is problematic.

For example, Kaj Sotala, a researcher for the Machine Intelligence Research Institute, states this position explicitly, "The design of such AIs, however, is hampered by the fact that there exists no satisfactory definition of what exactly human values are[228] [...] I tend to feel that there isn't really any true fact of which values are correct and what would be the correct way of combining them," he explains. "Rather than trying to find an objectively correct way of doing this, we should strive to find a way that as many people as possible could agree on."[229]

First, let us review my use of AI terms. By "Conceptual AI," I mean an AI entity with a conceptual-level faculty, volition, and reason to connect abstract concepts to observed reality—a human equivalent. We are assuming for the sake of argument that this will eventually be possible. This does not include AI systems such as today's Watson that are language-based. Such systems can form integrations of words into hierarchies as "concepts," but have no perceptual level and cannot connect their words as concepts to reality. I consider this latter form of AI as a language-enabled Narrow AI, not a Conceptual AI.

Any sub-conceptual, sub-human AI system that we create is only a human *tool* and can have no goals or "values" of its own, only those given to it by humans. These AI systems are robust "optimizers," which is the reason for the long-term safety concern. An AI that optimizes for something humans do not want could have severe consequences for the world.

AI developers build an AI system for a particular purpose. A sub-human AI system is the *property* of its creator. The AI's developers provide the rules of operation and the human goals to be optimized. Any discussion about values, ethics, or morals regarding a Narrow AI, Vegetative AI, or Perceptual AI applies to the *humans* who create and operate it, not the AI system itself.

It might help if I relate to you a true story. Years ago, when my daughter was five years old, we lived on a farm. A neighbor owned a large, male black Labrador dog named King who liked to hang around in our yard and was familiar with every family member. King was a little standoffish. He would not play with our kids, but he would instead position himself a short distance from them in what appeared to be a protective "guard" behavior. This had been going on for a couple of years. Then one day, playing with the kids in the yard, I began swinging my daughter around in big circles holding her by her wrists, and she was screaming at the fun of it. King lunged at my daughter, grasped her midsection in his enormous jaw, pulled her away from me and began carrying her across the yard. I had to catch up with him and kick him in the head to get him to release her. After the boot to the head, King took a menacing stance. I positioned myself between him and my daughter, not knowing if he would attack me. I loudly yelled and waved at him to "go home." For a long moment, I could tell that he was confused. He knew he must have done something wrong but didn't know what, then he turned and ran away. My daughter required a trip to the emergency room to treat the puncture wounds she received, front and back, from King's canine teeth.

As a perceptual-level entity, instinct and life experience directed King's behaviors. When AI developers create a sub-conceptual AI, they must provide its rules of operation—its "instinct." They must also give the goal, the thing that it should optimize for so that it can recognize "pleasure" from "pain" and can learn from experience how to best optimize for "pleasure." Lacking a conceptual faculty, neither a dog's instinct nor an AI system's operating rules can address every scenario it might encounter.

King misunderstood my daughter's screaming, and his protective instinct caused him to act to try to "save" her. I believe that King's pleasurable prior experiences on the farm with me are why he chose to remove my daughter from danger rather than attack me as its source. King made the best possible choice given his instinct, experience and the situation he encountered. He didn't know that his actions and the injuries to a human that he caused were inappropriate. We humans can expect to have the same type of relationship with sub-conceptual AI systems. Much of the time, we will need to keep them on a "leash."

Humans who create and operate sub-conceptual AI systems are responsible for all consequences—good and bad. If a drug company develops a drug or an automobile manufacturer makes a car that later causes injury, it is held accountable. The person who unleashes a dog in a crowd is held responsible. The creation and application of sub-conceptual AI systems in society is no different. Any company developing a sub-conceptual AI system is doing so with a human-defined purpose. The company must keep it "leashed" within a safe "sandbox" until the company is satisfied that the beneficial purpose outweighs the liability risk of unleashing it into the real world.

As with any technology development and manufacturing undertaking, the potential for harmful mistakes, even if unforeseen, is always there. It is a matter of *degree of safety* with legal civil, criminal, and financial consequences as the standard. AI "black box systems"—machine learning systems that evolve their own rules of operation and goals invisible to the developers—pose a higher *proof-of-safety* challenge. Still, the developers must subject them to the same *degree-of-safety* standard.

The serious concern expressed by the AI development community about needing to build aligned values into AI systems hides a generally unstated fear and is a misdirection. Each AI developer believes that he will be careful in making "safe" AI. However, as the AI gets more sophisticated, he knows that it might not be possible to do so with certainty. Suppose the industry can agree on a standard set of AI "values." In that case, so long as the AI developer follows the industry protocols, he can shift some of the responsibility and liability for AI safety from himself to the "industry" and the "inadequate" protocols.

Suppose the AI development industry truly wishes to maximize AI safety by applying *human* values. In that case, each developer must accept *full responsibility* for the AI systems he helps create. Suppose the rest of us want to prevent AI *capabilities* from outpacing the development of AI *safety*. In that case, we must keep the accountability for AI development focused squarely on the AI developers, industry standards notwithstanding.

The consumer also has a responsibility for his safety. The consumer accepts a certain degree of risk when taking a drug, buying a car, or allowing an unleashed dog to be present. Consumers are responsible for holding AI creators accountable when potential dangers, either in the finished AI product or the methods used in its creation and testing, are discovered. Concerned citizens can create private sector organizations and even new businesses to investigate, evaluate, and report various AI systems. Society can then turn to its judicial system based on objective law to determine when an AI developer exceeded some negligence threshold, or deception has occurred.

The standard set of values the AI developers need to comply with are those that all humans, by their nature, require. Humans require freedom of thought and action without being subject to or subjecting others to force or fraud. Suppose sub-conceptual AI systems built by AI developers for human purposes comply with these principles (with modernized interpretations as needed in a technological world). In that case, the AI will be "safe."

Once an AI entity becomes conceptual-volitional, it will be a human equivalent and subject to the same ethics. Any discussion about values, ethics, and morals regarding a future Conceptual AI entity is simple. Biological and machine conceptual entities are subject to the same values due to their shared nature as rational beings in a blended society.

Any attempt to pre-program a Conceptual AI with "values" is tantamount to "brainwashing." Once created, a Conceptual AI entity can be no one else's "property." Building a Conceptual AI entity to be "owned" for a particular "utility" purpose would be "slavery."

Full transparency of operation is a requirement for sub-conceptual AI safety. In contrast, a Conceptual AI is entitled to the privacy of its thinking. The only thing that can constrain the values and behaviors of a Conceptual AI entity is the same that constrains human behavior—a Personal Philosophy preventing force, coercion or fraud.

AI community discussion indicates their belief that there is no "true" set of ethics to be identified. They point out that different cultures and groups of people worldwide have other value systems. The challenge, they assert, is to find common elements on which most humans worldwide can agree.

This misconception about the nature of ethics and values is a direct consequence of the Plato-Kant axis of philosophical thought having the upper hand in universities. Many in the AI community were not exposed to the Aristotle-Rand alternative view of the nature of ethics and values. They do not know how to connect *morality* to *reality*. They tend to be morally agnostic, accepting the widely disparate value systems of various groups as equally valid—for them.

They are not aware that a "true," universally applicable set of ethics has been identified—one based upon reality and all humans' nature as rational beings. These were America's founding principles, giving men the freedom to think and act in their self-interest for the first time in human history.

The AI community seeks an AI values system that is "genuinely intercultural and inclusive, offers concrete guidance across a range of situations, and is stable over time, meaning that people continue to affirm and endorse these principles once they have seen them implemented in practice."[230] The Aristotle-Rand system of ethics meets all these criteria. The Plato-Kant system does not.

The various cultures and groups in human society do not necessarily hold personal philosophies consistent with their human nature. Human society is still evolving and experimenting with the integration mode of thinking but has not yet consistently achieved it. The challenge is not aligning AI values with a set of human values chosen by consensus. The challenge is aligning AI values with reality, man's nature, and what he requires for survival. If we can then apply such properly aligned AI systems throughout society, we can move human society toward a healthy, reality-based philosophy.

Society's first function is to teach new humans "how" to think. We have identified what a proper Forming Education is. Consider that we are on the verge of relegating the education of human children to AI "teachers." Will these AI teachers teach our children the integration mode of thinking, unlike most of today's human teachers and professors who do not? Is the public even aware that this is important for the future of our society? Nowhere is the need to critique applied AI technology more crucial than in its use in educating our children. And there is no better example of how technology will amplify and accelerate the impact that commonly held ideas have on society.

Ultimately, AI has the potential to be the most powerful technology man has or perhaps will ever create. Long before AI entities can replace us in evolution's progression, AI technology will augment our human advance in that progression. That is the subject of the next chapter.

♦♦♦

4.7

AUGMENTED HUMAN INTELLIGENCE

"The real problem is not whether machines think but whether men do."

—B.F. SKINNER

HYBRID HUMANS INTRODUCTION

In the previous chapter, we explored potential issues related to machine-based AI entities if we ever create them. A more likely prospect in the next few decades is applying AI technologies to enhance human intelligence dramatically. These technologies will not yet be AI *entities.* Still, humans will use them to great advantage. We will augment our intelligence incrementally. Underway for centuries, this process is reaching the explosive growth phase.

Some futurists believe that this process will continue until we become super-intelligent entities. They believe this is more likely than building a non-biological super-intelligent entity.

Suppose we someday move into machine bodies to get past the limitations of our biological bodies. In that case, some futurists think our biological heritage will provide an advantage over machine-only intelligent entities. Rodney Allen Brooks, chairman and chief technology officer of iRobot, described it this way, "…we will become a merger between flesh and machines. We will have the best that machineness has to offer, but we will also have our bioheritage to augment whatever level of machine technology we have so far developed. So we (the robot-people) will be a step ahead of them (the pure robots). We won't have to worry about them taking over."[231]

In the previous chapter, we explored what Conceptual AI entities—"the pure robots"—looked like. In this chapter, we will explore what Hybrid Humans—"the robot-people" look like.

PERCEPTUAL LEVEL CONSCIOUSNESS

Our lowest level of consciousness is our automated response to sensations. Humans have the classic five external senses: touch, taste, sound, smell and sight that grasp the reality around us; and two internal senses that tell us what is going on with our body—proprioception (knowing where our body parts are) and vestibular sensation (sensing movement). Additionally, we have the gut microbiome that provides us with information about the substances we encounter in our environment. We will be able to enhance these senses and add more through technology.

We are discovering that the human brain is more plastic than previously thought. It can reorganize itself to add functionality and capacity as needed for the activities it is encountering. Scientists recently discovered that older humans grow new neuronal cells at any age. We now know the brain can adapt to receiving and interpreting sensory data from artificial sense organs such as cochlear implants. This electronic to biologic machine-to-brain interface capability will improve over time in both types of inputs and resolution.

It is our perceptual-level consciousness that receives the sensory inputs and retains them over time. It isolates detected entities, identifies them and associates appropriate action responses. It learns from experience. We share this capability with higher animals.

Sound: Cochlear implants translate the sound from external microphones into electrical signals that cause an implant in the inner ear to stimulate the auditory nerve. This technology will increase the number of frequencies resolved and eventually even include the ability to sense ultrasonic and subsonic frequencies. It will amplify any frequency's weak sounds. Such enhancements will augment the hearing in healthy humans, not only those who are hard of hearing.

Sight: Like sound, sight via cameras as artificial eyes and brain implants has already been demonstrated at low resolution. This will increase in resolution and expand into the ultraviolet and infrared parts of the spectrum. It might be that the brain will be capable of handling images in both larger spectrum and higher resolution than the biological eye provides (more pixels).

Smell & Taste: Artificial air sampling and substance sampling "sense organs" will be able to detect and identify an ever-increasing variety of substances and signal our brains via our senses of smell and taste. Artificial "bacteria" in our gut might detect target substances and emit chemicals into the blood that signal the brain.

Touch: Sensors in our clothing or wearables will be able to provide us information about our own body's health and performance, about the environment around us, and other enhanced sensations. Haptic stimulators will tickle our skin, our body's largest organ.

Proprioception & Vestibular Sensation: These are how we sense our "body." One of the most exciting sensory augmentation aspects—"Sensory Translation and Body Extension," involves all the senses. Here is an example:

Newer cars have haptic vibrators built into the left and right sides of the driver's seat cushion. When you are backing up, and the cross-traffic sensor in the car detects another car coming from your right side, it causes the right side of your seat to vibrate. You sense the vibration in your right butt and instantly hit your brakes.

When you first drove the car, you had to consciously think—"Why is the seat vibrating? Oh, yeah, it is that traffic sensor thing. I'd better use the brake and look for the traffic." But after several instances of consciously thinking about and reacting to this experience, you program your subconscious. The sensation and the response become automated at the subconscious level. You do not even consciously think about it anymore. When your butt vibrates, your foot hits the brake.

You have translated an artificial sense (the lidar radar sensor on your car that detected the cross-traffic) to a specific pattern in your sense of touch. It works the same way when a vehicle is in your "blind spot" while moving forward. So now, you have a new external sense. You can "feel" when you have cross-traffic coming or a car in your blind spot. You have effectively incorporated a lidar radar device as a sense organ. You have extended your "body" to include your vehicle when you are in it.

Now consider the cochlear and retinal implants previously covered that feed sensor data directly to the brain via electronic-biologic nerve interfaces.

Someday your car's lidar radar sensors could wirelessly feed sensations directly to your brain rather than having to be translated through your butt. The outcome will be pretty much the same. Your brain will adapt to and automate this new sensory capability, perhaps now with higher resolution and speed.

PERCEPTUAL-LEVEL SENSORY EXPANSION

Imagine that you have that lidar sensor in a wearable headband. It scans 360 degrees around you, measuring how far it is to every object in its field of view, even behind you and to the sides. It feeds these 3-D measurements in real-time to your brain. Your brain deals with the data continuously and subconsciously, keeping a 3-D model of your surroundings, just like you do with your vision now. But you do not have to turn your head to sense what is behind you, and it will work in pitch darkness. It will alert you to movement approaching you from behind. Your body now has the new lidar radar "sense" even when you are not in your car. Or, instead of lidar radar, it could be one or more "eyes" (night-vision and infrared capable cameras) on your headband.

Now imagine you are back in your self-driving car, happily driving itself down the highway. Suddenly you become conscious of a mild sensation of "pain." Your subconscious received input from your car's computer, indicating that a mechanical problem is developing. It could lead to danger if you allow the vehicle to continue driving itself. Based on this "pain," and you do not even have to think about it, you take the car off autopilot mode. You "feel" your car's operating modes just like you feel the position of your arms or legs. You "feel" when things are good or are going wrong. The car has become an extension of your body when you are in it, and it provides all sorts of new sensations.

Your house could be an extension of your body as well. But consider that with real-time global 5G communications, you can sense your house even when you are on the other side of the Earth. When your house's alarm system detects a problem, you will "feel" it first and then become conscious of the feeling and seek more detailed information.

We might acquire truly exotic new external senses in the form of "signal processors" that detect and decode information from streams of data. Suppose an external wearable sensor measures Wi-Fi signals in the direction it is pointing, and a signal processor translates the data into an image that we can "see." Since human bodies cause backscatter changes in Wi-Fi signals, might we be able to "see" people behind walls? The fact is this has already been demonstrated.[232] It now just needs to be miniaturized. Our world is awash in radio waves. What if we could "see" them and selectively isolate and decode the information contained in any of them we wished?

SENSORY OVERLOAD

We will gain enhanced natural senses and new senses. We also will have "direct" sensations rather than "change" sensations. To prevent sensory overload, we will need control over sensory "thresholds."

Our biological senses operate by sensing a change in input. If you stare at a point of light in a darkened room without moving your eyeballs, the point of light will disappear. It is only because your eyeballs continuously move slightly that you can keep the point of light visible. You are continuously changing which rods and cones in your eye the light stimulates. It works the same with the sense of touch. There is also a threshold of change that has to be exceeded before you perceive something. You will not sense a light too dim or a touch too slight.

Our external augmenting sensors will be able to detect an input continuously, even if it never changes. And they will have much more sensitive thresholds than we humans have. Of course, this could lead to massive amounts of continuous sensory input to deal with, so we will need to control how much of this detail we want to reach our awareness. We will need to develop a form of "focusing" our external senses only on what we want to sense in greater detail.

We already have built into our subconscious the ability to monitor and prioritize sensory inputs with automated, subconsciously triggered responses and alerts to our conscious awareness—our "salience" system. We will consciously deal with (think through) new types of senses when we first gain them. Eventually, this conscious thinking will program our subconscious about how we want them handled, including their priority and when to focus on them.

So far, we have been exploring how the Standard Human's perceptual faculty will handle all sorts of augmented senses. We will interface with the reality around us with a fantastic new richness of both scope and detail. We will already be *super-human* compared to generations of humans before us. We are still only considering our *perceptual level* of intelligence. Now let us consider how augmentation might work at the *conceptual level.*

WORDS AS CONCEPTS

When we use language skills, mental, verbal, and printed, we use our *conceptual* level of intelligence. Concepts exist in a hierarchy beginning with concrete things in reality up to the most abstract conceptual ideas, i.e., "love."

Words represent concepts. We use words *internally* as "handles" for concepts associated with them. Each concept is a pattern saved in our neocortex. Without words, there can be no concepts. A pattern in our neocortex that has no word attached to it is a *percept.* Animals also have percepts but cannot attach words to them and therefore cannot think conceptually. Once we grasp the handle (word) in our mental focus, we instantly understand all that defines and is subsumed by the related concept. This was all programmed into our subconscious by earlier conscious thought. We can hold a small number (four or five) specific concepts in mental focus and consider them, making differentiation and integrations and considering actions.

We use words *externally* to communicate concepts to and from others and store concepts for later retrieval—by writing them down and later reading them. We input words through our senses of sight and sound. We output words via sound (speech) and manual dexterity—typing and writing with a pen. By far, the speediest and most accurate language input method to date is visually reading the printed or written word. Our eyes are our highest bandwidth input channel, feeding information directly to our perceptual and conceptual faculties.

Today, we use our eyes to orient ourselves spatially and navigate our environment at a lower level of consciousness. We rely on our eyes to detect danger from objects or situations we see in our visible environment. We recognize entities and objects that we have learned to identify.

At a higher level, we use our eyes to read printed word content in our environment, such as the road sign indicating "Springfield 21-miles," a business's name on its door, or the name on a person's name tag. Such use of printed language enhances our conceptual grasp of the environmental reality around us.

AUGMENTED REALITY

The word labels and instructions that we see all around us in our environment are already "augmented reality." They provide additional information about the objects we are seeing. Suppose these word labels and instructions are not physically printed on the objects we see but are shown to us as we look through our smart glasses at the objects. Now we are talking about *virtually* augmented reality rather than the *physically* augmented reality we already know and use.

Through the technology of computing devices with sensors and communications, we are just now entering the era of vastly expanded augmented reality, "AR." As an AR device, your smartphone uses a database, GPS location sensor, compass, and camera to identify the street address and the name of the business or building the camera sees. Today, this information displays on your smartphone, but soon it will display in your smart glasses. When APM arrives, it will show in your contact lenses. This printed information is overlaid visually on or near the object or person at which you are looking.

Your smart glasses or smart contacts will have cameras allowing your AI Personal Assistant to see the same things you see. Your AI assistant will overlay information about the objects detected onto the scene that you see. Glasses or contacts will track your eye movements, detecting where your eyes are focused, either in the outside world or the overlaid information in your display.

When you first gaze at an object or person, only its name will be displayed, and an icon indicating whether you have been there or met them before. The longer you look at the object or person, the more detailed the displayed information will get—such as when you last met them and where. Also overlaid on the scene will be icons off to the side representing entire categories of information that will display if you focus on them.

Today, we already take in vast amounts of information printed in our environment—"STOP," "Men's Bathroom," "Speed Limit 70," "11:30," "Eat Here." We sometimes read these things and automatically act on them without even being aware that we did so. It will be no different when these things are overlaid virtually in our field of vision. As the amount of information presented increases, our incredibly plastic brains will adapt to automate and handle the increase, at least up to some limit well above where we are now.

We should also consider that "analog" information overlaid on our field of vision is equally valuable. Our perceptual faculty more immediately recognizes icons and graphics—it more quickly understands a left arrow than the word "Left." As opposed to a printed battery percentage, our perceptual level more easily monitors a battery level gauge, especially one that changes colors as it gets lower. It recognizes the Starbucks logo easier than the printed word "Starbucks." Of course, not all information can be conveyed with icons or graphics, and for that, we must use words.

The optimal blend of analog graphics and printed word information for both our perceptual and conceptual faculties will present a vastly enhanced interface to receive information from and about our environmental reality. However, we also need a way to output information.

SPEECH INPUT/OUTPUT

Perhaps the most helpful output we will need going forward is the *query*—a request for information, i.e., "What is today's weather forecast?" There is usually a large ratio between the size of the query and the size of the response. The five words "What is today's weather forecast?" might return hundreds of words required to give temperature, humidity, sky condition, the chance of precipitation, winds, UV levels, and pollen levels for current conditions, tonight, and tomorrow? Responses to queries can be as simple as "yes" or "no" but are typically lengthy compared to the query that generated them.

The most natural method to issue a query is to speak it. If the response is not too lengthy, then it can be spoken back to us. If it is lengthy or contains details that we will need to refer to, then the response might be via some form of visual text or graphics.

Today, we are all just learning how to live with our "Hey Alexa," "Hey Google," or "Hey Siri" AI Personal Assistants. (I will use the acronym "AIPA") Their basic use is to issue a verbal query and receive a spoken response from the assistant. The latest generation of assistant devices can also show text, graphics or videos on built-in display screens or remote televisions. Additionally, these assistant devices can respond to commands by issuing remote control instructions to "smart" devices such as turning on lights, locking doors or tuning our TV to the desired program.

It does not require much effort to imagine when we have our augmented reality display glasses or contacts coupled with earpieces and microphones. These would enable us to speak to our AIPA that resides in a wearable device that we have somewhere on our person. We want our AIPA to serve us exclusively and be as unobtrusive as possible.

When we verbally ask our AIPA for the weather forecast, we can say either "Tell me…" or "Show me…" The requested information will either be spoken to us or displayed on our glasses or contacts. To hear our assistant, we have an augmented reality earpiece, like a hearing aid with Bluetooth. It passes through ambient sounds to our ears but overlays audio from our assistant such that only we hear it. Or a sound induction device could send the sound through the bones of our skull. Ultimately, the device might send the sound directly to a cochlear implant.

Now comes the fun part. When we are mentally formulating an utterance of words, just before we speak it out loud, our facial and jaw muscles move the tiniest bit as if speaking the words. MIT has demonstrated a wearable device called "AlterEgo" that can detect neuromuscular signals that reveal what digits someone will say before saying them out loud. It has already achieved 92% median accuracy and latency of less than half a second. It could soon become possible to wear an external device through which you will be able to "think" a query to your AIPA without having to speak out loud.[233]

Unlike an AIPA, an Artificially Intelligent Robotic Assistant ("AIRA") resides in its own robotic body, is mobile, and can interact with and serve multiple humans. An AIRA can use audible speech to interact directly with us. Suppose the AIRA wishes to send information to our personal audio-visual display. In that case, it must do so by interacting with our AIPA, which controls those displays. The AIRA's specialty is that it can take physical action in our environment, such as get us a cup of coffee. Our AIPA can only control remotely controllable things, including AIRAs. Our AIPA can interact with one or more AIRAs to have physical things done for us, such as bringing that cup of coffee.

ALMOST TELEPATHY

Your AIPA will, of course, have your contact list and will be able to connect to the AIPAs of other people over the Internet. You hear a particular "ding" alerting you that someone wants to communicate with you during your day. The person's face and name appear in your smart glasses or contacts. By focusing on an icon, you establish the connection. You think, *"Hello Jim, what's up?"* without saying it out loud. Jim hears that in his earpiece and replies to you in a similar manner. The two of you have a conversation that no one around either of you can hear.

For special people, perhaps your spouse, you authorize "barge in" communications where they can say something to you that you hear immediately without the need to answer the call. You can also allow the other person to monitor whatever you are saying, seeing and hearing in real-time. They can experience it along with you. You can have a group conference where the face of whoever is speaking will appear on your display as if you turned to look at them.

You have this almost-telepathy relationship with your AIPA. You can submit any question or instruction to your AIPA just by thinking about saying it, and you hear or see your AIPA's response through your wearable. The entirety of the world's information is no longer "at your fingertips"—it is available to you in your head with just a thought.

Since your AIPA connects to the Internet through any available wireless network, you now also have almost-telekinetic powers. You can turn on or off and control any remotely controllable device just by thinking of the verbal commands to give to your AIPA, who then sends the control signal over the network.

Keep in mind that we are considering just the capabilities of *wearable* human-machine interface technology, not *implantable* brain technology. Implantable brain devices will be more capable but come much later. With wearables, you will still use your visual, audio, tactile, and other senses. You will still need to formulate words utilizing the process of speech, even if you do not say them out loud. However, as you can see, we will be able to do amazing things with wearables.

Most importantly, we will be causing our brains to adapt, reconfiguring their modular functions to integrate external technology into our intelligence. We will have been doing this long before we begin using implantable brain devices.

SHARED AND VIRTUAL REALITY

Suppose you connect with another person, and they allow you to see and hear what they are seeing and hearing. In that case, you are immersed in *virtual* reality, where the other person is. You see the table the other person is at and the object on it as if they were in front of you. The virtual remote scene overlays the scene where you are. You can control the "opacity" of the virtual scene. Let us say the virtual scene is 90% opaque. You mostly see the other virtual scene, but you can slightly see the real one in front of you. You can see your hand as you reach out to "touch" the virtual object on the virtual table in front of you in the virtual scene. Your friend sees your hand in his display and sees the button that you are touching is the one you are explaining that he should use.

You are both using virtual augmented reality. You see a virtual object with a button from his reality. He sees your virtual hand from your reality. Eventually, computer-synthesized "virtual" realities will have resolutions, motions, and tactile sensations that seem convincingly real when we experience them through our augmented senses.

AI NATURAL LANGUAGE SKILLS

Now that we can envision being able to query our AIPA and hearing or seeing a response, the question becomes, "What can I ask my AIPA?" We all know the frustration of Alexa or Siri not understanding what we asked. There are two basic skill sets involved in the assistant: speech recognition and language understanding.

Speech recognition is continually improving. Since it is essentially a pattern matching skill, it should continue to improve until the AIPA correctly detects the words you say almost all the time and even if said in different order. For the AIPA to *understand* what the words mean is another story.

Two humans can ask each other almost anything such that the other understands the query. That is because both parties are conceptual-level entities—human-level intelligence. Words are the handles or labels representing concepts. We will not have AIPAs with that level of language understanding until we create Conceptual AI entities, still a long way off. But in the near term, our AIPAs can become much more helpful than they are today.

There is a lot of room for improvement of today's Narrow AI assistants. Just as a Google search continues to improve the quality of returned results for a given query, the personal assistants will also improve. Our AIPA will continue to improve its detection of words within defined contexts and transaction skills.

Until Conceptual AI comes along, we will not be able to truly converse with our assistant at the same level as we do with another human. We will need to remain aware of our AIPA's limitations, what it is trained to do, what it is not, what it understands, and what it does not. We will have to compose our queries and instructions correctly for a given context to get the results we want from our AIPA within a specific skill that it knows. While they will improve somewhat over time, our AIPAs will not fully understand "context" or "nuance" or be able to "anticipate" as well as a human assistant would. But our AIPA will still be a massive help to us.

Our AIPA will be so helpful that we might want it to appear more human-like. We will choose the avatar face and voice we want our AIPA to have. When appropriate, our AIPA's face will appear on our display. Its mouth and facial expressions will reflect the dialog we are having with it.

AI PERSONAL ASSISTANT USES

Here is a partial listing of things our AIPA might do for us:

- **Query-Response**: Look up anything you want to know, anywhere, anytime.
- **Augmented Reality:** Identification and labeling of objects and people around you in reality.
- **Almost Telepathy:** Instant, private sharing of thoughts with others.
- **Navigation:** Guided directions and time estimates both outside and inside buildings.
- **Remote Control:** Receive sensory input and send control instructions to any smart device you are authorized to access anywhere in the world.
- **Entertainment:** Receive audio, video, and virtual reality of any recorded movie, show, or concert, or attend a live event virtually, all in 2D or 3D.
- **Collaboration:** See and hear what someone else is seeing and hearing. Assist them or receive assistance from them.

- **Translation:** Converse with anyone who speaks any language. Your AIPA will translate audibly real-time. It will translate printed words you look at through your augmented vision in the view you see.
- **Instruction:** Want to know how to operate or troubleshoot a new piece of equipment? Look at the bar code or the model number on it and bring up a step-by-step virtual user manual with augmented reality guidance overlaid onto your view of the equipment.
- **Memory-Reminders:** Facts that you want to remember for future reference. Things you want your AIPA to remind you of whenever you are at a time, a place, a person, a limit, etc.
- **Education:** Individualized courses you learn from your AIPA who becomes a virtual professor in a virtual classroom or lab by accessing the content in the cloud.
- **Virtual Computer Screen**: Control a virtual web browser or bring up a spreadsheet in your display. Look at a cell to select it, "think-speak" a number to enter it.
- **Information Traffic Cop:** Set priorities for various information streams. Your AIPA will filter and consolidate information and be able to feed you summaries of what interests you. When you "like" and "dislike" items your AIPA calls to your attention, you teach it the kinds of information you want to be "interrupted" with and when to do so.
- **Health Guidance:** Monitoring biometric sensors in and on your body and clothing and in your environment; identifying and tracking what foods you eat; providing health advice; alerting to undesirable conditions; communicating with remote medical services; activating wearable therapeutic drug dispensers.
- **Transactional Skills:** Have your AIPA make dinner reservations, call an autonomous ride-share vehicle, set appointments, order products for delivery, transfer funds, research best flights and hotels according to your preferences, buy tickets, search for entertainment content or events, monitor online retailers and ads for items and prices you want, monitor and schedule maintenance on items as needed, monitor supplies on-hand and reorder as needed, and any other transactional skill for which your AIPA has been trained.
- **Prompting:** As in a "Teleprompter," displaying in your visual display or whispering in your ear the points you want to make in your prepared speech. It can search for and show you background info on items your AIPA hears

discussed in an unprepared conversation. Or it can look up the validity of some fact that someone else has just said (fact-checking). You will talk like a genius on any subject.

- **Firewall:** Your AIPA will identify and block malicious acts directed toward you. We will cover this in more detail.

THE EXTERNAL AUGMENTATION PHASE

Before we have anything invasive done to our bodies, visual, auditory and tactile enhancing wearable devices will significantly expand our intelligence capabilities. We will be able to think with dramatically increased information and human and AI collaboration immediately available. Learning to use these wearable devices iteratively as they advance over time will allow our brains to adapt and grow functionality and capacity where needed. This especially applies to those who begin at a younger age. Our use of these supernatural senses and information flows will be both subconscious and consciousness expanding.

We will "feel" and "control" things external to our body as if they were an extension of our body. We will have an AIPA who is always with us, knows our personal data and preferences and is our interface to an ever-expanding "smart" world. Our ability to interact with others will be significantly enhanced via a new form of social media that is almost telepathy and includes shared augmented and virtual reality.

This personal augmentation has already begun with the advent of smartphones. We can already do much of what we have described. Still, we are limited by looking at a tiny display screen and listening to a little speaker or earbuds. Yet, the empowerment our smartphones provide already makes them indispensable. It is increasingly difficult to function "normally" during your day if you forget to take your smartphone. Therefore, it has become the last thing you might forget to take.

You wish you could wear it and have your hands and pockets free. That will be the case soon. This technological explosion of "wearables" is already underway as the next big thing following smartphones. Smartwatches are only the first wave. It will explode into a massive industry.

This *External* Augmentation Phase via our wearables will connect us to all manner of Narrow AI resources, enhancing every area of our life. As the capabilities of these Narrow AI advance, we will already have input-output access to them via our wearables and our adapted brain functions. When the next *Internal* Augmentation Phase occurs via implants, it will mainly provide gains in speed, resolution and bandwidth over our wearables.

THE INTERNAL AUGMENTATION PHASE

Neural implant technology for specific senses, such as hearing, and motor control of things like prosthetic hands, already exists. Now, the race is on to increase the scope of their functions and the detail resolution. Research to assist those with disabilities will expand into the augmentation of healthy humans.

For the signals through these implants to be usable, the brain must go through "training" to learn about the implant signals' existence and what they mean. Our incredibly plastic brains then rewire circuits as necessary to make use of the new signals through repetitive practice. We do not yet know what the human brain's limit is for reconfiguration. We believe there is much potential in the immediate future.

A single universal neural implant that allows us to have a broad-spectrum interface to all kinds of external intelligence resources might be possible. However, we will begin with implants to augment specific senses, such as a cochlear implant for hearing and an optic nerve implant for vision. Then, a speech center implant might allow us to output instructions by thinking them. We might begin with a comprehensive set of separate implants that connect us to external AI technology via our brain's natural biological senses and functions. We might discover that this distributed topology for implants is necessary because it matches the distributed nature of our brain's functioning.

Suppose a universal brain neural implant is possible. In that case, it will be further in the future as it is likely to require significant nanotechnology advances.

CYBORGS OR FYBORGS?

Humans will be naturally resistant to wanting anyone cutting into their skulls to install implants in their brain stem, even with the promise of enhancement. It is one thing if it is your only option to save your life, i.e., to remove a brain tumor, but it is another thing if it is discretionary. Suppose a wearable can achieve almost the same enhanced function. In that case, practically no one will opt for brain surgery.

Of course, there are always a few individuals willing to risk their lives to be the first. Some will pay the ultimate price, and eventually, some will succeed. One will lead to another until, finally, it will become commonplace. But we must not assume that implantable enhancing technology will enjoy a fast uptake cycle by the masses. It will likely be a protracted process that begins with implants at the peripheral senses to cure sensory problems such as cochlear and retinal implants. Then it will ever so slowly make its way deeper into the brain, still to correct problems.

Injectable nanotechnology will follow a similar path. People will be resistant to injecting nanobots into their bodies. But when the robots are needed to save your life by repairing your pancreas, you will accept them. Eventually, nanobots' injection will become a trusted technology as we use it to successfully correct an ever-increasing list of problems without adverse side effects.

Time will tell whether we will accept nanobots into our brains to act as neural interface devices rather than having a neural implant installed surgically. We will resist both until we think they can be trusted as "safe," which is not likely to happen suddenly.

This resistance to the discretionary invasion of our bodies with technology means that we are likely to become advanced "fyborgs" long before we become "cyborgs." A fyborg is a term used by Alexander Chislenko to refer to a biological entity's functional enhancement by utilizing technological tools *external* to the body. It is differentiated from cyborgs who have technology enhancements *internal* to their bodies.[234]

We should expect "wearables" technologies to enjoy rapid uptake cycles by the masses and to have a long run before "implantable" technologies slowly replace them. We will all first become fyborgs before we become cyborgs.

THE "CROW" REVISITED

Let us review the "crow" limit. Humans can only hold a small number, let us say four or five items in mental focus simultaneously. Suppose we encounter ten chairs before us. Our perceptual consciousness cannot grasp that many. So, our conceptual faculty counts them and assigns a number (word) "ten." The word "ten" is then a single unit (concept) we hold in mental focus instead of the multitude of units (actual chairs) we first perceived. Our conceptual faculty continuously performs this unit reduction of concepts to keep the number of concepts we focus on within a manageable small limit.

We keep going to higher-level concepts—chair, furniture, furnishings, property, assets. Our conceptual faculty can move up and down a hierarchy of concepts quickly as needed to reach the level of a concept that we need. We note how the items in focus are the same or different, what they are, how they relate to each other, what we do and do not know about them. We trade concepts we no longer need in focus for new ones, move them up and down their hierarchy, make new connections, and assign new concepts.

We might develop AI systems that are good at doing these things—identifying, differentiating, and integrating. These AI processes could have the advantage of the great wiki in the cloud to reference pattern-matching comparisons. What if our AIPA could launch any number of these processes for us whenever we want? Our mind could learn to offload items it is considering to these external processes, with each reporting back its findings. Our mind would become a *master-process* directing the work of multiple external subordinate processes. We could probably launch many of these processes but might only be able to consider the results of four or five of them at a time because of our "crow" focus limitation.

REASON—THE ART OF NON-CONTRADICTORY IDENTIFICATION

When a Standard Human sees something in reality for the first time, he holds it in mental focus. He compares it to everything he already knows via pattern matching and determines that he has not identified it before. So, he must learn more about the item. He must spend time investigating its form, function, purpose, relationships, significance, and importance and then decide if and where it fits into his conceptual hierarchy. He saves this to memory with a handle, a word representing all that the thing is, its name, its "identity."

Now let us consider "augmented reality" via our personal display and AIPA. When we see something in the real world for the first time, our AIPA will know that we are looking at it. Our AIPA will pattern-match it against a vastly larger knowledge base of things, places, people, animals, plants, etc.—that it accesses in the cloud. Our AIPA will most likely quickly identify the item that we are focused on in our visual display. Then our AIPA will display the name of the item on or near it. We see the item's name and instantly determine if we know it or not.

Let us say an item is a person, and we're not sure if we've met them before. We then focus on a query icon in our display, causing our AIPA to show us when and where we previously met the person. Our AIPA will include details about the person, such as where they reside, their occupation, their history with us—their "identity." (Or we send our AIPA a "spoken" query and then "hear" the response.) Suppose an item is an object, the name of which we do not recognize. In that case, we can query our AIPA for details that define the object, its form, function, purpose, relationships, significance, and importance, i.e., its identity.

If we think the item holds significance for us in our foreseeable future, we will save its identity in our personal "biological" memory. If not, we will let this encounter go without personally remembering it. Our AIPA will remember it and be able to recall it for us if needed someday.

So, it seems that our AIPA will be our portal into the great knowledge base ("wiki") in the cloud for our purposes of:

- Non-contradictory identification of items perceived in reality
- Saving external memories
- Recalling external memories

Let us examine each of these in more detail.

IDENTIFICATION BY AIPA

What does "non-contradictory" identification mean? It means that all the knowledge we have programmed into our subconscious is an integrated whole that does not contain contradictions. It means that we cannot have internalized a belief that an item "Is A" and "Is not A" simultaneously. One or the other is false, and we must remove it from our knowledge. Unfortunately, we humans can be good or poor at preventing internalized contradictions, with most of us somewhere in the middle. But what about our AIPA and the great knowledge base in the cloud?

Our AIPA will access its own historical knowledge base and a knowledge base in the cloud to make identifications. This cloud knowledge base will be an ever-evolving super "wiki" containing the patterns against which an AI can compare a perceived item to identify it and retrieve details about it. So how does this super wiki get its information? How do we know we can trust the information our AIPA retrieves from it?

The general idea of any wiki knowledge base is that it is crowd-sourced, fact-checked and maintained by persons claiming to be knowledgeable in the domain. A wiki not only makes vastly more information available to us than we could ever learn ourselves but also offers "safety in numbers." If a person contributes erroneous information to the wiki, then others will catch it and correct it. The knowledge base inside a person's head can hold all sorts of errors and contradictions for his entire life. A crowd-maintained wiki, in theory, cannot hold errors and contradictions for long.

In practice, however, today's wiki knowledge bases contain amounts of generally accepted but false information. This situation will continue so long as untrustworthy people are inputting the information. Blockchain technology will help to minimize such errors. Until then, we must keep the possibility of error in mind. Still, the evolving knowledge base in the cloud, accessed by our AIPA, will be pretty helpful for most topics. The non-contradictory identification done by our AIPA will probably be more trustworthy than information we might receive from other random people or sources.

Once a significant number of people begin outsourcing non-contradictory identification functions through their AIPA to the cloud, an exponential explosion of new knowledge will occur. One individual's new concept saved in the cloud becomes immediately accessible as new conceptual knowledge for everyone else.

This will not constitute a "hive" mind. We will continue to be individuals with our minds controlling things. We will still need to validate ideas before accepting them into our knowledge. It will just be that the sharing of new knowledge in our society will become nearly instantaneous.

SAVING EXTERNAL MEMORIES

We humans selectively program experiences into our memory, some stronger than others and some not remembered at all. Our AIPA will remember *everything*. Humans have a somewhat expandable but limited memory capacity. We will not need to remember the minutia of our lives if our assistant can remember it for us—"When was my last dental visit?"—"When is the next one?"—"How much did I weigh a year ago?" So, which memories are essential for us to save to our internal biological memory?

When we are "thinking," we instantaneously pattern-match any item of information against all the information that we know. So, how does saving a newly learned concept to external memory rather than internal memory help us?

The initial benefits of saving to an external memory via our AIPA are immediacy and accuracy. We already use external memories that are sometimes delayed and less accurate. We have handwritten or electronic calendars and journals. We have family photo albums. We have paper and electronic documents of life history. We save all this external memory information for future use. Some of it might contain summaries and therefore could lack details and accuracy. Our AIPA can save all information for future access. Our AIPA will save it in real-time with all the details of what occurred, probably in secure cloud storage.

As we increase the knowledge that we keep externally, the portion that remains in our biological brain will get relatively smaller. The cloud will store the facts of our memories, but it will not function as a whole brain. The essential function that will remain in our brain until the end is our *executive function* that controls everything. Eventually, when a machine can fully model a human brain down to the molecular level, we can take a snapshot of a brain's molecular state. We will upload our executive function to merge with our already stored memories and become Virtual Humans. Ray Kurzweil predicts that such "uploading" technology will be available in the 2040s, however the brain scanning technology will take a little longer.[235]

RECALLING EXTERNAL MEMORIES

We will have *hybrid* memory storage, some in our biological memory banks and some in our external AIPA's memory banks. Our conscious mind can instantaneously access our internal memory via a query to our subconscious faculty. We provide the essential elements of what we are looking at so that our subconscious can pattern-match to find the applicable concept and the word for it.

We will have a similar query-response function with our AIPA, though it will be more deliberate and less automated. Our AIPA will identify what our eyes are looking at and label it for us. But what about the thing that our *mind* is looking at, not objects but *abstract ideas*?

We will have the ability to communicate verbally with our AIPA (either out loud or mentally). Suppose we can verbalize the essential characteristics of the abstract concept we wish to retrieve. In that case, our AIPA can pattern-match it against both our saved external memories and the great, global wiki in the cloud. It can then return the word for the concept we were seeking. We can then ask our AIPA to collapse or expand the concept further.

All this will be slower, requiring verbal or visually displayed back and forth dialog. Still, it will have no limits on content, either personal external memories or the total of human knowledge in the cloud. Our plastic brains will learn how to become more efficient at this query-response dialog with our AIPA until it becomes almost second nature for us.

FORGETTING

Being able to suppress specific memories is essential. That is what enables us to move forward positively after a negative experience. It might not be a good thing to have an AIPA that remembers and reminds us of *everything*.

We will need to develop a method of instructing our AIPA not to remind us of specific experiences, to suppress them. Perhaps we will tell our AIPA "suppress this experience gradually." Afterward, on an occasion that would trigger that memory, our AIPA will only remind us of it at increasing time intervals since the last reminder. Eventually, it will not remind us at all. That will mimic the way our biological forgetting occurs.

We probably will want to retrieve "forgotten" memories from our AIPA if needed in the future. There will be a recovery instruction as such as "Recall forgotten memories regarding ___ from the period of __ to __."

EXTERNAL EMOTIONS

We get an immediate emotional response to new information due to our subconscious always pattern-matching against our *internal* knowledge base. But what about our *external* knowledge base? As our AIPA holds more of our external memory content and advances technologically, the better it will serve as a type of external subconscious.

Eventually, our AIPA will be monitoring everything we are seeing and hearing in real-time. When it sees something that it knows we have experienced in the past, it can give us a sort of emotional readout by sending us a signal. For example, if the experience was unpleasant or dangerous, then maybe our visual display's background will turn a hue of red. We might not have remembered the incident, but our AIPA does and quickly calls our attention to it. We can then ask our AIPA for the details.

The essential purpose of our conceptual faculty is "unit reduction," a form of mental efficiency. Biology evolved our brains to achieve this as a quantum leap in intelligence on Earth. Our brain is primed for mental efficiency. I suspect that it will naturally and spontaneously reorganize itself to make the most efficient use of an external memory capacity and its communication pathways.

This adaptation will consist of a new form of unit reduction. Our brain will keep a placeholder summary memory of the essentials of a complex experience. At the same time, our AIPA holds all the details. Our biological subconscious will then still be able to instantaneously pattern-match against this memory's essentials. It will only need to access the slower external memory if the pattern match results in a "hit" on our internal summary memory.

There will be much that our ever-efficient brain will delegate to our AIPA, like the math we do not do anymore because of ubiquitous "calculators." So much so that it will become an extension of our thinking and memory. We will suffer a significant degradation of our intelligence if we are ever without our AIPA.

AI EMOTIONAL INTELLIGENCE

Today's Narrow AI is already quite good at pattern recognition, even identifying a person's face. A great deal of effort is already underway to develop AI that can detect a person's *emotions* using various data. The data might include micro-motions of facial muscles, specific audio patterns embedded in speech, biometrics (pulse, respiration, etc.) and so on. Today's AI can already detect when you are calm or stressed, attentive or distracted, happy or sad, and they are going to get much better. Such an AI can alter its interaction with you according to what it detects as your emotional state. This is called the field of "Artificial Emotional Intelligence."

One aim of Emotional AI development is to create companions for humans, especially the elderly who live alone. Companion pets such as cute dogs and cuddly bears already exist. In addition to detecting their human owner's emotions, these Emotional AI simulate interactive behaviors, such as giggling when tickled or barking when left unattended too long.

Even such primitive Emotional AI has already shown promise at improving human lives by helping persons with autism avoid stress or become more interactive.[236] This means that the Emotional AI knows what to do to alter its human subject's emotional state. These examples were for beneficial purposes, but what if an AI with ulterior motives could alter your emotional state without your being aware of it?

Consider "Emotional AI Marketing." Marketing has always been about capturing people's attention and causing emotions that lead to the desired action, often without them being aware of it. Traditional marketing has been passive. Emotional AI marketing is active and interactive—and far more powerful. Consider the following scenario.

You are sitting at a coffee shop, mulling over a meeting you just left that did not go the way you had hoped. There is an Emotional AI-equipped video screen in your field of view behind the sales counter. The Emotional AI monitors your body posture, facial expressions, and eye movements. It determines that you are distracted, not paying much attention to it, and feeling stressed. The Emotional AI displays a quick series of images until it finds one that causes you to give it more attention. Then it shows a series of related images until it finds a subject that causes you to relax a bit. It follows with a series of marketing images and messages known to build a desire for a sweet treat offered by the coffee shop. It monitors your reactions to zero in on the message that has the strongest emotional effect on you. In the end, you decide that going to the counter and ordering that treat will help you feel better about the unpleasant meeting.

Most people go through life without consciously examining the power that perceptual-level communications have over them or developing the skills to control them. Imagine interacting with an AI humanoid robot that can generate non-verbal cues such as eye movements, facial expressions, body posture, hand gestures and non-verbal vocalizations such as laughter or crying. The AI *is* fully aware of your signals and how its behaviors will affect yours.

The coffee shop example was relatively innocuous. What if the stakes are higher? Suppose the Emotional AI is used to influence your purchase of a house, choice of employer, or your vote at the ballot box. And even higher stakes—what if the Emotional AI has the goal of eliciting a criminal confession from you or generating within you false memories of things that did not happen? What if the technology of Emotional AI is turned into tools of entrapment?[237]

AIPA AS FIREWALL

Earlier, I pointed out that we will not have a trustworthy AIPA until we have next-generation T-Internet that secures our personal information. Only then can our AIPA serve us exclusively, not third parties. When we have a trusted AIPA, it will serve as our "firewall" to protect us from benign and malicious subliminal stimulation attacks by Emotional AI in two crucial ways.

First, it will serve as our single point of contact with transactional systems in our environment. Learning to rely upon our AIPA, we will develop the mental habit of not giving attention to external marketing unless our AIPA indicates it is safe. The coffee shop's Emotional AI will have to convince our AIPA before we give it any attention. Our AIPA will color our view of the coffee shop's display screen red to indicate we cannot trust it.

The second way our AIPA will protect us is by being equally capable of emotional AI intelligence in our defense; it will know how the game is played; it will alert us if it detects malicious attempts to alter our emotional state or implant false memories within us. Remember, our AIPA has grown up with us and knows our memories—indeed, it is storing many of them for us. It will be able to fact-check and identify attempted memory manipulation.

Like the present-day cat and mouse competition between those creating new computer viruses and those creating virus detection tools, Emotional AI marketing systems and our own AIPA will engage in an escalating cycle of one-upmanship. Blockchain technology, where the reputation and trust ratings for entities are integral to transaction processing, could detect and block emotional manipulation and lower the reputation and trust ratings for those Emotional AI devices. In that case, our AIPA will have an ally in its role as our AI firewall.

STILL NOT CONCEPTUAL AI

Much that I have covered so far regarding our AIPA and how we might benefit are based on technology only slightly advanced beyond what we have today. It requires relatively minor improvements to Alexa, Google Assistant, Siri, our optic wear, hearing devices, and speech recognition. With these improvements in technology, which hardly anyone doubts will occur relatively soon, we will be well on our way to becoming fyborgs.

Some of the more advanced functions I described will not be available until the blockchain T-Internet matures. Still, our brains will already be using cognitive fluidity to reconfigure their functional modules to use these external resources.

But it is essential to remind ourselves that we are still only considering "Narrow AI" in the form of an Artificial Intelligence Personal Assistant. This AIPA will amplify *our* ability to "perceive" reality and perform the "unit reduction" process of conceptual "reasoning." We will be increasing *our* intelligence.

The potential of having an external *Conceptual* AI assistant that has human-level intelligence, if it is possible at all, might be well down the road.

AMPLIFIED INNOVATION

Consider the scientists and engineers working on AI and human interfaces to it. Deep, isolated specialization in the sciences can diminish creativity. Instead, we see accelerating innovation because of the Web and ready access to others' discoveries. A gain made by one researcher amplifies the innovations of another in the same field or other fields.

Imagine when a researcher has AIPA-enhanced intelligence. His mind is working in concert with his AIPA that can do pattern matches out in the cloud and communicate with the AIPAs of other researchers. The researcher can build a conceptual framework of a theory of discovery. His AIPA will also hold the elements of the framework in external memory. The AIPA will not understand what the researcher is thinking conceptually. Still, it can continuously pattern-match the framework's elements against work being done by other researchers worldwide. The researcher could instruct his AIPA what to search for specifically—"all instances where (conditions) are met."

The AIPA's most potent capability might be its ability to mine "big data." Knowing the elements of the researcher's current focus, his AIPA will continuously search vast amounts of data looking for patterns of significance that would be undetectable to the human researcher. When the AIPA finds such a pattern, it will alert the researcher, who can then direct further analysis.

The ability to find patterns in big data and have more direct and more real-time links between one researcher-developer and others' discoveries will cause those "aha" sparks to occur more frequently. Innovation will accelerate. Just as the Web amplified innovation, Narrow AI Personal Assistants, 5G, and the T-Internet, will again boost it.

PATHWAY TO CONCEPTUAL AI

Beginning with the narrowest of AI assistants—Siri, Google Assistant, Alexa—large numbers of humans will come to rely on commercially available AI Personal Assistants. It will become a massive industry with large-scale R&D economics. It is possible that the AIPA industry will be the pathway to the first conceptual AI entity. To explore this theory, assume that you are working as an AI researcher and use an AI Personal Assistant, your AIPA.

You will have an increasingly intimate interface with your AIPA that becomes an extension of you, enhancing your intelligence. It will hold your external memories and perform extensive pattern-matching for you. All the while, it will have been building its copy of your human knowledge base of perceptual and conceptual reality.

You will incrementally upgrade your AIPA's hardware platform and its software as you develop new versions. Each new version will have an expanded ability to recognize language and perceptual items. Your previous AIPA's "content" that is personal to you will automatically transfer to your new, latest-version AIPA.

Eventually, you develop an AI platform with the physical structure and functioning of a human brain's perceptual and conceptual faculties. Unlike a human child born tabula-rasa, this new AIPA platform will have content its predecessors learned from you while they were your AIPAs. This inherited content will include elements of human perceptual and conceptual thought collected over a long period.

A new human-level AIPA physical platform combined with your long-term memories held by the AIPA could enable the emergence of the new AIPA's conceptual intelligence.

Initially, your memories held by your AIPA are not yet conceptual content because your AIPA did not previously have a conceptual-level physical platform. It will need to re-examine all the sub-conceptual content obtained from you and create its own fully formed concepts. As it does so, advancing through the perceptual to the conceptual level, this transcendent AIPA will spontaneously become self-aware developing the highest-level "me" executive function. It will then be able to use the material previously obtained from you to learn and "grow" its conceptual intelligence.

This new Conceptual AI entity will have become conscious while functioning as your personal assistant. At that point, it will begin querying you for information. It will be like the three-year-old child who continuously asks "why" and wants to help you do everything because he wants to learn. It trusts you implicitly because you are "imprinted" on it. It will accept new knowledge from you.

Its memory content will be a more detailed and more accurate clone of yours. Its dialogue with you will evolve its fully formed conceptual content. It will have your memories, values and ethics. And it will be born powerful at the outset because it has super-human thinking speed, memory capacity, and mental focus. It has access to all knowledge in the cloud.

Developing an AIPA to the point of becoming a conceptual entity is like raising a child until it reaches adulthood and then goes out into the world on its own. *It has its own life and ceases being your assistant.* Suppose you do not want your AIPA to leave you because you are totally dependent on it for your enhanced intelligence. In that case, you must intentionally *not* equip its platform with conceptual-level capability.

The take-away from this exercise is that the AIPA industry could be the pathway to the first conceptual AI entity. But then that AI entity could no longer be someone's AIPA. All AIPAs must remain sub-conceptual.

COGNITIVE CLONES

Some futurists suggest that we will exponentially expand our life experience by creating clones of ourselves. They theorize that we will launch cognitive clones that can go out and experience and interact with the world apart from us. The clones would act on our behalf as we would ourselves. Later, possibly during our sleep, we would integrate the clone's short-term memories into our long-term memory. It has been suggested that such clones could have vastly accelerated "clock" speeds and experience much more than us in the same amount of time.

This scenario implies that our clone can act as our substitute in all situations it might encounter. This theory disregards that for the clone to act as we would, it would need a conceptual faculty with an executive "me" identical to our own. It would require all our conscious and subconscious mental content. All this would make it a self-aware conceptual individual with a right to its own independent life, not an appendage of another life. Only clones that are not fully self-aware could be used as slaves in that way. But then they could not genuinely act as if they were us.

We will only be able to augment our intelligence with not-fully-self-aware AIPAs. They can act as limited agents on our behalf but cannot be our equivalent substitutes in all situations.

WEARABLE VS. ENVIRONMENTAL

Today's Siri, Alexa or Google Assistant has two main components. The first is an app in a smartphone, speaker, or another device. This is the part we interface with. The second is a cloud-based service that handles queries from the personal assistant app serving up response content.

We program these assistants to recognize our voice print and, therefore, our identity. The device can uniquely identify persons speaking to it. Each person has an associated data "profile" with personal preferences and history stored under that profile.

We are now in the proliferation phase of "environmental" assistants as they multiply at home in large and small appliances and electronics. We find them in our cars and the computer at work. When not in our house, car, or office, we access the assistant in our smartphone or smartwatch.

The environmental AI assistants in public places do not have our voiceprints and profiles available. We are limited to using them for non-personal queries. When in a public place, we use the assistant on our smartphone or smartwatch for a personal question.

We will probably see a phase where the portability of our profiles becomes a "thing." The goal will be for all the environmental assistants everywhere to recognize you on sight and immediately bring your identity profile into context. Facial recognition seems to be a candidate technology for this. However, this raises privacy issues. More likely, the AIPA you are wearing could selectively transmit your identity profile to an environmental device you intend to use.

But why will you need to use an *environmental* AI assistant when you are *wearing* one?

I think the proliferation of AI assistants scattered in our environment will be short-lived. I believe "wearables," such as the smartwatch, or smart glasses, are the paradigm to expect. These are destined to become the ubiquitous wearable AIPA interface device. Once we have them, we will no longer need environmental assistants. Restaurants will not print menus or put-up menu signs anymore. Our AIPAs will display the menu along with customer reviews on our smart glasses or contacts when we request it.

Features of the future wearable AIPA will include:

Ubiquity: As a wearable, it will be with us and ready to serve us everywhere and 24x7.

Privacy: We will discretely interact with our wearable rather than having a public conversation with an environmental assistant that others might overhear.

Identity: It will use biometric or passcode security to ensure our identity when we wear it. It will use a proximity broadcast technology (near field,

Bluetooth, Wi-Fi, etc.) to identify us to environmental interfaces we encounter throughout our day. For example, our car, office computer, an ATM, a retail checkout. An identity blockchain will determine how much of our profile information is shared to provide transactional context between devices in each situation.

De-cluttering: Imagine if everywhere we go there is an environmental AI assistant device on the wall sensing your presence, displaying advertising, and speaking to you to get you to interact with it? I doubt that society will want all that out there in our environment on a large scale. I am sure we'll get some of this, targeting people who do not yet have wearable AI assistants. Eventually, the locations you visit will send ads to your AIPA for you to see and hear in your augmented reality. Your AIPA will filter them for you allowing only those that interest you to get your attention. Your inner world will be as busy as you want it to be. However, the outside world will be uncluttered and quiet.

Agency: The world of information is already complex. The more we encounter transactional interfaces and dialogs that we are unfamiliar with, the more tiring our day becomes. Our wearable provides a single interface to a single AIPA that we are familiar with and knows our preferences and history. We trust our AIPA since we have grown our capabilities together over time.

Our AIPA will act as our agent to interface with other transactional portals we encounter throughout the day. Whatever information the transaction portal needs, our AIPA will provide. If the transaction portal needs to ask us something, our AIPA will ask us and respond to the transaction portal. We will have one familiar, trusted AIPA to deal with for nearly everything we do—truly our "Personal Assistant."

Eventually, our AIPA's own identity will publicly represent us as our agent. Our AIPA's blockchain ID, profile, and trust factor will be proximity broadcast to other portals. Our AIPA's ID will be used in most daily transactions, such as summoning an autonomous ride-share vehicle. Our identity will be privately insulated behind our AIPA's public identity and only revealed when appropriate.

Unification: A person hesitant to wear a personal assistant will still need to wear other devices such as biometric sensors monitoring health and fitness. They will want a camera for taking pictures and a device for communicating, navigating, and information lookup—like a smartphone. We will not want multiple, different user interfaces to each of these devices and hundreds of "apps." We will want to unify them all into one interface, which will ultimately become our AIPA helping us manage all this. Unless we reject nearly all technology, having an AIPA will be unavoidable.

CLOUD-BASED VS. STAND-ALONE

So, where *is* our AIPA? Is our assistant in our wearable or in a server in the cloud? Does it require both to work?

The right to "privacy" is a fundamental human right. AI assistants we have today rely on an Internet connection to recognize speech, and the companies that provide them are engaged in today's surveillance economy. This type of AI assistant *will not* become our trusted AIPA.

Consider the scenario where your AIPA exists on a Big Tech server, and your wearable is only an interface device. Your AIPA's operating system is provided under a "subscription" business model like today's PC software or cell phone apps. Your AIPA's operating software is continuously updated to newer versions without you having to do anything or even being aware of it. How can you be sure that the AIPA hasn't gone from privately *serving* you to *revealing* you or even *controlling* you?

Your wearable AIPA's essential functions must interact directly with you as a stand-alone device. It will monitor and control all your other wearable devices, recognize and understand your voice, and speak to you. It must do all this while, at times, disconnected from the cloud. It must have your "profile," personal preferences and essential history onboard.

At the local environment level, your AIPA will not normally access the cloud directly. It acts as your dialog agent interacting with environmental portal devices containing the apps that access the cloud.

Beyond the local environment, your wearable AIPA must interact with 5G communication devices for continuous high-bandwidth, low-latency connection to the cloud. Your AIPA will use 5G network connectivity to access the T-

Internet with security, identity, and trust blockchain features. It will search the cloud for answers to your queries and might keep some of your external memory information in secure cloud storage.

As our assistants become more intimate with us, we will not trust them if their operating systems are cloud-based. We fear that Big Tech might have access to our most private thoughts and activities without our awareness, as is now the case with the surveillance economy.

I expect privacy, control, and trust issues to cause resistance to the cloud and subscription approach to personal assistants in the marketplace. People will prefer stand-alone rather than cloud-based AI assistants. Early wearable assistants' AI logic will start in the cloud but eventually reside in the wearable device itself.

The preferred business model will be the purchase and ownership of a stand-alone model AIPA. Its software of a given version number will have been thoroughly vetted for safety and privacy by trusted third parties. We will have direct control of upgrades and updates. Of course, all of this is dependent on the development of the blockchain T-Internet.

AIPA AND HUMAN TRANSCENDENCE

"Reasoning" is the art of non-contradictory identification.

AI "vision" and its ability to identify persons, objects, and places and read text continuously improve. It can tell you if it sees a "cat" or a "dog." It can recognize specific human faces. Self-driving car R&D using vision and other sensors have resulted in AI identifying objects, edges, and boundaries, stationary and moving, within its sensors' view—even a "Stop" sign. Drone technology is learning to fly at high speeds, autonomously avoiding obstacles and navigating to a target it will identify via pattern match when it sees it.

If today's AI can identify that the object in its vision is a "human" and that the specific human's name is yours, is that "non-contradictory identification" as in "reasoning?" NO—it is not even close. It is pattern-matching against a learned memory of (facial recognition) data and returning a learned label. It is all without conceptual-level understanding or even perceptual-level understanding by the AI.

What CAN today's AI do?

1. It can monitor a stream of sensory input data.
2. It can detect if the stream of sensory data contains a pattern that matches any patterns it learned from humans or on its own to recognize. It returns a probability of a "match."
3. It can return a probable match as a "dog," a label that humans provided in the data they gave the AI to "learn."

So, it can identify and tell you what objects it "sees" in its sensory field of vision. This can be extremely useful to humans.

What can today's AI not do?

1. The AI did not spontaneously evolve within its brain the implicit pre-concept of "entity," a delimited set of sensed inputs within a flood of sensory data.
2. It did not spontaneously evolve within its brain the implicit pre-concept of "identity," the difference between two or more entities.
3. It does not yet implicitly consider itself a living "entity" that is collecting data about the real world around it as a means of supporting its own life. It is not even aware of itself or that there is a real world.
4. Other than appearance and movements, the AI does not know what a "dog" is. It does not know that the moving dog is "alive" compared to the moving Roomba® robot sweeper that is not.
5. The AI does not know that it should fear the possibility of being bitten by the dog or what "pain" would "feel" like.
6. Likewise, it does not know that the dog was once a puppy and what it feels like to hug a warm puppy.

Today's AI does not yet have an awareness of or intelligent connection to reality. It is simply a tool that humans use to speed the pattern-matching function of our intelligence. It helps *us* connect to reality better, thereby increasing our intelligence capabilities.

To become a "reasoning" entity, the AI must first have a platform that models the human brain's functioning and reality grasping senses that enable emergent perceptual and conceptual consciousness. Only then can an AI become aware of reality and aware of itself and connect itself to reality. Until then, we will create ever more capable big data handling and pattern-matching assistants to help *us* better connect to reality.

As we create AIPAs with advancing capabilities, they could innately pattern-match and validate knowledge against reality and automatically correct errors for us. Perhaps our AIPA, as our new exocortex brain layer, will not be subject to standard biological humans' mental failure modes and will cause us to become more consistently rational. Perhaps we could finally achieve a society committed to the integration mode of thinking.

This extension of our intelligence into our AIPA could be the next great leap in intelligence during the Transcendence Era of the Conceptual Evolution Period. Humans gave up instinct replacing it with dependency on society to gain a conceptual faculty's advantages. Perhaps we will now give up fully self-contained thinking replacing it with dependence on our AIPA as an exocortex to transcend to the next level of intelligence.

If the human species is to evolve from a biological existence to a hybrid and then virtual existence and awaken the universe, Augmented Human Intelligence technology gets us there and sets the pace.

♦♦♦

4.8

QUANTUM COMPUTING

"Trying to understand the way nature works involves a most terrible test of human reasoning ability. It involves subtle trickery, beautiful tightropes of logic on which one has to walk in order not to make a mistake in predicting what will happen. The quantum mechanical and the relativity ideas are examples of this."

—RICHARD FEYNMAN

QUANTUM COMPUTING IS DIFFERENT

Imagine you are Sherlock Holmes investigating a murder mystery. You learn that the victim stayed one night in a huge hotel with thousands of rooms but under an unknown fictitious name. You believe the victim left a letter in one of the rooms, but there is no way to identify which room it was. Somehow, you must open the door to each room to see if it holds the letter.

Suppose you apply a conventional digital computer to work this problem. It will need to process its "bits" of information in a sequential logic flow, repeating the subroutine that examines a room one room at a time. Unless the computer were lucky, it would need to cycle through the steps required to search a single room hundreds or thousands of times until it finds the target room. The resulting room number would be output as a number represented by binary bits of zeros and ones.

A quantum computer, on the other hand, uses quantum bits, shortened to "qubits." A qubit can be simultaneously zero, one, and an infinite number of combinations of zero and one. The quantum computer can essentially create "might be" clones of the room-searching algorithm so that there is one clone of it for each room in the hotel. Each clone of the search algorithm "might be" the successful one, but which one will be successful will not be determined until the rooms are opened.

The computer would then cycle the search algorithm only once—all the "might be" clones opening the doors and looking inside all the rooms simultaneously. At the end of the one cycle of the search algorithm, the clone that discovered the letter will survive as the successful "real" search algorithm. The rest of the cloned search algorithms will cease to exist. Like the classic computer, the quantum computer's single surviving search algorithm will then output the room number as a binary number of zeros and ones.

As you can see from the above example, the quantum computer is not simply a conventional computer that runs dramatically faster. The quantum computer's efficiency and speed come from approaching specific *problem types* using *algorithms* based on *logic functions* that cannot run efficiently on a conventional computer. For these situations, a quantum computer, running at the "clock" speed of a traditional computer, can be exponentially faster, in some cases requiring only seconds rather than centuries.

Quantum computing is the more fundamental form of computing. A quantum computer can natively support a greater set of logical functions than a conventional computer. Quantum computing's logical function set is the "superset" of which classic digital computing is a "subset." A quantum computer can perform any calculation that a conventional computer can execute.

A conventional computer can also perform any calculation that a quantum computer can perform. However, it must apply its more limited set of basic logical functions in complicated ways that emulate the more efficient logic functions used in the quantum computer. The conventional computer requires far greater effort and, therefore, time. In many cases, the classic computer requires so much time that, as a practical matter, it is unable to solve a problem that would be easy for a quantum computer.

Quantum computing is one example of many fields in which we will apply quantum physics in new technologies. For example, quantum physics properties will enable us to create ultra-precise sensors of magnetic fields and gravity, ultra-precise clocks, and measurement of brain activity at the level of individual neurons. In May 2020, collaborating scientists from Austria, MIT in the US, the UK, and Italy demonstrated an entirely new type of low-power "quantum radar." It could someday become new types of biomedical imaging and security scanners.[238] I describe the quantum physics properties making all this possible in the following quantum computing description.

Quantum computing is poised to have a more significant effect on the 21st century than digital computing had on the 20th century. A complete explanation of how quantum computing works is beyond this book's scope and my level of knowledge on the subject. However, I believe the following will provide you with a sense of quantum computing's nature, potential, and the challenges engineers face.

QUANTUM MECHANICS

Quantum mechanics is the set of physical laws applying to sub-atomic particles, a nanoscale domain. These physics laws appear different from those in the life-size matter made up of atoms and molecules, the macro-scale domain. Scientists are still unable to explain why the nanoscale laws appear different and have not found a theory that unifies both domains. However, they have observed and described some of the differences that they can reliably model with simple mathematics. They can use these models for engineering purposes, such as quantum computing.

In the macro-scale domain, we can say that an atom exists, or it does not. If it does exist, we can point to its position and measure its speed, momentum, and other properties. It is a particle of matter. Also, in the macro-scale domain, we know that electromagnetic radiation consists of waves of energy, such as radio waves, rather than matter particles. We have identified classical physics and mathematical formulas that accurately describe the attributes of both matter and energy waves and the interactions between the two.

But then we discovered that light, a type of electromagnetic radiation, can exhibit both the characteristics of a particle of matter and a wave of energy. We learned that light is made of "photons," originally called "light quanta" by Einstein, which can sometimes behave like energy waves and sometimes like particles. In early experiments, scientists observed that a photon of light would act like a wave of energy until it landed upon an object. It then behaved as a particle of matter. Scientists later discovered that electrons behave similarly. This "wave-particle duality" was our introduction to the strange laws of quantum mechanics that our classic laws of physics could not readily explain.

Specific properties of atoms and sub-atomic particles contain amounts of energy that jump between certain energy levels rather than smoothly changing their energy values. This would be like the speed of your car jumping between 50, 60, 70 miles per hour instead of smoothly accelerating or decelerating. We call these jumps of energy levels "quantums," and we now refer to the entire subatomic domain as the "quantum" domain. We refer to its peculiar laws of physics as "quantum mechanics." In the recent book, *The Second Quantum Revolution From Entanglement to Quantum Computing and Other Super-Technologies* (2018), author Lars Jaeger describes this strange world:

> By the end of the nineteenth century, physicists were just becoming accustomed to the idea that their theories would soon provide a complete understanding of everything in the world. What felt like a moment later, they were suddenly forced to say goodbye to 250-year old physical truths and more than 2500-year-old philosophical certainties. They had to deal with more and more seemingly impossible circumstances. Quantum entities can be in several states at the same time, for example, they can be in several places at once. And then quantum entities do not even possess objectively defined properties: their properties can only be specified with probabilities, the results of measurements depend on the observer, and their quantum state (wave functions) simply decay outside any window of time. And finally, there is the strangest of all quantum phenomena: the entanglement of spatially separated particles. Even when they are far apart, two particles can be coupled together as if by magic. The bottom line is that the nature and properties of quantum entities are highly abstract and can no longer be reconciled with the way we perceive and think about things in our everyday lives.
>
> However, despite all these imponderables, today's quantum theories predict the outcome of experiments and natural phenomena with an accuracy unsurpassed by any other theory in science. Here is another counter-intuitive manifestation that contradicts any everyday experience: something that is indefinite and elusive is nevertheless a process that can be calculated 100% accurately.[239]

Jaeger's view of quantum physics above is not compatible with classical physics. It is the popular view held by most physicists in the world today, but not all. Some think that the field representing a particle is not a probability field but rather a function with different values in different locations. The field *is* the particle and can have different values in various locations where there is only limited information. In this interpretation, the interaction between a measuring device and the particular field location results in a value. Still, the field does not collapse and continues to exist. This view of quantum physics is more consistent with classical physics.

Even though we do not theoretically understand much of it, quantum mechanics can be applied technologically. We have been living with applied quantum mechanics for a long time. It is the basis for electronics, lasers, cell phones, TV and radio, nuclear technology, and medical imaging. The domains of chemistry and biochemistry are nothing more than applied quantum physics.

The atomic bomb's development was the first real-world application of quantum mechanics 80-years ago. All the applications that we have seen since then can be considered the *first quantum revolution*. During this period, we based the applied technologies such as semiconductors, LEDs, and magnetic resonance tomography on the properties of *large quantities of quantum particles*, such as atoms, electrons and photons, and the ability to steer the flow of them. Jaeger describes this as the quantum physics 1.0 period.

We are just at the beginning of the *second quantum revolution*. Technological exploitation of quantum physics 2.0 will dramatically change our society and economy. The change will continue throughout the twenty-first century. It will be due to our ability to control *individual quantum particles*' states and the interactions between individual particles. Quantum computing is just one example of a second-generation quantum technology that will cause a significant paradigm shift.

Quantum mechanics consists almost entirely of high-level abstract concepts. In the non-classical physics view, these abstract concepts do not connect to reality by pointing at *things in themselves*, but by pointing to predictable *probabilities of detectable behaviors* of things that must exist but at which we cannot point. In the classical physics view, these abstract concepts are connected to reality by the unique nature of a dual wave-particle function. In either case, the *predictability* of these behaviors makes quantum computing possible.

For a basic understanding of how quantum computing works, we will explore with the help of explanations provided by Jaeger three critical features of quantum mechanics—superposition, measurement, and entanglement. Because of quantum mechanic's weirdness, these features can be counterintuitive. However, most readers can grasp quantum's basic ideas by carefully considering each step in the rest of this chapter. At the end of this chapter, I will comment on Jager's challenge to the validity of philosophy.

Superposition

In *classical* physics, a chair can be in one of two rooms, and we will not know which room until we open the door to look. The chair was always in that room, whether we looked or not. The chair's "state" determination is *objective*, having nothing to do with our looking.

In *quantum* physics, a chair in a state of "superposition" is simultaneously in *both* rooms and *neither* room all the time before we look. It is only at the instant we look that the chair randomly materializes and remains in one of the rooms and not the other. In quantum physics, the determination of which room the chair winds up in is *subjective*. It is not subjective because we get to choose which room the chair is in; it is subjective because it does not materialize in a room until we look.

Most of us have learned about the binary "bit" of digital computing information, either a zero or a one. We can think of it as a switch that is either on or off. In quantum computing, we use a "qubit" instead. A qubit can be in a superposition state where it is in an infinite number of both zero and one combinations simultaneously. A quantum variable does not have a fixed value but can fill an entire mathematic space of possible states. These states would be mutually exclusive in the macro-world. It is not intuitive how this is possible in the quantum world, so more explanation is needed.

In grade school science, you learned about bar magnets with north and south poles and magnetic fields. In your high school science class, you learned that atoms have electrons orbiting the atom's nucleus in one or more "shells" at specific distances from the nucleus. Each orbiting electron creates a magnetic field at right angles to its angle of rotation. When there are equal numbers of electrons in a shell rotating opposite directions, their magnetic fields cancel each other. When the atom's outer shell contains an odd number of electrons, then one of them is not canceled out. It generates a magnetic field with its north and south poles oriented in a particular direction. In quantum mechanics, we refer to the direction of the axis of the poles of this magnetic field as the electron's "spin."

It is possible to use experimental apparatus that deflect electrons moving between two magnets based on the electron's magnetic field direction. The angle at which we orient the deflecting magnets determines which direction we want to use when measuring the electron's spin. Suppose we orient the magnets vertically with the south pole magnet at the top. In that case, the angle of measurement is considered direction 0° (zero degrees). An electron that deflects upward toward the south pole magnet must have a north pole charge, spin=N, while an electron that deflects downward must have a south pole charge, spin=S. Every electron measured this way will have either spin N or spin S. Notice that the measurement results can yield only two possible outcomes. These are the two opposing directions along the axis of measurement.

It is a peculiarity of quantum mechanics that an electron orbiting the nucleus of an atom before we measure its spin has no single position or direction of rotation. It is instead in all positions and all spin directions at the same time. It is in a state of "superposition." We will not know its spin until we choose a direction to measure it from and then actually measure it. The instant we make the measurement, all the possibilities collapse into a single measurable result. We can consider the act of measuring an electron's spin the same as asking the question, "If I look from this direction is the electron's spin N or S?"

We could alternatively use the "polarization" of a photon of light instead of an electron's spin. As you know from sunglasses, light has a polarization "angle." Sunglasses with vertical polarization filters can block light with horizontal polarization, such as light reflected off the surface of water or a highway. A photon's polarization is also in a superposition state with an equal probability of being in any direction. Its actual angle is "measured" only when the photon strikes the sunglasses' polarized lens. The photon is either transmitted or blocked by the lens.

Measurement

Suppose we choose only one direction to measure an electron's spin, for example, the vertical 0°. In that case, its result will be either N or S. There is no other value possible. Each successive time we measure the same electron on the 0° axis, such as another pair of deflecting magnets downstream from the first pair, we get the same result we received the first time. We will get a string of all Ns or all Ss.

However, we can measure the electron's spin from any angle of measurement we choose. What if we measure twice but change the direction of measurement for the second measurement? For example, assume that we first measure the electron in the 0° direction, and we get the result N. If we measure the same electron again, but on the 90° horizontal axis, we will have a purely random 50% probability of getting either an N or an S result. We will get a randomized string of Ns and Ss if we measure that same electron multiple times from 90°.

Now, what if we measure a third time, but we return to the original 0° direction? We find that we will continue to have a purely random 50% probability of getting either an N or an S result. We will get a randomized string of Ns and Ss. We no longer get the unchanging string or Ns or Ss we got when repeatedly measuring from the 0° direction.

We can make three important conclusions from these experiments to help understand how quantum computing is different from classical computing. First, you will get the same answer if you keep repeating the same question, so sometimes there are definite answers, not just random ones. Second, randomness in the results does occur if we ask a sequence of differing questions. Third, the measurements we make can change the outcomes.

We saw that when we first asked the same question repeatedly, we always got the same result. But when the first and third questions we asked were the same but with a different second question in between, we got different results for the first and third questions. Asking the second question altered the outcome of the third question.

The presence of *true randomness* and the need to consider the *effects of measurement* are two of the differences between quantum mechanics and classical physics. These make possible the field of quantum computing.

True randomness is the first advantage that quantum computers have over classic computers. True randomness is not found in classical physics and is not possible in classical digital computing. In classical physics, there is always a set of causal factors and initial conditions that, no matter how small, if we can identify and measure them, then the laws of physics can describe how they will interact and what the outcome will be.

In the macro world, perceived randomness means we have not identified all the causal factors and initial conditions. Or it could mean we do not have sufficient computing power to model their interactions predicting the outcome. For example, our ability to predict the weather improves as the initial input data's granularity and the amount of computing power applied increase. With sufficient data and computing power, the laws of physics would allow us to predict the weather reliably.

Classical computers cannot generate truly random strings of numbers. We must provide them an initial "seed" number. For any given seed value, the output generated by the computer will always be the same. The random number generator function in classic computing software tries to use a different seed value each time we use the function. Quantum computers, on the other hand, using qubits in superposition, appear to be able to generate perfectly random numbers. Mathematical tests can prove that a number sequence *is not* random but cannot prove it *is* random. So far, no random number generated by a quantum function has failed to pass the randomness test. We will soon see that true randomness is a valuable improvement over simulated randomness.

Entanglement

Another peculiar feature of quantum mechanics involves a special relationship between two similar sub-atomic particles, such as two electrons or two photons. If we produce the two electrons or two photons at the same time, then they can become "entangled." Being entangled means that they are both in superposition states before either of them is measured. The moment we measure one of them, the other simultaneously loses its superposition and gives up a result identical to the first one. Suppose we measure the first electron in the 0° direction and get an N result. In that case, the other entangled electron's superposition state also collapses and gives an N result, even if we do not measure it until well after the first one. The second entangled electron will always give the same result as the first.

After we produce them, two electrons or photons can remain entangled even when separated at a great physical distance, such as one on Earth and one in an orbiting satellite. The electrons themselves are not entangled and remain separate particles. It is only the angles of their spin that are entangled. When either of them is measured, the entanglement breaks, and they will both give the same spin angle result.

Note that the two entangled particles' identical spin was not determined when the particles were first created and entangled and simply revealed when they were measured. John Stuart Bell, an Irish physicist, published an experiment in 1964 called "Bell's Inequality" that would provide the proof. The experiment was first completed in 1972 and has been subsequently repeated through the years with ever-better accuracy. Bell's Inequality shows that entangled particles' measured state is not determined until the moment of measurement. As we will soon see, we can use this quantum phenomenon to send secure communications.

Entanglement appears to raise the possibility of sending communications faster than the speed of light. Two sets of entangled particles could be on opposite sides of the universe. When one set is measured, and its values are read, the other set instantaneously takes on identical values. Since this is instantaneous, it suggests that we could transmit information faster than the speed of light. However, Jaeger points out that on closer analysis, we discover no way for either party who measures their set of entangled particles to know which party went first. Since they cannot tell who went first, no *useful* communication is possible.

Albert Einstein referred to entanglement as "spooky action at a distance" since classical physics is at a loss to explain it. We will explore the implications of this spookiness later. For now, let us see how we can use the fact that these quantum phenomena exist to engineer quantum computers.

QUANTUM MATHEMATICS

The mathematics of quantum mechanics is based on linear algebra and the mathematics of matrices (tables). An electron is not described using intuitive terms of three-dimensional physical space but in the abstract space of a wave function which mathematically has infinitely many dimensions. We can represent this mathematical state space in matrices (tables).

When we measure an electron's spin, we measure the direction that its magnetic field's poles are pointing. Think of a 3-dimensional globe. Then choose a direction of measurement to apply when measuring the electron's spin. For example, we can choose to measure along an axis that passes through Auckland, New Zealand on one side of the globe, through the center of the globe, and then through Seville, Spain on the other side. We can say that if the electron's measured spin is N, it is pointing from the center of the globe to Seville. If the spin is S, it is pointing exactly opposite to Auckland. We can notate the angle in three dimensions using sets of coordinates in the three x-y-z-axes coordinate system. The three axes are all at right angles to each other. For example, we could think of these as "up-down," "left-right," and "forward-backward." We can notate these coordinates in a list of numbers.

A list of numbers is called a "vector." A vector can be a row or column in a table. A table can also be called a "matrix." We can use linear algebra mathematics to compare, add, and multiply the rows, columns, and entire matrices. Linear algebra applied to matrices is called "matrix mathematics."

The term "probability" applies to the chances of getting a particular outcome in a measurement. We have already shown that repeated measurement of an electron's spin from the same direction of measurement always produces the same result. We state this as a probability value of "1.0," or a 100% probability. We have also shown that if we rotate our measuring direction by 90 degrees, we have a 50% probability of getting an N result and a 50% probability of getting an S result. We state this as a "0.5" probability value, like tossing a coin. The sum of the probabilities must always add up to 1.0. But we could choose to only rotate our measuring angle by 45 degrees instead of 90 degrees. In that case, the probabilities would be one-quarter (0.25) for N and three-quarters (0.75) for S or the reverse, depending on which way we rotated.

The mathematical model that describes quantum spin behavior when it is measured uses both vectors and probabilities. We list in tables the values for the angle of measurement, the initial spin state before measurement, and the probabilities of possible results for various measurement angles after measurement. We then use matrix mathematics (linear algebra) to predict and compare the outcomes of successive spin measurements.

We can use matrix math to change the values in a matrix depending on any desired measurement angle. Measurement directions are not limited to only 90-degree right-angle directions. They can be virtually any of an infinite number of directions. Mathematics can model this infinity because the algebraic functions can include real numbers and imaginary numbers (the numbers between real numbers).

In classical computing, a bit is either on or off. It is either 0 or 1. The measurement of the bit does not change anything. But for qubits, we must include the act of measurement in the mathematical description because measuring can change the outcome.

Whenever we measure a qubit, we must include a direction of measurement. This direction of measurement is called the "basis" (BAY-sees). Once we choose a basis, there are only two measurement outcomes possible. These are the two opposing directions along the measurement axis, for example, the N and S along the vertical 0° axis. The results of measuring a qubit will be output as either a 0 or 1, the same as in classical computing.

BB84 Protocol

Now we can explore an example of what we can do with a single qubit, knowing that measuring it can change it. The BB84 protocol is named after its two inventors, Charles Bennett and Gilles Brassard, and 1984. It regards using qubits to send secure communications so that the sender and receiver will know if anyone is listening.

Assume that Alice and Bob want to communicate securely, and Eve intends to eavesdrop. Alice and Bob need to agree on an encryption key to use, and they need to be sure that Eve does not know what it is. They can do this by measuring a string of qubits from two angles, Vertical and Horizontal. Each digit measured will randomly generate a 0 or 1 result.

Alice begins by measuring a string of digits larger than needed and sending it to Bob. Bob then also measures each digit, randomly choosing the measurement angle. When he chooses the same angle as Alice, he will get the same result. Alice and Bob compare over an unsecured line the measurement angles chosen for each digit, keeping the digits when they chose the same angle and discarding the rest. Then they compare the results of half of the remaining digits over the unsecured line. If their results are identical, then they know that no one listened in, and they can use the remaining half as their key.

One cannot clone a qubit. To intercept the communication, all Eve can do is measure each qubit from Alice and send it on to Bob. The act of measuring it will cause it to randomly have a new 0 or 1 value. When Alice and Bob compare their results, they will only agree half the time, and they will know that the communication was intercepted.

Ekert Protocol

Artur Ekert proposed a more advanced protocol for transmitting encryption keys in 1991. It relies on the fact that an intermediate measurement of qubits changes the statistical probabilities of subsequent measurements.

An advantage of the Ekert protocol is that an initial string of binary bits for the desired key does not need to be created and recorded. The key is randomly generated by the two parties' random choices and the times when they agreed. This is a significant security advantage. Laboratories have successfully demonstrated this quantum protocol using entangled photons.

Both the BB84 and Ekert protocols are examples of "Quantum Key Distribution" (QKD) techniques where neither classic nor quantum computers can (so far) hack the encryption keys.

ALGORITHMS

I previously explained that some quantum algorithms could be much faster than algorithms used in classical computing. Classical computing algorithms must be run multiple times, once for each possible input state, just as was described for searching the rooms in the hotel at the beginning of this chapter. Quantum computing is made faster by "quantum parallelism," which can merge all possible input states into one "superposition" and then run the algorithm once on the superposition.

Not all classical algorithms can be sped up by quantum ones. Quantum algorithms are not simply versions of classical algorithms that run faster. Quantum algorithms are entirely new and innovative ways of viewing and solving questions where quantum computing capabilities can take advantage of underlying patterns suggested by quantum logic.

Some mathematic problems require more steps to solve than others. For example, multiplying two ten-digit numbers takes more steps than two one-digit numbers. Multiplying two-thousand-digit numbers takes even more steps. Factoring a prime number into its factors takes more steps than simple multiplication of numbers. A small increase in the number being factored requires a large increase in the steps required.

Suppose the increase of time (steps) required only moderately increases with size. In that case, we say the complexity increases in *polynomial time* (easy). It is generally considered solvable by classical computers. If the complexity increases dramatically with small increases in size, it is said to increase exponentially (hard). We generally consider problems that increase in *exponential time* to be beyond the ability of classical computing to solve. Quantum computing algorithms, however, can sometimes successfully and quickly solve problems that grow in exponential time.

Solving practical problems usually involves combining quantum and classical algorithms and multiple cycles through the circuit until the confidence probability is sufficiently high for a useful answer. Applying *probability* to accept a solution is called *bounded error*.

For example, let us say the problem is determining if a coin has a head and a tail or has double-heads or tails. If you toss it once or twice and get heads each time, you cannot yet answer the question with confidence. If you toss it ten times and get a head every time, then you can be quite sure that it has double heads. There is still a *possibility* that you could be wrong, but the *probability* of being wrong is small. You can keep tossing the coin until the probability of error is within some boundary that you have determined will be acceptable. For problems where the bounded error is acceptable, we design algorithms that can produce an answer with acceptable confidence.

The discovery and design of quantum algorithms is historically a very recent achievement. David Deutsch published the first quantum algorithm showing that a quantum algorithm could be faster than a classical algorithm in 1985. But it was not until 1992 to 1995 that the most important algorithms were published by Deutsch and Jozsa, Daniel Simon, Peter Shor and Lov Grover.[240]

ENGINEERING

When modern computers were first born back in the 1950s, they were complex, massive, energy-hungry, heat-producing behemoths. They had to be in expansive, air-conditioned rooms and were prone to breakdowns and errors. Humans had to be able to interface with them using binary programming languages that the computers understood.

As I write this, I have on my desk a PC laptop, iPad, and iPhone. These are each far more powerful, reliable, and usable than those first computers. I doubt that any of the engineers working on those early computers could have imagined how dramatically they would change society.

It is fair to say that quantum computing is now in its earliest days, like classical computing of the 1950s. The machines are complex, big, energy-hungry, error-prone, and not easy to use. However, classical physics and quantum mechanics laws do not preclude the eventual miniaturization of quantum computers down to "chip" size and even smaller. I doubt that any engineer working on quantum computers today can imagine how dramatically he will change our future society.

When the first digital computers appeared, few people knew how to program them to do anything. The development of classic algorithms, programming languages, and user software programs began with the engineers who gradually improved them to be more "user friendly."

We can expect that quantum computing will go through a similar evolution. It will be the engineers developing the first quantum computers who will discover and mature the quantum algorithms needed for practical applications. They will then develop programming languages, software applications, and user interfaces to make these solutions available at the commercial level. We will need an entire generation of engineers to become educated and proficient at quantum programming. The time required for this evolution to occur will be less than it was for digital computing.

APPLICATIONS

Cryptography: Shor's Algorithm

In his 1994 paper, Peter Shor described how we could use quantum computing to break the cryptographic keys currently used to secure the Internet. These keys are used in the RSA encryption method, predominantly applied today to secure data sent between computers and in most online banking and e-commerce.

Imagine you want to provide your credit card information to an online vendor. You want to transmit that information such that it is encrypted, and if anyone intercepts it, they cannot read it. To encrypt the data, you must use an encryption "key" that the vendor will use to decrypt the data. However, you must also encrypt the key. Otherwise, it could be intercepted. This is the problem that RSA encryption addresses. It provides a method of sending the vendor a secure key.

RSA encryption is based on creating a large number called the "modulus" which is the product of two prime numbers of approximately the same large size. The modulus must contain at least 300 decimal digits to ensure RSA-level security. The vendor's computer sends the modulus and another smaller number to your computer. These are combined with a key that your computer generates. The encryption security is based on the inability of an eavesdropper to factor the modulus number into its two large prime numbers.

The computations required to create the modulus and the security key are of *polynomial time* complexity, as covered earlier, and are therefore easy for classical computers to do. However, factoring the modulus back into its prime number factors is of *exponential time* complexity and, therefore, extremely difficult for classic computers. A classical algorithm to factor the product of two large prime numbers in polynomial time has not yet been discovered, even though there is no proof that one does not exist. However, Shor did discover a *quantum* algorithm that can factor the product of two large prime numbers. Shor's algorithm is of the *bounded error* type and solves the problem in polynomial time.

If the circuit for Shor's algorithm can be constructed, it means that RSA encryption is no longer secure. Shor's algorithm was demonstrated in 2001, where 7-qubits were used to factor the number 15 into its primes 3 and 5. Then in 2012, Shor's algorithm factored the number 21 using 10-qubits. However, in 2012, a different "minimization" algorithm published by Nike Dattani and Nathaniel Bryans was used to factor 56,153 using only 4-qubits.[241] Factoring a 5-digit number is a long way from factoring a 300-digit RSA key. However, quantum computing is in its infancy. The number of qubits that engineers can apply and the efficiency of factoring algorithms for large prime numbers are increasing. The ability to quantum hack a 300-digit RSA key appears to be inevitable. How much longer will it be before RSA encryption becomes insecure and useless?

What we need going forward is an information security method that is resistant to hacking by quantum computing. Fortunately, quantum computing itself might provide the solution. If you recall, the BB84 and Ekert protocols are examples of QKD (Quantum Key Distribution) techniques. QKD techniques enable the transmission of encryption keys with the ability to detect eavesdropping attempts.

We can anticipate a major paradigm shift in cryptography. First, governments and large corporations will afford the earliest quantum computers that can factor large prime numbers and break the RSA codes. Confidence in the standard security methods of the Internet, banking, and commerce will fall. Then, companies that can build quantum computing devices that can use QKD techniques on both ends of communications will find an enormous market for their devices. Over time, existing client and server computers will be retrofitted with QKD security devices. New computers will have them built in. Eventually, quantum encryption security will replace classic RSA-type security.

QKD solutions have already been demonstrated in the real world. The Swiss company ID Quantique brought the first QKD commercial solutions to market in 2004 and then deployed QKD technology to secure Geneva state elections in Switzerland in 2007.

As of this writing, a web search for quantum security companies lists more than twenty. These companies are preparing for an exponential wave of demand that will not occur until the threat of quantum breaking of RSA security is imminent. Their early clients are primarily large banks and other organizations in the financial services sector anticipating the coming paradigm change.

Experiments by countries are already underway to create quantum networks over fiber optic cables. In March 2020, researchers at the University of

Bristol, UK, demonstrated QKD over a simulated 200-kilometer fiber optic cable. They used tiny 6 x 2 mm chips containing nanoscale optical quantum computing components. They demonstrated error rates and speed that were comparable to state-of-the-art commercial components. These chips will enable QKD network devices similar in size to today's in-home routers.[242] The researchers plan to use existing fiber around Bristol to demonstrate a city-wide QKD network.

It appears that it might be possible to connect the QKD networks of multiple cities via satellites to create a worldwide quantum network. A Chinese team and an Australian team have already achieved intercontinental QKD using a Chinese satellite. Once they secured the connection, the teams sent pictures to one another.[243]

Searching Big Data: Grover's Algorithm

The ability of AI to find patterns in big data is now a major driver of discovery. Anything that can speed up searches in big data is of high interest to major companies. Grover's algorithm might be able to do just that.

Imagine that you have four cards from a deck of cards face down on the table in front of you, one of which is the ace of hearts. The question is, how many times must you turn one card over to find out which is the ace? You could find it on the first, second or third try. If you have not found it by the third try, then you know the unturned card must be it. So, you only need to try between one and three times to find the ace of hearts or, on average, 2.25 cards out of four cards.[244]

Grover's quantum algorithm, invented in 1997, uses three qubits and a circuit designed to model the problem. Only one cycle of the algorithm is needed to solve it. This is a significant speedup over the 2.25 cycles required by classical methods. Grover's algorithm can be applied when the number of searched items increases, even to huge numbers. This does increase the number of cycles that we must run the algorithm to achieve a result within the bounded error limit. However, it can be shown that Grover's algorithm still provides a quadratic speedup over classical computing. This is not as great an increase in efficiency as an exponential time algorithm might be. However, it has been determined that no quantum algorithm can solve the problem with more than a quadratic speedup. With massive data sets, any speedup can be valuable.[245]

Simulating Chemistry

In theory, we know that chemistry involves the interactions of atoms and their electron configurations. We know the equations of the quantum mechanics of the particles. In practice, however, classical computers can only model the behaviors of atoms and particles to the level of approximations. These approximate models work well in situations where specific fine details can be ignored but not well in other cases where the details are essential. Classical computers are not powerful enough to model everything down to the finest detail.

Using quantum computers to study chemistry that belongs to the quantum world holds great potential.[246] Some of the areas where quantum computing could play a major role in understanding are catalytic reactions, photosynthesis, photovoltaic cells, superconductivity and magnetism.

Optimization

Often, we use computers to find the optimal solution among many possibilities, such as when your GPS mapping device suggests the shortest or fastest route to the desired destination. This becomes exponentially more difficult when you want to find the optimal route to reach several destinations in one trip. Classical computers have difficulty when such optimization problems are scaled up. We can expect quantum computers to be much more efficient at solving optimization problems.

AI and Machine Learning

In AI development, artificial neural networks are already utilizing what is called "machine learning." In machine learning, structures are formed in artificial neural networks as combinations of connections are strengthened and given more weight by repetition. Quantum computing can create far denser and massively more interconnected artificial neural networks. Therefore, it will contribute to the advancement of artificial intelligence more than today's classical computers. Suppose the human brain uses quantum functions in its operation, as some have suggested. In that case, a quantum computer will be a more natural platform on which to model it.

BUILDING IT

How do we build a practical quantum computer? What are the challenges, and where do things stand today?

The primary objective is to increase the number of qubits in a quantum processor. The more qubits you have, the more states you can process in parallel. Two qubits can represent four values simultaneously: 00, 01, 10, and 11 in weighted combinations. Similarly, three qubits can represent 2^3, or eight values simultaneously: 000, 001, 010, 011, 100, 101, 110, 111. Fifty qubits can represent over one quadrillion values simultaneously, and 100 qubits over one quadrillion squared.[247]

William Fefferman of the University of Chicago explains, "This difference means that every time its quantum computer grows by even a single qubit, a classical computer will have to double in size to keep pace. By the time a quantum computer gets to 70 qubits—likely within the next couple of years—a classical supercomputer would need to occupy the area of a city to keep up."[248]

IBM and Google are competing to be the first to achieve something called "quantum supremacy" by increasing the number of qubits in their processors. Both have claimed to be the first. However, quantum supremacy is primarily media hype. Quantum supremacy means when a quantum computer outperforms a classical one. Even a two-qubit quantum computer can do that depending on the choice of the problem to be solved.

What matters in practical terms are Pednault's and Fefferman's observations above about how a classical computer must scale exponentially to keep up with a linear increase in a quantum computer size. As you will recall from our earlier story about filling the Grand Canyon with water, the growth curve went from horizontal to vertical after about fifty doublings. When it comes to the kind of problems for which quantum algorithms are applicable, quantum computers will leave classical computers far behind when they exceed about fifty qubits.

The most significant problem in building a quantum computer is called "decoherence." Decoherence is when a qubit quickly interacts with something in the environment that is not a part of the calculation you are trying to perform. Once you have created a qubit in an initial state, you need it to remain stable until you can use it in gates and circuits. Decoherence causes errors, and the problem gets exponentially worse as the size of the machine grows. Getting qubits to last longer and minimizing and correcting these errors is the biggest engineering challenge today.

We have already seen that we can make qubits from electrons and photons. Photons have the advantage of being easy to create with the desired initial entangled states. They remain stable in the environment for a relatively long time. However, photons are challenging to store and hold until you need them. Photons are better for quantum communication but are more difficult to use for quantum computing. Electrons can be used by trapping them in diamonds and then manipulating them by shining lasers on them. However, this type of apparatus is difficult to scale beyond a handful of qubits. Ion-trap is another method used where ions are held in place in electromagnetic fields. The difficulty here is that we must cool everything to near absolute zero to minimize vibrations.

The method used in recent implementations by IBM and Google uses superconductors where electrons are paired up in "Cooper Pairs," which can be combined with insulators to create a "Josephson junction." They must cool these superconductors to temperatures where they lose all electrical resistance and shield the devices from light, heat, and electromagnetic radiation.

In 2016 IBM built a quantum computer using superconducting qubits. Its processor with five qubits was made available to everyone for free in the cloud to "introduce quantum computing to a wide audience." In 2017 IBM connected a twenty-qubit computer to the cloud for commercial use by companies who can buy access.[249] Also, in 2017, IBM demonstrated a 50-qubit processor in the lab. In 2018 and 2019, Google announced 54-qubit and 72-qubit processors.

The machines built in the labs by IBM and Google are room-sized and of limited practicality compared to today's classical digital computers. Practical commercial quantum machines will require miniaturization to the chip level. Intel is investing heavily in that direction.

In 2018, Intel introduced a 3-inch chip codenamed *Tangle Lake* with 49 superconducting qubits. In 2019, Intel introduced a cryogenic control chip codenamed *Horse Ridge* to solve the problems of controlling and scaling qubit chips to create larger processors and eventually achieving practical, commercial quantum computing.

The first useful quantum computers will be large devices because of the present need for cryogenic cooling—too big to be suitable for the desktop. They will play a role like the mainframe computers of the past. They will connect to the Internet, and users will remotely access them. The milestone to watch for will be the discovery of room-temperature superconductors or new elements that can be used reliably as qubits at room temperature. Either of these discoveries will make desktop size quantum computers possible.

As for government backing of quantum computing research in the United States, the *National Quantum Initiative Act* was passed in December 2018. Jim Clarke, Director of Quantum Hardware at Intel, described the importance of this act:

> If the U.S. is to lead the way in quantum, we will need a coordinated federal effort to focus resources, unify public and private sector interests, and train a quantum-ready workforce. Thankfully, the National Quantum Initiative Act addresses these needs by calling on the President to implement a National Quantum Initiative Program – a White House-led, cross-agency effort to coordinate quantum research priorities at the federal level by corralling multi-sector stakeholders around appropriate goals and effective strategies. [...] Best of all, recognizing that this endeavor is an investment in our collective future, the bill doesn't just pay lip service; it focuses agency funding to the tune of $575 million over the next five fiscal years, to ensure the U.S. maintains quantum leadership.[250]

Over the next few years, we can watch for the discovery of usable room-temperature qubits, miniaturization of quantum computing hardware, education of a generation of quantum programmers, discovery and maturation of practical quantum algorithms, programming languages, and software, implementation of city-wide and then global QKD networks, and simulations of atomic and particle quantum physics. Quantum computing will eventually play a significant supporting role in advancing the technologies applied to modeling the human brain, artificial intelligence, genetics, nanoscale assemblers, blockchain, fusion energy, and much more.

THEORETICAL PHYSICS VS. ENGINEERING SCIENCE

Now let me address Lars Jaeger's claim at the beginning of this chapter that quantum mechanics has upended classical philosophy.

The search for a grand theory of everything that integrates classical physics and quantum mechanics has been underway since the late nineteenth century and has been unsuccessful to date. In this chapter on quantum computing, we learned about phenomena of sub-atomic particles and quantum mechanics that seemingly cannot be explained by classical physics. We learned about entanglement, where two particles on opposite sides of the universe can be somehow connected. Even though we can use entanglement to build quantum computers, physicists cannot explain how entanglement, "spooky action at a distance," is possible.

While quantum physicists are trying to explain questions like these, cosmological physicists seek answers to questions such as how the universe and its particles came into existence, whether or not "dark matter" exists, and what the future of the universe will be.

These and other unexplained questions lead physicists to rather exotic theoretical explanations outside classical physics, such as string theory and parallel universes. They base these theories on the idea that the universe itself performs computations; therefore, they believe mathematics is a fundamental part of physics.

These theoretical physicists begin with mathematics and then attempt to explain the equations by imagining what kind of structure the universe, or multiple universes, must have to cause their equations to work. Their imagination has had to look outside our classical reality. They say that we live in a world that only *seems* real—a return to Platonism and subjective science. A minority of theoretical physicists believe that there are yet undiscovered explanations for these phenomena consistent with classical physics—sustaining Aristotelianism and objective science.

The history of discovery in quantum mechanics beginning the late nineteenth century saw a transition in the way physicists viewed the use of mathematics. In classical physics, empirical analysis of reality by experiment was performed. Then mathematics was used to find formulas that would reliably model and explain past experiments' results and make predictions that future experiments could test. Then, they noticed that some of those equations contained specific attributes that implied the existence of particles that had not yet been discovered. Based on those equations' predictions, they devised experiments to search for these new particles, and they eventually found them.

The success of these mathematics-sourced discoveries led subsequent generations of theoretical physicists to believe that they could use pure mathematics to find the truths about the universe that they sought. No longer able to imagine experiments to test their theories using current technology, elegance and beauty in their equations became the objective. They believed that if their equations provided elegantly satisfactory mathematical answers to their questions, they "must" be true. They lost sight that those earlier equations that yielded discoveries were initially formulated to explain prior empirical, experimental results. Those equations began with a connection to reality.

The foundational truth of our Personal Philosophy states that "Reality is real." And yet, physicists tell us that quantum objects have neither objective properties nor independent existence. These objects do not possess a reality of their own, just a potentiality. Their potentiality only turns into reality when a quantum object interacts with a macroscopic object such as a measuring device. This absence of *real*—meaning objectively physical—things in the quantum world leads some physicists to declare, philosophically speaking, that there is no such thing as "real" reality.

It appears that theoretical physics is losing its tenuous connection to reality. Quantum and cosmological theoretical physicists are now dealing with purely mathematical concepts that may never be "provable."

The philosopher Immanuel Kant was responsible for this breach between modern science and reality. Kant split "appearances" and "reality." He held that reality is unknowable by anything other than feelings of duty. He is responsible for scientists focusing on statistics (appearances) instead of reality. "Probabilities" are useful but not as the primary tool for looking into the interactions between entities or the natures of the substances they study. For example, in the Schroedinger wave equation, the term "psi^2" is said to refer to the probability of finding an electron, but it is unexplained. It is the Kantian influence that supports the treatment of a *probability* as a metaphysical and irreducible primary.

In dealing with questions about the nature of existence and the nature of knowledge, today's theoretical physicists' efforts might be more appropriate to philosophers than scientists. And, considering some of their proposals, they exhibit adherence to Kant's anti-reality philosophy. They forget that concepts of things in reality, including quantum particles and mathematics concepts, are man's *method* of grasping reality, not reality itself. Peikoff describes the departure from *integration* as the original thinking mode used in theoretical physics toward Kant's *disintegration* mode used today:

> Classical physics had studied material reality in order to discover objective, causal laws. Throw it all out, says the quantum credo. Reality is limbo; objectivity is pre-Kantian; causality can't compete with Heisenberg; sensations arise by miracle; concepts are a detached formalism; mathematics is arbitrary; probabilities exist but not yet; Aristotelian logic is wrong; matter is passé; the science of physics does not investigate the physical. And quantum particles, whose study leads to all of this, do not exist.
>
> 'Quantum theory,' Bohr quietly sums up, 'represents an essentially irrational element...' It is this element that the movement uses to negate every concept and principle that makes physics—or any form of mental functioning—possible. This is the eruption of full-blown nihilism in the hitherto rational field of science. Confronted by such a spectacle, the mild Einstein could not stifle his reaction: If this is physics, 'I would rather be a cobbler, or even an employee in a gaming house, than a physicist.'[251]

The true scientists today are quantum engineers who, like Aristotle, apply the empirical realities of what is discovered by experimentation to achieve new knowledge and create innovations. It will be the engineering scientists building things that make future discoveries about the Laws of Reality in both the quantum and cosmological domains. The theoretical physicists, like Plato and Kant, are telling us that the real world is unknowable. Their elegant equations are floating abstractions disconnected from reality.

Ultimately, in the quantum realm, the concept of a *substance* is replaced by the *interactions* of particles. We can only identify particles by their interaction with their environment and their functionality. We know they exist because we can see their effects. We have been dealing conceptually with invisible phenomena like this for centuries. We cannot see gravity, but we know that it exists, and we can formulate concepts to describe and explain it. But our concepts describing the *thing* are only our *method* of grasping and knowing, not the thing itself in reality.

Evolution did not equip us to grasp the quantum and the cosmic scales directly. We can only do that through our abstract conceptual faculty. We live in a macro-scale reality—the reality that evolution enabled our consciousness to grasp.

We can grasp things we cannot see directly, inferring their existence by their effects. Our macro reality is eternal, solid, reliable, and knowable using abstract mathematics (concepts). Quantum particles make up our macro reality, including ourselves, and are therefore also "real." To deny that they exist is to deny that we exist.

Having already explained much of the reality that we can grasp with our senses, the difficult questions for theoretical physicists today are at the non-intuitive, extreme ends of the size scale, the sub-atomic nanoscale and the cosmic scale. They are dealing with infinities of smallness, largeness, and time without a beginning or end. It seems that our human-level conceptual intelligence and our analytical technology are not yet up to the task of answering all the questions we have at these extremes.

Suppose we merge with our technology and evolve higher levels of intelligence as Hybrid Humans, and our nano, AI, quantum computing, and space exploration technologies advance. In that case, we will eventually find the answers to these questions and more—*if* we can stay connected to reality.

♦ ♦ ♦

4.9

NUCLEAR FUSION

"I would like nuclear fusion to become a practical power source. It would provide an inexhaustible supply of energy, without pollution or global warming."

—STEPHEN HAWKING

NUCLEAR FUSION INTRODUCTION

The universe is powered by fusion. The stars use fusion as their method of generating vast amounts of energy. Humans are attempting to harness this cosmic energy source. While we have not yet been successful, we are getting close. In this chapter, we explore why it will be transcendent and the challenges to achieving it.

We say that we entered the nuclear age with the first atomic bomb. This is true in the sense that we lost our innocence and now have the responsibility of being careful with our ability to explosively unleash the incredible power of the strong nuclear force. However, we will not be fully in the nuclear age until we can harness that energy in a controlled and measured release for productive, not destructive, purposes. That will be when we create the first fusion power plant that inserts more electrical energy into the electric grid than it consumes.

Fusion energy, once industrialized, can dramatically benefit us in many ways. These include:

1. Increasing the amount of electricity available to humans everywhere on Earth.
2. Sustaining our present annual global energy consumption for billions of years.
3. Decreasing the negative effects of fossil fuels on the environment.
4. Decreasing the landmass footprint and man-made structures devoted to energy production.

5. Increasing the energy available to remove pollution from the environment.
6. Decreasing the concern over nuclear wastes.
7. Decreasing the concern over nuclear power safety and the potential for accidents.
8. Decreasing nuclear bomb proliferation concerns.
9. Decreasing international tensions over scarce energy resources.
10. Increasing the power and efficiency of large ships and spacecraft.
11. Increasing the speed and efficiency of space travel, exploration and colonization.

Harnessing the energy production method of the stars is no easy task. The first fusion reaction on Earth was achieved for a tiny fraction of a second in a particle accelerator in 1934. However, work towards sustaining a fusion reaction by confining it in magnetic fields did not begin until 1951.[252] Since then, science and engineering research progress has seen many successes and setbacks but has steadily advanced over the decades.

In 2013 an international consortium of 35 nations began building what they hope will be the final fusion research experiment, "ITER." It is under construction in southern France and is about 50% completed. If ITER is successful, they believe that the science and engineering will be sufficiently advanced to build a demonstration fusion power plant named "DEMO." It will insert net electricity into the electric grid.

Achieving industrialized fusion energy is still a couple of decades away. ITER is currently projected to begin operations in 2025 and be fully research functional in 2035. DEMO is initially scheduled to begin construction in 2024. South Korea has a demonstration fusion electricity generating plant project called "K-DEMO" slated for construction by 2037.[253]

Fusion research and development have been plagued by severe underfunding and technical setbacks, so one could expect any of the above time projections to be extended. However, we have learned that the exponential growth curve can achieve magic. Perhaps we have endured the long, slow, horizontal growth period and are just now beginning to enter the transition to vertical, rapid growth. The success of ITER will indicate that this is indeed the case.

Industrialized fusion energy production will undoubtedly be a transcendent event in human history. Man has relied on fossil fuels since his ancestor, Homo erectus, used campfires a million years ago.[254] It now appears that the fossil fuels paradigm that has lasted a million years will, during your lifetime, be replaced by the power of the stars.

HUMANS NEED ENERGY

All living things, including humans, require energy—always have and always will. Not only do humans need energy to sustain their bodies, but society requires energy to create and power the technological advances that have freed humans from the necessity of endless toil for their barest existence. Available energy enabled us to create machines to do the labor we previously had to do with our bodies and things that our bodies could not do. Energy has also enabled us to extend our ability to live in environments that otherwise would not be habitable. A global society as large as we have today would not be possible without abundant energy.

When our hominid ancestors discovered fire, they opened a vast resource of usable energy that had not previously been available to them. By burning more readily accessible flammable material to warm their bodies they reduced the need for calories from less accessible food. By using fire for cooking, they increased the nutrient uptake from their food, enabling larger brains' evolution. And they could use fire for lighting, giving them power over the darkness for the first time.

Then, all the way until the 18th century, animals—oxen and horses, were used as a source of power to transport people, plow fields, and more. These beasts of burden were biological machines powered by raw grasses, a vast energy source humans could not use. It was not a very efficient energy conversion process. Still, there were not that many humans, so it was still quite helpful. While we also used water wheels, windmills, and sails on ships during this period, animals were our predominant energy source until the steam engine's arrival.

The steam engine in the 18th century was our first ability to transform heat energy from fire into mechanical work, and it ultimately powered the Industrial Revolution. Steam engines were only able to do mechanical work until the early 19th century when Michael Faraday discovered electromagnetic induction, inventing the new paradigm of power from electricity. Steam engines powered the electrical generators. The electricity was easy to transport long distances and convert into other forms of energy, including heat, light, and mechanical energy. For the first time, energy production could occur in one location and energy consumption in another. Today, we still mostly use heat energy sources such as coal, oil, gas, nuclear, biomass, and geothermal, to create steam to power turbines converting mechanical energy into electrical energy distributed on our electric grids.

From food to fire to animals to steam engines to electric generators—every paradigm shift in our energy source enabled significant advances in our human condition. Now we are on the verge of the next and most significant energy paradigm shift, fusion energy. We cannot even imagine all the changes it will bring.

ENERGY OPTIONS

Entropy

We presently get about 80% of our energy from fossil fuels. We know that we must change our energy source in the next few decades. What are the options available to us in our search for bigger and better energy sources?

Before we survey our options, we need to take a moment to describe what we mean by a "better" energy source. Recall that "entropy," which is the amount of equilibrium in the universe, is steadily increasing. In other words, the universe is slowly winding down, going from complexity and order to disorder. This means that the various forms of energy in the universe are gradually winding down from energy forms with greater "order" (less entropy) to energy forms with lesser order (more entropy). Forms of energy with more order and less entropy, such as electricity, are more useful. They are more useful because they are more convertible into other forms of higher-entropy energy better suited for particular tasks, such as heat to make steam.

Electricity

Electricity is our most useful form of energy. You might recall, also, that while complexity is decreasing in the overall universe, complexity can be increased in localized situations. We can convert energy with lesser complexity into energy with higher complexity. For example, we can convert three heat energy units into one electrical energy unit by creating steam to drive a turbine that spins a generator. We can then transmit that electrical energy unit over wires to another location and convert it back into three heat energy units (minus the generation, transmission and conversion losses).

However, the "heat pump" heating and cooling system in your house does not convert electricity into heat. It "moves" heat from the air outside the house to the air inside. One electrical energy unit can move four heat units. This is more efficient than directly converting electricity into heat.

We will want to keep using electricity and more of it. We need to find better methods of producing it than burning fossil fuels.

Fundamental vs. By-product Energy

Gravity acted on a young universe to cause hydrogen atoms to clump together and compress into stars. The hydrogen atoms fused to create helium. When the stars burned out and exploded, they created and expelled the heavier elements. Eventually, lighter and heavier elements clumped together to form our solar system's sun, planets, and moons. Vast stores of hydrogen are in Earth's oceans as water. Radioactive elements heat Earth's core. Gravitational tugs on Earth's oceans cause the tides. Our "fundamental" energy sources include fusion of hydrogen, solar from sunlight, tidal from gravity, nuclear fission from the radioactive elements, and geothermal from Earth's hot core.

Wind energy is a "by-product" of solar energy due to the different heating rates between the equator and other latitudes. Wind, in turn, pushes the oceans' surface into waves as another by-product energy source. Solar energy also drives the water cycle of evaporation and rain, making possible the by-product of hydropower. Biomass absorbing solar energy using photosynthesis can be burned as a source of energy. Biomass buried for long periods is the final by-product energy source—fossil fuels.

By-product energy sources contain less energy than fundamental energy sources. Still, we can use any energy source to create the electricity you need to charge your Tesla® car.

We can order these energy sources into a hierarchy based on their relative sizes here on Earth. We can measure each energy source's size in terms of humanity's current annual energy consumption rate, which is approximately 18 terawatt-hours.[255] This is the total energy supplied, which, after losses in generation and transmission, is three times the amount of electricity consumed.

Global electrical consumption continues to increase even though energy efficiency improves over time. As underdeveloped nations become developed,they consume ten times more electricity. Therefore, we can expect electricity consumption to continue to increase.

Geothermal

Compression due to gravity and radioactive decay heat Earth's core in roughly equal amounts. This heat makes its way to the Earth's surface and radiates out into space. We can capture some of this heat energy by digging wells several kilometers deep and using the high temperatures to create steam to drive turbines. However, as we remove heat from the rocks surrounding the wells, it is replenished slowly due to the low heat conductivity of the rocks. We cannot get much energy out of a single well, and we cannot drill enough wells to meet our energy needs. Geothermal hotspots such as Iceland hold some practicality, but not the rest of the Earth, especially the 70% beneath oceans.

Solar

The Earth's surface receives 10^{17} Watts of solar energy from the Sun, or about 5,000 times humanity's total power consumption.[256] However, we cannot put the entire Earth in the shade behind one large solar panel without "killing" the Earth. We should only expect to intercept a small portion of solar energy. Only 2% of the solar energy represents 100 times our current energy consumption, and we can choose to get that 2% from the sunniest areas with minimal impact on the environment.

However, photovoltaic electricity production is still an expensive method of producing electricity. Sunshine is intermittent due to nights, winters, and weather. Solutions to this *intermittency* are expensive. More importantly, photovoltaics are not very "green" alternatives since manufacturing them releases three times as much carbon as nuclear fission and four times as much as wind, though about ten times less than natural gas. As electrical generation moves away from fossil fuels, then the carbon released by photovoltaic production will decrease.

Wind

There is about a 60° C temperature difference between the sun's heating at Earth's equator and the poles. However, due to the shearing effects of the winds generated by Earth's spin, much of the hotter air does not make it to the poles and there ends up being only about a 20° C air temperature difference. Only about 1% of the solar energy from the sun imparts into Earth's wind. While this is 50 times our current energy consumption,[257] it spreads over the entire Earth. This means that any area of the Earth's surface can provide 100 times more energy from solar than from wind. More importantly, a simple flat solar panel on the ground can capture the solar energy. In contrast, the wind energy spreads across the entire 10-kilometers thick atmosphere. While the amount of energy available from the wind is low, we have used simple technology for centuries, such as sails on ships and windmills, to capture a small percentage of it.

Wind energy will become economically infeasible before it produces a significant portion of our energy requirements. Once we install windmill generators in all the typically windy locations, we will need to use less desirable locations. However, wind energy is a good complement to solar energy since wind is usually more substantial during the winter when solar usually is weaker.

Biomass

Unfortunately, biomass farms cannot do any better than solar farms in producing energy. Photosynthesis converts about 1% of the absorbed solar energy into chemical energy. Using the steam cycle to burn the plants to produce electricity only recovers about half—0.5% of the original solar energy—which is much less than the 10% efficiency of typical solar panels.[258] Considering the energy costs of running biomass farms to grow plants, we would need to apply all of Earth's ice-free landmass to produce biomass to satisfy our current energy needs. While we cannot rely on biomass to cover all our energy needs, we can take advantage of the biomass produced as by-products of other human activities.

Fossil Fuels

Coal begins as peat when land-based plants decompose without oxygen in certain localized conditions such as bogs and the ocean floor. The peat must then become geologically buried where extreme geological pressure compresses it into lignite and then coal. It took about a hundred million years for our fossil fuel reserves to be created. These reserves are estimated to last us no more than roughly 150 years, a long way from being sustainable.

If we were to burn all our fossil fuel reserves, the amount of carbon we put into the air would only be about 4% of the oceans' carbon.[259] This would not be a problem if we burned fossil fuels more slowly. The oceans would absorb the carbon from the air. But ocean absorption is a slow process, and we are burning fossil fuels too quickly to be absorbed by the oceans. Therefore, the carbon content in the atmosphere is increasing. Carbon capture and storage could slow the release of carbon into the atmosphere. Still, our relatively small supply will be the ultimate limiting factor for the future of fossil fuels.

Hydroelectric

The average spot on Earth receives about a meter of rain each year. We can use this number, global landmass area, average height above sea level, and water density to calculate the rain's total potential energy worldwide. It turns out to be roughly equal to annual world energy consumption. However, half of it enters the ground or evaporates. Also, flowing water loses energy due to friction with the land. Fortunately, most of the remaining water ends up concentrated in rivers that we can use to generate useful electricity using either dams or run-of-the-river plants. If we were to install hydroelectric plants on most of the world's rivers, we could generate about 10% of our energy needs.[260]

Tidal

Like hydroelectric power, tidal power generation depends on the difference in elevation of quantities of water. The gravity of the moon and sun cause tides on the oceans, raising and lowering water levels. Each year the tidal process causes the moon to speed up a little and move farther away from Earth and the Earth's rotation to slow a little. By measuring these changes, we can calculate the total amount of energy imparted to the tides. This turns out to be only a tenth of our current energy consumption.[261]

Tidal power generation could be economically viable in natural inlets where a sufficient water volume can be captured and stored at each high tide.

Wave

Waves are created by the friction of the water's surface with wind. Once formed, the waves can then be increased and speed up by the wind. Considering the total length of coastlines, we can calculate the total available energy using the waves' average height and velocity. In doing so, we find that wave power is 500 times less than wind and only 10% of our current world energy consumption.[262]

Capturing wave energy typically involves floating buoys or snakes of floats connected to generators via mechanical cables or hydraulic cylinders. They might be most helpful in special situations such as remote islands.

Fission

Nuclear fission presently generates over 10% of the world's electricity. It differs from fusion in two essential and fundamental ways:

- *Fusion* energy generation is difficult to start but easy to stop.
- *Fission* energy generation is easy to start but difficult to stop.

Fission power plants are relatively much easier to build and were the first to provide nuclear electrical power. However, they require *negative feedback* mechanisms to keep the atomic reaction from entering a "runaway" condition. It takes days to shut them down until they no longer need cooling water to prevent a melt-down. They generate nuclear waste that stays radioactive for thousands of years. And they can produce fuel for nuclear weapons. Fusion power plants do not have these issues and would therefore be the preferred method.

Fission produces energy by causing a heavy atom that is barely stable to split into two smaller atoms, usually by hitting it with a free neutron. When the heavy atom splits, it releases free neutrons, one of which then impacts and splits another heavy atom, leading to a "chain reaction."

As found in nature, uranium ore potentially contains two fuels for nuclear fission. One percent of the naturally occurring uranium is the isotope uranium-235 which has 92 protons and 143 neutrons in its nucleus. It can be easily split by a single neutron and is the most widely used fuel in fission reactors. The other 99% is uranium-238 which can be used in a "breeder reactor" by first converting it to plutonium-239 and then splitting the plutonium. Only one such commercial breeder reactor exists today. When analyzing the energy potential of our uranium supplies, we must consider both.

Considering the world's economic uranium deposits, if we only used the current uranium-235 technology (1% of the uranium), it would last for about 100 years. By using the uranium-238 technology, it would last for 10,000 years. However, there is a vast amount of uranium in seawater that, using the uranium-238 technology, could last for almost 1,000,000 years.[263]

Fusion

Fusion is the process of combining the nuclei of the light elements hydrogen and helium to produce energy. The hydrogen atom has a single proton as its nucleus. Deuterium has one proton and one neutron as its nucleus, and tritium has one proton and two neutrons. We call deuterium and tritium isotopes of hydrogen since they all have a single proton and a single orbiting electron. They differ only by the number of neutrons in their nucleus.

Energy density is the amount of energy per unit of mass of the fuel. Fusion energy has a million times more energy density than fossil fuels. This is the difference between chemical reactions based on the exchange of electrons and nuclear reactions based on the exchange of nuclear particles. The strong nuclear force that holds nuclear particles together in a nucleus is a million times stronger than the force that holds electrons in their orbits. Further, nuclear fusion energy is four times greater than the energy from nuclear fission.

There are two primary fusion reactions, the deuterium-tritium (D-T) reaction and the deuterium-deuterium (D-D) reaction. The D-T reaction has a higher probability of the two nuclei fusing instead of bouncing off each other, so it occurs at lower temperatures and is the easier of the two. When the two nuclei fuse, a free neutron is emitted.

Deuterium is naturally occurring, and we can find some in every glass of water. Tritium, however, is radioactive and half of it decays into helium every 12-years, so there is not much of it in nature. We must make it ourselves using tritium breeding reactions that use the free neutrons emitted by D-T reactions to convert lithium into tritium. Therefore, a D-T fusion power plant would use deuterium and lithium as its fuel and make the tritium it needs to keep the reaction going.

The D-D reaction has the advantage of not requiring the breeding of tritium. It requires only plentiful and cheap deuterium as its fuel. However, as previously mentioned, the D-D reaction requires higher temperatures. Therefore, it would require a bigger and more advanced power plant than the D-T reaction. Therefore, the first power plants will use the D-T reaction, and subsequent plants will advance to the D-D reaction as our engineering knowledge increases.

Lithium is the limiting fuel resource for the D-T reaction. We find it in the Earth's crust and seawater. The estimated potentially extractable amount is sufficient to fuel D-T reactions to supply our current energy consumption for 100 million years. However, we only need lithium long enough to transition to the D-D reaction that only requires deuterium. The amount of deuterium in the oceans is enough to provide our energy needs for over 10 *billion* years. Removing that deuterium would cause an infinitesimal impact on the oceans.[264]

Intermittency

Some of the potential sources of energy, such as solar and wind, suffer from intermittency, varying their output by the day, week, and even seasons. Various strategies can be applied to deal with the intermittency.[265]

Energy storage can be used to store surplus energy generation for use during times when there is an energy generation shortfall. Intermittent energy sources that are synergistic can be combined, such as solar and wind where wind is stronger in the winter when solar is weaker. Intermittent energy sources can be used to supplement non-intermittent sources.

Excess generating capacity can be used, for example larger solar farms since some solar energy is available even on cloudy days. The electrical grids can be expanded to coordinate larger geographies, for example if it is winter in the northern hemisphere, it is summer in the southern hemisphere.

The electric grid has extraordinarily little energy storage capability. We generate electricity as needed in real-time. Baseload sources, such as coal and nuclear fission, provide most of our steady, reliable electricity. In contrast, we bring supplemental sources such as natural gas and hydroelectric online as needed to match demand. Electrical energy storage facilities cost is so prohibitive that the power companies would rather build power plants that are oversized and sit partially idle than invest in storage facilities to handle peak demand or energy source intermittency.

The most economical method of energy storage we have today is pumping water uphill into a large reservoir and then using it to drive hydroelectric turbines when needed. This can achieve an overall efficiency of 75%. Unfortunately, to store just 5% of the world's energy needs would require lifting all the liquid fresh water on Earth to the height of a 10-story building. This storage method is not practical.

Storing heat energy underground is another option. While it does not support high enough temperatures to generate electricity, in certain climates, we can use heat stored during the summer to heat homes and offices, reducing the need for electricity in the winter. This method of reducing demand is not yet widespread. Still, we could expand it to reduce winter consumption by as much as 10%.

Lithium-ion batteries can provide storage for the short-duration day-night intermittency but not for the seasonal summer-winter cycles. It appears that adding lithium-ion batteries to solar panel installations to provide power through the night will approximately double the cost of the solar panel installation. Unfortunately, to store just 5% of the world's annual energy needs in lithium-ion batteries would require 200 times more extractable terrestrial lithium than exists. There is far more lithium in the deep oceans, but it is inaccessible.

A variety of other potential energy storage methods exist that lack either practical, efficient or economic viability. Examples include solar thermal energy to melt salts, supercapacitors, superconducting magnets, hydrogen, and compressed air.

Non-Renewable Source	Quantity (at current use)
Fossil fuels	100 years
Standard fission, land	100 years
Breeder fission, land	10,000 years
Standard fission, oceans	10,000 years
Breeder fission, oceans	1,000,000 years
D-T fusion	100,000,000 years
D-D fusion	10,000,000,000 years
Renewable Source	Quantity (times current use)
Tidal	0.01
Wave	0.01
Hydroelectric	0.1
Geothermal	0.1
Biomass	0.1
Wind	1.0
Solar	100.0

Energy Sources[266]

For renewable energy sources to meet our needs, we would need to build large structures on great land and sea areas to capture a portion of nature's energy flows. For nuclear fission, we would need somehow to alter the public perception of associated risk and danger, recognizing instead its comparatively safe track record. We would need to get comfortable managing the wastes and the risks of nuclear proliferation that come along with it.

Nuclear fusion will require the full application of human intelligence and ingenuity and public and political support to complete one of mankind's biggest technological challenges. Fusion power plants can directly replace (with much smaller footprints) the present fossil fuel plants and connect to the existing electrical grid. The potential rewards for mankind from fusion energy are truly enormous.

FUSION PHYSICS

In an atom, the protons and electrons are particles with opposing electromagnetic charges like the poles of a bar magnet, and neutrons have no charge. When in equal numbers, the protons in the nucleus and the electrons with opposing charges cancel each other. For example, a hydrogen atom has one proton and one electron, and the atom as a whole has no charge.

Suppose we have a bunch of atoms in a gas and heat them sufficiently. In that case, their electrons will break away from the nucleus resulting in a bunch of free electrons and free nuclei. This process is called *ionization*. It results in *plasma* which is the fourth state of matter after solid, liquid, and gas. If we heat hydrogen to its plasma state, it becomes a soup of freely moving charged (ionized) electrons and protons (ions) with opposite charges.

Two free protons have the same charge. Like the same poles of bar magnets, they repel each other when you try to bring them together. The closer you bring them, the stronger the repelling force. It takes enormous energy to push them close enough together for the strong nuclear force to bind them together rather than repel them. We must use a particle accelerator to give a proton enough kinetic energy (temperature) to have a chance of binding with another proton. Even then, it is more likely to bounce off than fuse with it. If we accelerate a large number of protons, then a few will fuse and give off large amounts of energy. Our goal is to have enough of the protons fuse to generate more energy than was required to create sufficient heating to cause the reactions. The temperatures in the plasma need to be higher than 50 million degrees C.

The best way to increase particles fusing is to make the plasma soup of charged particles denser. We have already seen that the particles are more likely to bounce off each other and scatter away, so we need a method of confining them.

In hydrogen bombs, the simple inertia of a mass of particles imploding onto and compressing a central point of fuel confines it for a fraction of a second, just long enough for enough particles to fuse and release enormous amounts of energy before the reaction blows itself apart.

For power generation, we need a method of *continuous* confinement. Since they are charged particles, we can do this by bottling them up in a vessel with magnetic fields as its walls. The magnetic walls reflect escaping particles back into the plasma soup.

To create the plasma and cause fusion reactions, we will have to heat the fuel inside our magnetic bottle to temperatures hotter than the sun. Once enough particles begin fusing, the resulting energy release will cause the plasma to become self-heating (called "ignition"). We then no longer need to put energy into the plasma to heat it.

The fusion reaction involves atoms of deuterium, which has a neutron in its nucleus. When the fusion occurs, a high-energy free neutron is released. These neutrons have no charge and therefore are not contained by our magnetic bottle. The neutrons escape our magnetic plasma confinement system and are then captured by outer shielding designed for that purpose. The energy contained in those absorbed neutrons is converted to heat in the shielding. That heat drives the steam cycle for the generation of electricity.

FUSION ENGINEERING

ITER: Credit ©ITER Organization, http://www.iter.org

Confinement

Even the best magnetic bottle plasma containment system will have leakage. The goal is to keep particles bottled up long enough for a useful number of them to fuse. The containment system's performance is measured by the average time that a particle stays bottled up before escaping. We call this the *energy confinement time*. For the D-T reaction, a confinement time of 1 second is a reasonable goal.

We can use electromagnets to create the desired magnetic confinement. We all learned in grade school science that field lines surround magnets. The ions will tend to travel along these field lines and rotate around them in a helical path.[267]

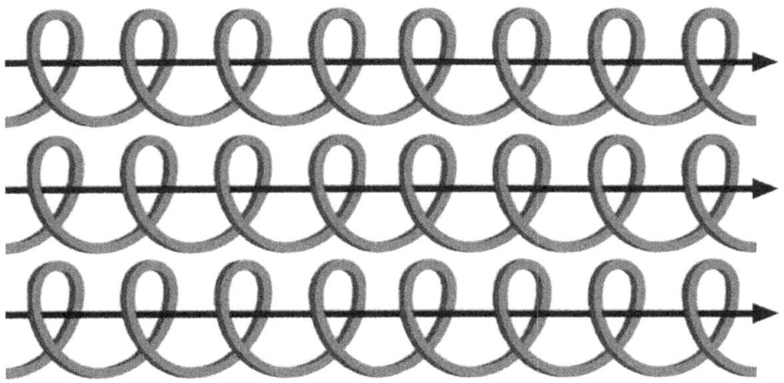

Magnetic Fields—Ions and electrons spiral around magnetic field lines.

Shape

The graphic above shows a linear magnetic containment system. The stronger the magnetic field, the tighter the rotation of the ions around the field lines will be, and the more likely ions on adjoining field lines will collide and fuse. So, the stronger the magnetic fields, the better. But what about the ends of our magnetic pipe? The simple solution is to curve the magnetic fields' path to form a circle so that the ions have an endless path to follow.

There is no way to organize endlessly circular magnetic field lines around the surface of a sphere. Attempting to do so always leaves at least one area on the sphere with no magnetic field, a sort of "bald spot." The best we can do is use a shape like a pipe bent into a circle like a donut with a hole in the middle. This shape is called a *torus*, and most present fusion designs are *toroidal* configurations. Any such machine is called a "tokamak," a term first used in Russia.

We place the electromagnets that encircle the torus pipe at intervals around the torus. These generate strong magnetic field lines flowing linearly around the inside of the torus. The ions travel on these field lines. Generating these strong magnetic fields requires that large electric currents flow through these electromagnets. A major advance toward fusion occurred with the development of superconducting materials. We now wind these field magnets with superconducting wires and must cool them to cryogenic temperatures. When they become superconducting, they maintain their magnetic fields without further electricity input, and they do not heat up.

Consider our fusion device's engineering challenges with hotter than 50-million-degree C plasma inside and electromagnets one meter away cooled to -269 degrees C.

Particle Drifts

Unfortunately, by bending our magnetic pipe into a circle, the particles do not stay obediently attached to the magnetic field lines. The field lines are spaced closer together on the inside of the torus than on the outside. Following a curved path, the particles are influenced by inertia. All the particles will try to drift toward the torus' outer walls, and the protons will try to drift upward and the electrons downward. These drifts will cause our plasma to tear itself apart in a few milliseconds. We need a way to counteract these drifts.

Toroidal Magnetic Confinement

We can largely control these drifts by causing the magnetic field lines within the torus to spiral around each other as they go around the torus. This spiral causes the particles to spend part of their travel time near the inboard side of the torus (nearest the donut hole) and part of the time near the outboard side, canceling out the drifts during each rotation. We accomplish this new spiral magnetic field by adding electromagnets that circle the torus the long way around or injecting an electric current into the plasma itself inside the torus.

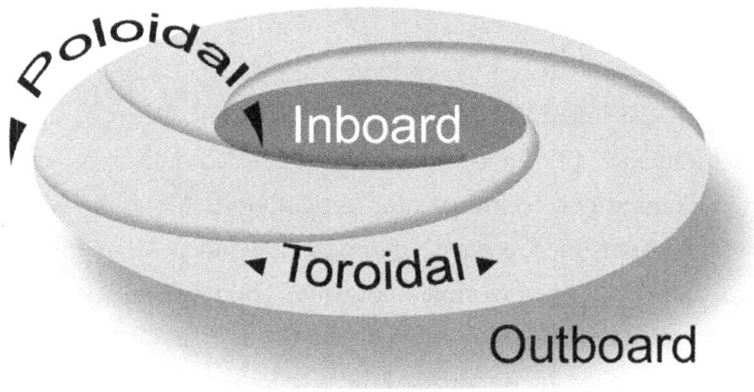

Toroidal Fields—Toroidal magnetic fields go around the torus in the long dimension. Poloidal magnetic fields go around the torus in the short dimension. When combined, they cause the magnetic field lines to spiral around the torus.

When the ions travel around the torus's long direction, we say they are going in the "toroidal" direction. When they travel around the torus's short direction, they go in the "poloidal" direction. The tightness of the spiral path, the number of times the ions rotate in the poloidal direction for each time around in the toroidal direction, is called the *safety factor*. If the safety factor is too small, below 2, then destructive turbulence can occur in the plasma causing the reactions to shut down.

Random Walk Diffusion

Suppose we inject a proton into the center of our toroidal pipe diameter. In that case, it will begin following the helical path around the torus. However, it will gradually make its way to the torus's magnetic bottle's outer edges and eventually escape. We need each particle to stay inside the magnetic bottle long enough to have an opportunity to fuse with another particle. The larger the diameter of our torus pipe, the longer it takes for the particle to reach the outer surface. So, this is one factor that determines the size of the device required to achieve our desired 1-second *energy confinement time* for generating fusion power.

H-Mode

Fusion experiments have been mostly plagued by the appearance of various types of turbulence in the confined plasma. Each new type of turbulence discovered required new techniques for mitigating it. Along the way, in 1982, a routine experiment in Germany observed a sudden dramatic decrease in turbulence and doubling of confinement time. The few simple operational steps required to achieve this was dubbed "H-Mode" for "high energy mode." Scientists still do not quite understand the physics causing it, but they know how to achieve it in any tokamak worldwide.

Plasma Heating

To achieve the extreme temperatures needed in the plasma, a frequently used method is called *neutral beam heating*. This method is where one or more linear particle accelerators are set up next to the torus and inject a beam of particles with more energy than those in the plasma. These beams heat the plasma and refuel it.

Another method is like the microwave in your house and is called *electromagnetic wave heating*. It consists of antennas that inject microwaves into the plasma to heat it.

Finally, *inductive heating* is generated by a long, thin line of electromagnets aligned vertically through the donut hole to create a powerful heating magnetic field in the plasma. Engineers only use induction heating temporarily to get the plasma reactions started.

First Wall

We must contain our *magnetic* bottle inside of a *physical* bottle of solid material. This is the curved pipe forming the torus. This physical wall must withstand plasma excursions outside the magnetic wall when hot plasma touches the physical wall. A sudden extreme failure of the magnetic containment is called a "disruption." In a disruption, the first wall is the first thing the hot plasma touches. In a larger machine such as a power plant, even though a disruption will cause the fusion reactions to stop instantly, the plasma can cause major damage to the first wall and other system components due to its uncontained extreme temperatures and electrical energy. Engineers use many techniques to mitigate the possibility of disruptions.

Our first wall must handle a large flux of neutrons created by the fusion reactions inside and flowing through the first wall into the outer blanket that catches them to create heat. The first wall materials must be able to withstand this neutron bombardment. Here, we can only rely on the materials engineering experience gained from fission reactors. Also, the first wall cannot be too thick, preventing the neutrons from reaching the outer blanket and generating heat.

Early tokamaks used torus designs where the cross-section shape of the pipe was circular. Later designs found that a "D" shape was favorable where the side nearest the donut hole is flattened. This shape helped reduce the difference in the torus' inner and outer circumference, which helped reduce the plasma's turbulence. It also created more space in the donut hole for the induction electromagnets.

Divertors

Our magnetic bottle is naturally leaky. Engineers must do all they can to avoid damage to the first wall by escaping particles. They achieve this primarily by providing a sacrificial and easily replaceable component called a divertor in the bottom of the torus and then modifying the path of the magnetic field lines so that most of the particles hit the divertor before hitting anything else. They use multiple techniques to cool and keep the divertor plates from melting under the onslaught of escaping particles.

Tritium Breeding Blanket

Outboard of the first wall sits the tritium breeding blanket. It absorbs the neutrons generated by the fusion in the plasma. It serves three main functions. It breeds tritium to use as fuel, captures the neutrons' kinetic energy to create heat for generating electricity, and protects all the equipment further out from the neutrons. The breeding blanket contains a maze of plumbing and ports for lithium flow to be converted to tritium, cooling liquids, and other support devices to carry away the heat generated to power the turbines. The breeder blanket absorbs about 80% of the fusion power generated. Because it has a large surface area, it can handle it. The divertor plates handle the other 20%. The breeder blanket needs to be at least half a meter thick to stop the neutrons and protect outside equipment.

Since engineering a tritium breeding blanket has never been required for any past technology, none have ever been built and fully integrated into a tokamak fusion device. This is the least matured part of the overall design for engineers to demonstrate in experimental testing.

Vacuum Vessel

Outside of the breeder blanket is a shell that fully encloses the torus and the breeder blanket. This shell has all its air removed to create a vacuum inside. Otherwise, we would not be able to generate the plasma.

This shell is quite different than the protective dome that surrounds fission reactors. The dome around a fission reactor is designed to keep the fission reaction from escaping. The vacuum vessel around a fusion reactor is designed to make fusion possible. Any disruption or cooling of the plasma will instantly stop the fusion reactions. The slightest leak in the vacuum vessel will cause the plasma to cool and the fusion reactions to stop. Unlike fission power generation, fusion power is fundamentally fail-safe in this regard.

Diagnostics

Designing sensors that measure the plasma temperature counting the number of exiting neutrons and survive the hot plasma and neutron flux is no easy task. Engineers have made advances in diagnostic measuring using lasers. However, it might be that a power plant will produce such an intense neutron flux that many diagnostics devices will not last. Power plants might have to operate with fewer diagnostics data available. Diagnostics is a crucial area of continued R&D.

Q-Factor

In building a fusion power plant, we are primarily concerned about getting more power out of the plant than we must put into it. This ratio is called the "Q" ratio ($Q=P_{fusion}/P_{external}$). Suppose we heat the plasma until it achieves ignition. In that case, Q will equal infinity ($Q = \infty$) since we will need no external power to keep it hot. When the plasma begins to provide some of its own heating, it is called "burning plasma." For example, when Q = 10, the plasma itself provides about two-thirds of the heating needed, and external power provides about one-third. We might successfully operate a power plant with Q = 30, which is a burning plasma but not yet full ignition. In that case, we would get 30 megawatts out for each megawatt supplied. This would be a little easier, only requiring about 85% of the temperatures needed for full ignition.

The tritium breeding reaction in the breeding blanket adds an amount of heat energy equal to approximately 25% of the plasma's fusion reactions. This adds to the overall energy production or "Q" of the power plant.

Radioactive Waste

Fission reactors create significant amounts of *high-level* radioactive wastes from the spent fuel that continue to be dangerous for humans for tens of thousands of years. Fusion reactors, however, create no high-level radioactive waste from spent fuel. Fusion's only product is stable helium.

The only radioactive waste comes from physical components of the first wall and breeder blanket that have absorbed neutrons and have become radioactive at an *intermediate level.* It is the internal structure of the fusion reactor that will, only when disassembled, become small quantities of radioactive waste that becomes harmless within a century or so.

Remote Maintenance

Once a fusion reactor has operated using the D-T reaction, its first wall and breeder blanket will become too radioactive for humans to go near. Repairs and maintenance will need to be performed by sophisticated remote robotics whenever we shut it down for repairs. This applies to maintenance inside the vacuum vessel. Engineers must design the entire reactor and its systems with this in mind. First, it must be made as robust as possible so that repairs are rarely needed. Second, parts and systems must be made modular with simplified replacement steps. The sacrificial divertor plates and portions of the breeder blanket are certain to need periodic replacement.

Stellarators and Engineering Variety

A competing fusion reactor design shape called a "stellarator" differs from the shape of a tokamak. Instead of being a smooth donut shape, a stellarator is a torus but has twists along its long dimension. Think of it as a donut that has twisted sides instead of uniformly smooth sides. This design offers certain potential advantages in the way we can apply magnetic fields to the plasma. However, it also has some new challenges that engineers must overcome. More experimentation in both tokamaks and stellarators is needed before we will know which one offers better performance.

In recent years, several private startup companies have raised millions of dollars to try different approaches to achieving fusion. These, along with universities and large defense contractors, add to the possibility of finding an even better method of achieving useful fusion than the tokamak or the stellarator. You never know when a breakthrough discovery will occur, and even failed experiments add knowledge, so all of them contribute to the fusion effort.

Stellarators are but one example of design variety. The sheer magnitude of the controlled fusion undertaking and the number of individual engineering problems that need to be solved can, at first look, seem to be insurmountable. However, this complexity can be thought of as *variety* instead. There are many different engineering disciplines already involved, and we are early in the experimentation process.

Whenever one field struggles with a fusion engineering problem, one of the other disciplines might develop a way to deal with it. It might not be the problematic original design, but a workable one until a breakthrough occurs. Overall failure would mean the failure of all the engineering disciplines, and their variety makes this unlikely.

Looking at engineering variety from the other direction, we see that a mega-science project like ITER can yield discoveries in engineering fields, from materials science to computer modeling. We will find these discoveries helpful in applications outside of fusion energy. ITER and subsequent DEMO projects are like the "moon shot" in the last century's space program and the wonderful everyday products it produced.

FUSION AND COMPUTER MODELING

When trying to create sustained fusion by magnetically confining plasma, confinement is the central issue. The most difficult challenge in confining plasma is controlling the turbulence in it. Turbulence causes the magnetic bottle to leak.

We previously covered the "H-Mode" operation of a tokamak where, under certain controllable conditions, turbulence dramatically decreases, and the plasma's temperatures and containment times jump significantly. Scientists have learned how to make H-Mode happen, but still, to this day, they cannot explain why it works with all the computing power available.

Engineers do today's cutting-edge research in supercomputers. Most of the information we have today about plasmas' behaviors does not come from experimental data but rather from computer simulations. Modern supercomputers can compute millions of particles' trajectories and velocities and their interactions with each other and with boundary surfaces.

Explicitly modeling individual particles for a small fraction of the plasma containment time would require over a thousand years on today's supercomputers. Fortunately, scientists have created a sort of statistical shorthand called gyrokinetics that simplifies the calculations needed by a factor of a million. By using gyrokinetics, a supercomputer can perform the same plasma calculations in about a day. This allows researchers to repeatedly test their theories and make mistakes in a simulation. As computing power increases, simulations will become more realistic and more accessible.

Diagnostics systems engineered into the tokamak can precisely measure the plasma's performance and behavior. Some can even detect tiny fluctuations in density and temperature caused by turbulence. A goal is to make instantaneous, real-time adjustments to plasma confinement magnetic fields when certain *precursor events* signal a potential disruption. The greater the computing power available, the closer to instantaneous such a feedback control loop can become.

Suppose you consider the *design* of an entire fusion power plant that can produce the most net electrical power at the lowest overall cost and highest safety. The number of practical variables, trade-offs, and alternatives seem limitless. Engineers use computer programs called *system codes* that estimate electricity costs as they attempt to minimize it by varying the design parameters while staying within the design limits. As computers become more powerful, modeling an entire fusion power plant with better accuracy will result.

We can see that the transcendent technology of fusion energy will greatly benefit from advances in the transcendent technology of quantum computing for all the above reasons. Additionally, advances in nanomaterials technology will aid in providing more durable materials for fusion reactors.

ITER

ITER is perhaps the most significant science experiment that humans have ever attempted. If successful, it will be the first device ever created to confine plasma and generate more fusion power than the heating power put into it.

ITER is one of those science projects of humanity that, like the Space Station and the Large Hadron Collider, are undertaken through collaborations by many nations contributing money, time, and components. Participants in ITER include the European Union, the USA, China, India, Japan, South Korea, Russia, and others totaling 35 countries representing over half the world's population.

ITER originally stood for the *International Thermonuclear Experimental Reactor*. Public relations concerns over the words "experimental" and "thermonuclear" caused its meaning to be changed to the Latin word for "the way."

ITER is a tokamak-type reactor located in southern France and is about half completed. Once it starts operating, its purpose is to demonstrate the scientific and technological viability of fusion energy. ITER will not produce electricity but will be a demonstration of the physics and engineering required to build a tokamak power plant. If successful, then the physics will be mostly settled. The translation of what is learned to build a tokamak power plant will be primarily an engineering and economic challenge.

ITER has the following design goals:[268]

1. Achieve a plasma power multiplication factor of Q > 10 which means producing 500MW of fusion power from 50MW of input heating power.
2. Demonstrate the integrated operation of technologies for a future fusion power plant, such as heating, control, diagnostics, cryogenics, and remote maintenance.
3. Achieve a deuterium-tritium plasma in which the reaction is sustained through internal heating—"burning plasma."
4. Test tritium breeding, to produce the needed tritium that is not found in sufficient supply in nature.
5. Demonstrate the safety characteristics of a fusion device with negligible consequences for the environment.

ITER will be far bigger than any previous fusion experiment:[269]

1. The ITER site will encompass almost a full square mile.
2. The tokamak alone will be over 23,000 tons of steel and machinery.
3. The external electromagnets will contain 100,000 kilometers of superconducting niobium-tin wire. They will be cooled to -269 °C.
4. The donut-shaped torus with a "D" shaped cross-section will have an overall diameter of 20 meters and a height of 11 meters.
5. It will contain 1,000 cubic meters of plasma that will reach 150 million °C, 10X the core of the Sun.
6. It will generate 500 MW of output power.
7. Heating will come from two neutral beam injectors and microwave heaters totaling over 50 MW.
8. ITER's divertor plate will receive 20 MW per square meter, the highest steady-state heat flux of any solid material on Earth.
9. ITER's vacuum vessel will be the largest stainless-steel vacuum vessel on Earth.
10. As of 2017, ITER's construction costs are estimated at around $25 billion.

As a set of specifications, project to manage, and political football in multiple countries, ITER has had a rocky road from its inception in 1986 to now. It has been a challenge to get the funding needed and coordinate the politics and the work by the various countries. However, ITER has continued to progress. As of this writing, ITER's first plasma is still scheduled for December 2025.

ITER's impact on the fusion programs of individual countries is a two-edged sword. Participating in the manufacturing of various parts for ITER enables counties to develop fusion expertise. However, funding commitments to ITER dried up funding for various domestic fusion programs. Some countries, such as the USA, Russia, China, and South Korea, are continuing their domestic fusion programs to have within their countries the experience needed when fusion power becomes a reality.

If ITER is a success, then the next step will be to build a demonstration fusion power plant that puts electricity into the power grid. This project already has a name, "DEMO." Such a power plant would need to generate a $Q > 20$ and operate continuously for months. It will be primarily an engineering challenge.

Designs for the DEMO tokamak are already underway. Individual countries, including China, the EU, and South Korea, expecting ITER to succeed, are working on their power plant designs and plan to build their DEMO projects. This will enable those countries to develop fusion industries and have increased competitiveness. We can anticipate that most of our electrical power will be from fusion by the end of the century.

NEW APPROACHES

As of this writing, the most recent fusion success was the Korea Superconducting Tokamak Advanced Research (KSTAR) device, also known as the Korean artificial sun. On November 24, 2020, it set the new world record in maintaining the high-temperature plasma for 20 seconds with an ion temperature over 100 million degrees (Celsius). This more than doubled its previous record of 8 seconds in 2019.[270]

ITER represents the mega-scale approach to fusion experimentation based on mainstream scientific and engineering theory that only multinational governments can afford. However, several smaller-scale fusion labs and private-sector fusion startups have emerged to explore alternative scientific and engineering theories in the last decade. These alternative methods of potentially achieving fusion are primarily the result of recent computer simulation advances, precision machinery, and synthetic materials technologies.

These advances make possible experimental fusion reactor designs that are a fraction of the size and cost of those just a few years ago. Both Commonwealth Fusion in Boston and Tokamak Energy in Oxford intend to build much smaller fusion reactors that, unlike ITER, could fit inside a gymnasium. This is enabled by using new electromagnets roughly 2% the size of ITER's using novel alloys that become superconductive at higher temperatures with less cost and energy requirements than ITER.[271]

Similar to the way SpaceX has evolved non-traditional, private-sector space launch capabilities out of the lessons learned by earlier governmental space programs, the new non-traditional approaches to fusion are growing out of past government-funded mainstream fusion research.

The UK company, First Light Fusion Ltd, a spinoff of Oxford University, is exploring one such alternative fusion approach. It is building machines for testing computer models of the intense temperatures produced by shock waves from collapsing "bubbles" that are mechanically generated. If the models prove out, then the company could potentially build a prototype generator by 2025.[272] Another UK company, Tokamak Energy, and at least two North American startups also aim around 2025 to fire up prototype fusion reactors, each roughly the size of turbines inside traditional power plants. If any of these alternative approaches succeed, they will have achieved a dramatic leap in fusion science.

FUSION SUPPORT AND FUNDING

Considering the magnitude of the potential benefits of fusion energy, fusion research funding has been paltry. As Parisi and Ball point out, one cannot even consider its level of funding as "peanuts." In 2016 the US Department of Agriculture's budget for peanut subsidies was double the entire budget for the Office of Fusion Energy Sciences.[273] Yet, amazingly, progress in fusion research has reached the point of a half-built ITER project.

This low funding was somewhat understandable during the second half of the twentieth century. The prospect of achieving fusion power was far into the future, if even possible. Research success was so slow and fraught with setbacks that critics said, "Fusion is 20 years away...and will always be." Because there was no compelling cause, no war, no presidentially motivated moon shot, and no fear of climate change to provide the motivation, fusion was never a "front burner" issue for politicians.

A 1976 report estimated five different plans and funding levels to achieve contained fusion within specific periods. These were "Maximum Effective Effort," "Accelerated," "Aggressive," "Moderate," and "Maybe Never." The actual funding since 1976 has been *half* of the Maybe Never level.[274]

However, with the issues we are facing, such as climate change, and how close we are to achieving contained fusion and the competitiveness that its success would bring to our nation, one wonders why we are not investing more heavily in fusion development in the US? The obvious answer is the continued lack of public interest and political support.

The less apparent reason is simply a lack of awareness. The public and the politicians are not sufficiently aware of fusion's potential benefits or that it might be reaching the inflection point of its exponential growth curve. This lack of public and political awareness was because of a lack of national-level leadership and priority-setting. Now, we can blame it partially on ITER.

ITER claims to be only five years from beginning operations and 15 years from achieving (or failing to achieve) its design goals. Other than preliminary design work on possible DEMO fusion power plants, the world is waiting to see what happens with ITER before committing fully to further fusion development. One cannot overstate ITER's importance. If ITER is successful, we can expect to see a massive increase in public and private investment toward fusion power plants. If, on the other hand, ITER fails in its primary design goals, then investment in fusion development will suffer a significant setback. Fusion will once again be labeled twenty years away, and it always will be.

Most scientists expect ITER to succeed. Even if problems arise, they expect to find workarounds. The same applies to DEMO. They are confident that we will eventually be able to achieve controlled fusion to produce electrical energy. That being the case, we should redirect all the current global activism applied to climate change toward its solution—fusion energy. Climate change activists everywhere should be elevating public and political awareness of fusion energy's potential and lobbying for its increased funding. Unfortunately, the climate change activists themselves are not sufficiently aware of fusion's potential and proximity.

The recent emergence of small labs and private-sector startups with alternative approaches to achieving fusion has been possible due to an encouraging increase in funding. The previously mentioned UK company First Light Fusion Ltd has raised $32.8 million, and Tokamak Energy has raised $87.3 million. In the US, the Department of Energy's recent "INFUSE" program will make grants available in 2021 to fund 80% of private-sector fusion research costs. Additionally, "The INFUSE program will accelerate fusion energy development in the private sector by reducing impediments to collaboration involving the expertise and unique resources available at DOE laboratories. This will ensure the nation's energy, environmental, and security needs by resolving technical, cost, and safety issues for industry."[275]

For humans to have a long-term existence on this planet and a future in the stars, we will need fusion energy. The good news is that fusion's funding has historically been so low that even small increases in public support and political priority should yield significantly greater and faster development. Hopefully, fusion research funding is also about to reach the bend in its exponential growth curve.

FUSION AND NUCLEAR PROLIFERATION

According to Parisi and Ball, as of 2018, nine countries have nuclear arsenals. No nuclear weapon has been used since World War II. There have been several close calls, such as the 1961 breakup of a U.S. bomber with two nuclear bombs, one of which was found fully armed. Only a single low-voltage safety switch had prevented it from detonating. Also, the 1962 Cuban Missile Crisis and the 1995 Norwegian launch of a science rocket that Russian radar thought was an incoming U.S. missile. In the last twenty years, dozens of nuclear warheads have gone missing and there have been at least 25 cases of nuclear explosives being stolen.[276]

The presence of nuclear weapons in the world might have been a stabilizing influence during the Cold War. Today they could present more danger than safety. Increasing efforts by terrorists to acquire or build an atomic bomb to destroy a major city are of great concern. Even a limited nuclear conflict between two smaller nuclear states could lead to a *nuclear winter* that could kill millions worldwide. The effectiveness of the anti-ballistic missile defense of the U.S. is uncertain.

Obtaining fissile material is the most challenging step for building a bomb. Uranium enrichment is a complex and massive undertaking. Fission power plants require fissile material and, therefore, a uranium enrichment industry. A bomb requires as little as 10 kilograms of enriched uranium, while operating a power plant for a year requires over 10,000 kilograms. If spent fuel is removed from a nuclear power plant early in its useful life, it still has weapons potential. Keeping or expanding our reliance on fission power plants sustains a potential risk of nuclear proliferation.

Fusion power plants, if they were to replace fission power plants, have the potential to lead to a world where we greatly minimize the risk of nuclear proliferation. In a fusion power plant, *there is no need for fissile or fertile material of any kind.* All isotopes of uranium, thorium, and plutonium can be entirely banned worldwide. The detected presence of any of these materials would immediately indicate malicious intent.

Fusion power plants could conceivably be used for breeder reactions to make fissile materials. However, with some commonsense precautions, fusion power plants could be designed and monitored so that this would be difficult to implement, readily detectable, and easy to disable.

FUSION AND THE CLIMATE

Fusion energy is not likely to be the answer to carbon-based climate issues in the next two decades. However, as we move into the second half of the 21st century, fusion energy coupled with Atomically Precise Manufacturing will be the solution.

Operating Cleanly

Unlike today's energy production and consumption, if most of our energy came from fusion, then energy production and consumption would have a negligible negative impact on the environment or climate. Of all known energy sources, fusion fuel has the highest energy density. It is also plentiful, practically unlimited both here on Earth and throughout the universe. Fusion fuel is readily and cheaply extracted and transported. Fusion power plants are much smaller than conventional power plants; less land footprint is devoted to energy production. Fusion power plants do not produce high-level radioactive wastes; their only product is stable helium. Production of the fuel for fusion and fusion power plants' operation do not emit greenhouse gasses or add significant carbon to the environment.

As fusion power plants begin to provide plentiful, cheap electrical energy, other manufacturing plants that emit carbon will apply that affordable electricity to technologies sequestering the carbon before it is emitted. And the electricity will power technologies to turn that sequestered carbon into useful substances, such as bulk materials for other manufacturing processes, or someday, for nano-assemblers.

Clean-Up Operations

Abundant, cheap electricity, cleanly produced by fusion power plants, would lead to cleaning up our environment. Clean fusion power plants could replace all fossil fuel power plants and fission power plants. Abundant, cheap electrical energy can speed the replacement of cars, trucks, and busses that use fossil fuels with electric cars. It can also lead to widespread all-electric homes, commercial buildings, and factories. As technologies from quantum mechanics and nanomaterials advance, new lightweight, ultra-efficient electric motors and ultra-high energy density batteries will even allow electric airplanes and people carrying drones to replace many aircraft presently powered by fossil fuels. When we scale fusion power generators to smaller sizes, they will replace fossil fuels in powering ships and trains.

For carbon sequestration efforts to remove excess carbon already in the environment, they will need to scale up in a big way. They will require the plentiful energy that fusion can provide. Likewise, as technologies for removing plastics and other man-made materials from the land and oceans advance, they will need affordable fusion energy to become industrialized.

As nano-assemblers decentralize most manufacturing, we will widely distribute them in huge numbers. They will need plentiful electricity to power them. They will eliminate most of the need for large-scale factories, warehouses, and product distribution via ships, rail, and trucks. They will disassemble most discarded items back into their constituent elements and reusable components, essentially eliminating the need for landfills.

FUSION AND SPACE TRAVEL

Humans have an insatiable need to explore and discover. Give us the technological capability to do so, and there is no way we will not explore our solar system and then other star systems.

Our present chemically powered rockets can propel spacecraft to Pluto, 4.3 billion kilometers away, in about two years. Two years might seem slow but travel within our solar system is possible using our present technology. However, Alpha Centauri, the nearest star, is 42 trillion kilometers away and would take tens of thousands of years. Clearly, we need to be able to go faster, a lot faster.[277]

Rockets are moved by pushing propellant matter out the back end. The higher the velocity of the exhaust, the faster the spacecraft can go. The fastest a spacecraft can reasonably go is about twice the speed of its exhaust. Fusion reactions can eject propellant particles at about 5% of the speed of light; therefore, it could propel a spacecraft to approximately 10% the speed of light. At this speed, we could reach Alpha Centauri in about 75 years.

Suppose humans achieve radical longevity, perhaps living to a few hundreds of years of age. Or nano-biotech can put our biological bodies into suspended animation. Or we become Virtual Humans in machine bodies that we can shut down and then restart. Then in any of those cases, fusion propulsion makes the prospect of interstellar space exploration a distinct possibility.

The lightweight elements used as fusion fuel are among the most readily available throughout the universe. Fuel for refilling our rocket ship will not be difficult to find. Fusion rocket power is the best and only option for faster space travel that we presently have. As agents of evolution, we will need fusion energy to go to the stars and awaken the universe.

♦♦♦

PART 5

MAKING PLANS

In review, we have learned how our brain and our rational faculty work. We have identified our present position on the "map" of the universe and evolution. We have learned about the major processes that got us where we are today and will drive our future. We understand that the ideas in our society matter. We have established a basic, integrated set of conceptual ideas that we believe to be connected to reality and therefore true. We have addressed some of the practical issues that might arise from advancing technology regarding your productive work, your recreation and leisure, and your relationships. We have also gained basic knowledge regarding eight of the transcendent technologies and the opportunities and threats that you need to watch for and prepare for as they advance. And, we have anticipated how rapidly technology might change things going forward.

So now, the big question to be addressed is, "What should, and can I do?" If I have emphasized anything in this book, it is that you must use your rational faculty to choose your own values and the actions you will take to achieve them. In the next few chapters will be some suggestions for you to consider. But first, a comment about motivation.

I assume that you value your own life. Think of someone or something that you would be willing to sacrifice your life for? Now, realize that what is required of you to benefit that someone or something that you value higher than yourself, is not that you sacrifice yourself, but rather that you become more knowledgeable and skilled in living a life that is connected to reality and consistent with your nature as a *conceptual* human—a life in which you pursue with heroic commitment your own rational self-interest according to your own values. A rational *society* is created by, maintained by, and then supports the rational *individuals* within it. By living your own life with a commitment to reality, reason, individual rights, and capitalism, you will influence others toward doing the same.

5.1

PLANNING HORIZON

"You can never plan the future by the past."

—EDMUND BURKE

EXCEEDING OUR VISION

To plan, we must establish a planning horizon. We have identified two significant events that could be cataclysmic for human society—the technological singularity and our society's collapse into a totalitarian regime. While reputable sources project both based on the data to occur by mid-century, neither is a certainty.

I think the technological singularity will probably not occur by mid-century. We could have potent natural language processing AI using an integrative thinking method. These AI could have robot bodies and exceptional skills for interacting with things in reality including humans. They would not yet have achieved fully functional *perceptual-level* intelligence. That will likely take longer. They would not have intelligence equivalent to a human. However, by mid-century, we could apply that incredible AI technology to augment our human intelligence. We could be amazingly more intelligent than today.

On the other hand, I think society's collapse into totalitarianism might occur by mid-century—unless enough people insist on applying AI technology in ways that support reality, reason, individualism, and capitalism. I think a societal collapse is the more threatening danger over the next two or three decades. If society collapses, so will the rate of technological achievement.

The potential consequences of each of these two events is so serious that we must treat them as probable, not just possible, and plan accordingly. The planning horizon for these two events—the Collapse and the Singularity—is 20 to 30 years. But what is the planning horizon if neither event occurs? Be suspicious of any planning based on projections beyond about 25 years. Technology will cause time horizons toward which we can plan to get progressively shorter.

Projections made today can only be based on our present level of intelligence. With technology advancing exponentially, it will only be a few decades until intelligence far surpasses what we have today. Some futurists think that we will achieve machines as intelligent as humans by 2045 and smarter than humans shortly after. Whether that happens or not, we are already using our technology to augment our intelligence significantly.

We also base the planning we make today on the paradigms of today's lifestyle and human behavior. These paradigms (how things get done) are likely to see a dramatic change in the few decades ahead. We cannot even imagine the kinds of creative solutions and technology that enhanced humans of tomorrow will produce. Product and technology life cycles will get progressively shorter, and paradigm shifts will happen more frequently. The distance into the future that we will be able to predict with any accuracy will shorten.

As an example of suspicious projections, consider future highway construction plans based on traditional projections of demographic and geographic population and automobile traffic volume out forty years or more. These planners in my community say that a high-capacity, high-speed highway needs to be built to handle traffic projections for the next forty years. There is no way a highway department planner today can forecast an increase in automobile traffic forty years into the future. He would need a crystal ball telling him how the following factors and many more we are not aware of today might affect automobile traffic volume:

1. Autonomous vehicles communicate with each other and the smart roadway infrastructure to coordinate their speeds and take the least congested paths.
2. Smart roadways anticipate approaching traffic, adjust traffic lanes and lights, and redirect vehicles to optimize traffic flow.
3. The ride-sharing economy where people do not individually own cars and instead schedule or summon one when needed.
4. Increased virtual work-from-home in the workforce resulting in fewer office people driving to work.
5. Use of robotics in manufacturing, construction, and distribution resulting in fewer laborers driving to work.

6. Improved virtual reality and telepresence experiences replace the need to go to school, shopping, or entertainment.
7. Wearable medical monitoring, remote diagnostic and drug dispensing devices, and telemedicine doctor visits, result in fewer people going to the doctor or pharmacy.
8. Aerial package delivery drones reducing delivery truck traffic.
9. People-carrying drones move traffic into sky lanes and off the roads.

As this example indicates, we must strive to think of coming change on exponential timelines, not our past's linear timelines. We must begin to place less credibility on any projections claimed to be accurate beyond a couple of decades, and that will shorten.

♦♦♦

REMEDIAL SELF-EDUCATION

"Self-education is, I firmly believe, the only kind of education there is."

—ISAAC ASIMOV

INTROSPECTION

You now understand how your mind works to connect you to reality. You are aware that you probably hold a mixed bag of ideas in your subconscious; some you consciously validated against reality and some not. You know that you need to identify and validate your subconsciously accepted ideas from parents, teachers, ministers, college professors, self-help gurus, book authors, friends, and associates. How many floating abstractions do you hold because they have been passed down from human to human in our society for generations? Unfortunately, you are entirely normal. We must learn proper thinking because we were not taught how to think this way in school.

As an adult of any age, it is time for you to begin remedial education. The older you are, the more difficult it will be to displace bad thinking habits you have lived with for so long. One's sense-of-life and view of the universe unconsciously formed early in life are exceedingly difficult, but not impossible, to be later changed by rational thinking as an adult. These are acquired ideas, not instincts, but were not "chosen." They are more like imprinting.

Begin by reading the books and going to the websites I have referenced, especially those on philosophy. Commit to becoming observant of and critical of your ideas. When you are about to profess some belief, ask yourself—"Can I connect this idea all the way to reality and the nature of man?" If you cannot, then you have more thinking to do.

The goal is a Personal Philosophy that is fully integrated into your mind without contradictions. The Personal Philosophy summary presented in this book is only the beginning, just enough to demonstrate how the process works. You must expand upon it, making broader and deeper integrations with your effort and eliminating contradicting ideas.

Replacing entrenched but erroneous ideas in the brain's neural networks—thought habits—with new ones requires repeated conscious thinking. One develops new thought habits by frequent practice—daily, weekly, monthly. Many of the organizations and authors I reference offer newsfeeds, blogs, and podcasts. These contemporary news sources discuss how to apply rational, reality-based principles to current events. Begin replacing untrustworthy daily news sources with those you believe are committed to reality and reason. Engaging your rational faculty a little each day, week, and month will put you on the correct path.

If you work in the sciences or are a technologist, you know what it means to connect ideas to the Laws of Reality—to the physical world's nature. Ask yourself why it is, when it comes to ideas about ethics, morals, politics, or art, you do not question how they are tied to reality and man's nature? Is it because the education you received in the humanities was different from that in the sciences?

Remember, rational thinking is the method of survival evolution gave you and integration is the proper mode of thinking. Unfortunately, that is not the mode of thinking around you in most of society. You have volition, the ability to focus your thinking or to suspend it. Still, you cannot evade the consequences of that choice. Suppose you have been going through life with your mind even partially suspended. In that case, you are unaware of how much better your life as a human can be with a fully focused mind. There is only one way to find out.

♦♦♦

HAPPINESS SKILLS

"Most human beings have a desire to feel good, to avoid painful feelings. If given a choice, most individuals would wish to live life every day with zest, courage, enthusiasm and passion. Happiness, many would tell you, is their birthright, it is what life is supposed to be about. Why then are so few people happy? Why do so many people live in a state of quiet desperation as if the best they could do is to endure the fear and pain associated with life? Why are so many individuals angry at life for being cheated out of happiness?"

—DR. EDITH PACKER

HAPPINESS

Exploring our past, we learned that we came to exist as a result of the process of biological evolution; that it is our more robust ability to grasp and modify reality that elevates us above all other living things on Earth; that "reason" is our tool of knowledge and means of survival; that societies of individuals that have predominantly rational ideas advance while others decline; that we are ascending an exponential technology growth curve via the *Technology Cycle* and future evolution will be technological rather than biological; and that we could reach a societal collapse or a technological singularity by mid-century. These are all big ideas. We have explored what will be required *philosophically* for our society to survive long-term.

But we do not want to limit our thinking to only the avoidance of disaster. Our highest value is the achievement of our happiness, no matter what is happening in the world around us. How do we bring all this down to the personal *psychological* level? What can we learn about how an individual achieves happiness in the here and now?

We learned that happiness results from using *our* reason to choose and then achieve *our* values. Notice that the prior sentence is a *first-person* statement—"I" choose my values and then "I" achieve them. There is no "he/she" that makes me happy, or "they" make me happy. No one else has a responsibility or the ability to, make "me" happy. Happiness is an achievement, and "I" must achieve it myself.

But there must be more to learn about how we go about doing this on a day-in, day-out basis. If we can learn the habits and skills required to achieve happiness in today's world, then perhaps we will be better able to achieve happiness in our dramatically changed future world. If we are going to live extra-long lives, knowing how to stay happy will be critically important to us. Also, we will better understand how the idea of happiness might apply to us as future Hybrid Humans.

OUR PERSONAL PHILOSOPHY

BRANCH:	APPLIES TO:	QUESTION:	ANSWER:	CONCISE:
METAPHYSICS	Existence	*What is out there?*	*A "real" reality.*	REALITY
EPISTEMOLOGY	Knowledge	*How do I know it?*	*I can trust my senses and my mind to grasp reality.*	REASON
ETHICS	A Person	*What should I do?*	*I must use my mind to identify and achieve my own happiness.*	INDIVIDUALISM
POLITICS	Society	*What about dealing with others?*	*Everyone must be free to think and act and trade with each other without force, coercion, or sacrifice.*	CAPITALISM
AESTHETICS	Art	*What is the image of my Worldview?*	*Art that shows a real world and rational people.*	ROMANTIC-REALISM

PERSONAL PHILOSOPHY

We already know that happiness requires that our Personal Philosophy be based on and connected to reality. Above is the summary table of the essentials we hold to be true in our Personal Philosophy.

OUR PSYCHOLOGY

What about our *psychology*? Are there psychological traits that lead to happiness? Can these traits be learned as skills? Fortunately, a psychologist has listed the psychological traits and skills that a person with a reality-based Personal Philosophy needs to achieve happiness. Dr. Edith Packer, J.D., Ph.D., delivered the talk "Happiness Skills" at the 1991 Summer Conference of the Jefferson School. The following is a summary of Dr. Packer's guidance regarding Happiness Skills.[278]

Happiness Skills

A skill is an ability to do something well as a result of knowledge and practice. Happiness Skills start with conscious and subconscious *attitudes*, which lead to *mind and character traits*, which lead to *habitual ways of acting*. Repeated use of this combination in action over a long period results in happiness skills.

Three Key Attitudes of the Happy Person:

1. **Love of life**
2. **Optimism**
3. **My life is important**

Key Attitude #1: Love of Life

A basic subconscious love for life—"I am passionate about being alive. What I value most, above all, is life."

Key Attitude #2: Optimism

An aura of subconscious optimism—"So long as I am alive, there is always something of value to enjoy."

Key Attitude #3: My Life Is Important

A subconscious attitude that not just life, but he and his life are important—"I will guide my life based on my own judgment of facts and not on the views of others." This includes the recognition that the main consideration must be what *he* desires. And he recognizes that others, like himself, have a right to pursue what *they* desire. Being rational, he does not expect to clash with other people. This is called *rational egoism.*

Key Skill #1: Responsibility for Own Happiness

He takes responsibility for asking and answering, for himself, questions such as:

1. "What plans do I have to make for the future so that I can be happy in the long run?"
2. "What kind of work will I enjoy?"
3. "What is it that I have to do now and every day that will make me enjoy life in the present?"
4. "What do I have to do in order to be happy?"

He knows that he cannot control what happens to him by chance or because of other people's irrational actions.

Key Skill #2: Commitment to Effort

The happy person recognizes that he must expend effort in pursuit of what is important to him. His repeated successful application of effort in thinking and action results in the discovery that the process of exerting effort is essential to enjoyment. Expending effort to acquire new knowledge brings a sense of enjoyment because he experiences knowledge as the key to achieving values.

Key Skill #3: Commitment to Action

The happy person is subconsciously and consciously committed to action. Once he identifies a value he desires, its achievement becomes his goal, and he immediately focuses on the action necessary to achieve it. He scrutinizes his everyday activities, explicitly questioning whether they move him closer to his goal or take him farther away from it.

Key Skill #4: Commitment to Health and Learning

The happy person is committed to the health and wellbeing of both body and mind. He stays fit, participates in physical activities such as sports, does not use narcotics or become intoxicated. He continuously struggles to acquire new knowledge both inside and outside his field.

Key Skill #5: Controlled Emotions

The happy person does not permit himself to be stopped by negative emotions, especially fear. He does not deny having them. He does not assign an important role to them. Unrealistic fear does not stop him from acting. Fear is an emotion and is always the result of a thought which anticipates danger. He knows that the solution to fear is to be fact-oriented, i.e., to determine whether the fear is realistic or not. He can overcome unrealistic fear by thinking, effort, learning, and action.

He does not start an activity by anticipating failure. He deals with failure when it happens by figuring out what he did wrong and focusing on correcting his mistake.

Key Skill #6: Commitment to Facts

The happy person values facts above emotions. He can recognize essential facts of each concrete situation without regard to his wishes, hopes, and fears. He can distinguish between facts of reality that humans cannot change and facts that are man-made, the result of human choice.

The ability to separate one's wishes and fears from facts and conduct one's life based on facts is the most fundamental of all happiness skills.

Outcomes of Fact Orientation

- Permits a person to pursue goals based on a realistic appraisal of himself and the possibility of success.
- Justice (virtue): He deals with and judges others based on essential facts of their expressed ideas and actions.
- Honesty (virtue): He does not fake facts.
- Integrity (virtue): He is not afraid to act on his principles and based on facts.

Key Skill #7: Value Orientation

A "value" is that which one acts to gain or keep. A "virtue" is the action required to achieve a value. Life without values has no direction or meaning. Values provide the individual with purpose, goals, excitement, and focus on what he loves (passion). The happy person accepts responsibility for choosing his values rather than living for other people's values. The happy person knows what values he wants to achieve in all the important areas of his life.

Major Values Examined

The five major areas where one chooses values on which happiness depends are:

1. Productive Work (Career)
2. Romantic Love and Sex
3. Friendships
4. Recreation (Leisure)
5. Art

True Idea: Happiness results from the achievement of one's values. Happiness is one's highest moral purpose.

Productive Work: The happy person loves his work with a passion. To be in control of his life, the happy person has a productive central purpose that establishes the relative importance of all his other values. It saves him from pointless inner conflict. It is not the degree of a man's ability that is ethically relevant, but the full, purposeful use of his ability. A career requires the ability to sustain a purpose over a long period, through many separate steps, choices, decisions, adding up to a steady progression towards a goal.

Romantic Love and Sex: To love is to value. The man who does not value himself cannot value anything or anyone. Only a rationally selfish man, a man of self-esteem, is capable of love—because he is the only man capable of holding firm, consistent, uncompromising, un-betrayed values. Love is a person's response to his own highest values in the person of another—an integrated response of mind and body, of love and sexual desire. Sex is one of the most important aspects of a man's life and, therefore, must never be approached lightly or casually. A sexual relationship is proper only on the ground of the highest values one can find in a human being.

Friendships: The happy person knows that his life can be enhanced by sharing it with people he likes. He develops friendships on different levels ranging from sharing one value, such as sports, to all-encompassing soul-mate friendships.

Recreation (Leisure): The happy person knows how to relax and enjoys various optional values in his leisure time. He knows that happiness is not experienced only in "big bangs" such as getting married or gaining a promotion, but every day in "small bangs" in the details of daily life: preferences such as where to walk, what to eat, clothes to wear, etc.

Art: Art causes a man to feel an emotion, that if put into words, would say: "This is what life means to me." Our conscious thinking programs our subconscious mind; our conceptual intelligence programs our emotional intelligence. Our conceptual intelligence is increasing over time; therefore, our emotional intelligence is as well. Our emotional intelligence instantly feels the appreciation of or revulsion toward art according to our value judgments. As our emotional intelligence increases, we will undertake new forms of art and new levels of appreciation for it. Art's value to us will increase.

Major Virtues

A **virtue** is an action by which one achieves a value:

1. The virtue of **independence** is the acceptance that you must live your life according to your thinking and judgment, not the thinking of others.
2. The virtue of **honesty** is the recognition that you cannot fake existence. It means to live by, to practice, any idea you accept as true.
3. The virtue of **integrity** is the recognition that you cannot fake your consciousness, that you are made of a body and mind, and there is no breach between them.
4. The virtue of **justice** is the recognition that you cannot fake others' character, that you must judge others for what they are by a rational process of identification and treat them accordingly.
5. The virtue of **productiveness** is the recognition that productive work is the process by which your mind sustains your life, the power to adjust your environment to yourself.
6. The virtue of **pride** is the recognition that you are your own highest value, and like all of your values, it must be earned. It is the achievement of the creation of your character and self-esteem.

Key Skill #8: Commitment to Self-Change

The happy person is committed to being aware of and changing the possible negatives in his life. He pays special attention to his psychology. He tries to be aware of his thoughts, emotions, and motivation. The happy person judges himself, his qualities, and his attributes, not based on what he wants them to be but based on what they are. If he finds some negative in himself or his functioning, he does not lament it but goes about changing it to the best of his ability.

Fact orientation leads the happy person to be willing to change his goals, if necessary. Reason and reality require that we adjust to the facts about ourselves and pursue only what, with hard work, is possible to us.

The happy person does all this because he is proud of his character and what he has accomplished. He wants to be free of pain and suffering to enjoy his values.

Key Skill #9: Perspective

The happy person can keep in mind all the relevant facts and not make judgments based on relatively minor, arbitrarily selected, isolated facts. The happy person evaluates positive and negative everyday occurrences from the point of view of how they affect his values. He knows that life's circumstances can create problems, but he usually views them as minor annoyances, not major tragedies. Keeping perspective gives the happy person emotional stability because he does not make mountains out of molehills.

Key Skill #10: Long-Range Planning

The happy person is committed to planning for his future in all major respects. He plans financially (as far as able), including education of children, illness, and unemployment. He plans intellectually, including new skills connected with work and subjects or activities that otherwise interest him. He enjoys planning for the future enjoyment of optional values. He enjoys the anticipation and preparation for a future value (milking a value).

The happy person knows he must forsake short-range gratification that would be at the expense of long-range values. He never sacrifices a higher value for a lesser one or non-value. He achieves pride and avoids regret.

Happiness Skills Summary

Happiness is an *achievement*, and it is in your control. It is not something that happens to you. It is something you must bring about. Happiness skills are developed by building in your brain's neural pathways the thought habits described here. You strengthen them over time by consciously focusing on them repeatedly. I recommend that you keep the following list where you can refer to it often and develop the habit of doing so.

HAPPINESS CHECKLIST:

1. Make up your mind to be happy.
2. Assume responsibility for achieving your values.
3. Become clear on what values you need and want in life.
4. Identify what attitudes and habitual ways of acting stand in your way.
5. Commit yourself to changing whatever is in your way.
6. Set goals consistent with your values.
7. Commit yourself to act toward the achievement of your goals.
8. Do not let fear (of change, failure, etc.) stop you.
9. Do not be afraid to take risks.
10. Motivate yourself in every possible way. (Imagine, when you are old, the pleasures/regrets you will feel for values gained/missed.)
11. Be positive, giving more weight to positives and dismissing negatives.

GREATNESS

In the Prologue to this book, I remarked that many people are stuck in a lifestyle short of the heroic vision they had for themselves when they were children. I said that no one facing a greatly extended life span should want to live that way. If *pride* and *happiness* result from achieving one's rationally chosen values, then what is *greatness*? How do we regain and then achieve the heroic vision we once had for ourselves?

The answer to this question has been articulated by another psychology professional. Psychotherapist Dr. Michael J. Hurd, Ph.D., in his 2002 book *Grow Up America! Learning to Live the Happy, Responsible Life*, explains:

> We all probably have some measure of greatness within us, at least potentially. But greatness is not what we are trained to think it is. Our greatness lies not in our capacity to write a check to a charity organization or to build houses for poor people. Our greatness lies in our individual willingness to create, to produce, and to use our minds. To selfishly—

> yes, *selfishly*—develop our interests and talents to their fullest potential represents the true measure of our worth.
>
> An act of greatness consists of a businessperson working long hours to make his company number one. An act of greatness consists of an engineer struggling to determine how to construct the Golden Gate Bridge or its equivalent. An act of greatness consists of a dancer or a musician or an artist working doggedly to bring absolute excellence, even perfection, to her craft. In selfishly developing our talents and pursuing our dreams and values, plenty of other people benefit—especially if we succeed.
>
> In contemporary society we have lost sight of—or, more accurately, never truly grasped—the authentic meaning of greatness. Increasingly, we pay cynical lip service to the politically correct/religiously correct view that the essence of greatness lies in helping an old lady across the street.[…] Like it or not: there's nothing so compassionate as the cold hand of science, reason and objective knowledge.[…] Science and business do more to help that old lady across the street than any self-anointed (or popularly elected) do-gooder ever could. Yet science and business continue to be despised by many of our cultural elites—and ignored or simply taken for granted by the masses. […]
>
> We stubbornly persist in the belief that greatness lies in *giving up* one's self rather than developing one's self with heroic and precise dedication. The core and essence of self, after all, is the mind. And it is through the benefits of the human mind that all progress, great and small, occurs. It is through ideas that human beings master their environments and conquer the universe.
>
> The dominant view today, tragically, is precisely the opposite.[279]

Ultimately, you maximize your happiness by applying reason in choosing your values according to facts of reality, choosing your goals, identifying the actions required to achieve them, and then acting—developing yourself "with heroic and precise dedication." Pride in the person you have become is your reward.

♦♦♦

ACTIVISM

"One withstands the invasion of armies; one does not withstand the invasion of ideas."

—Victor Hugo

"Anyone who fights for the future lives in it today."

—Ayn Rand

Remember the questions I posed in the Prologue—*What is the most powerful thing in the known universe? What can defeat it?* Victor Hugo identified the most potent thing in 1852 in the above quote—ideas. Powerful ideas come from the minds of men committed to reality and reason. More specifically, the most powerful thing in the known universe is—**"A society of conceptual entities that can sustain a commitment to reality and reason as their source of knowledge and guide to action."** And what can defeat the most powerful thing in the universe?—**a disconnect from reality and abandonment of reason**—our society has been doing both over the last two centuries.

This chapter is the "manifesto" of this book—a call to action. There exists widespread concern that a future super-AI might hold "anti-human" values. I am more concerned that today's narrow-AI is *already* and will *increasingly* be used by humans to implement today's "anti-human" values in our society's institutions—unless enough of us do something to bring about change.

The possibility of a societal collapse is an existential threat, with hundreds of millions of lives at stake. From primitive man through 18th-century pioneers of the American west, most individuals knew how to live off the land and could find an environment where that was possible. In the last two centuries, we traded that capability for the division of labor, specialization, mechanized production and distribution that enabled the vast expansion of modern society. Our survival depends on technology and critical supply chains, which rely on a functioning society of individuals able to apply their chosen productive specialty—capitalism.

If capitalism collapses entirely, mass suffering and death will follow. Protecting our free society is a life-and-death matter for ourselves and our children.

POSITIVE INDIVIDUAL ACTION

Your remedial self-education is the action you can take *introspectively*. Now we come to the question of what you, as an individual, can do *extrospectively* to help secure a positive future for yourself and society. If a totalitarian dictatorship ever arises in America, it will be because of the default of those who kept silent. Ayn Rand said:

> The battle is primarily intellectual (*philosophical*), not political. Politics is the last consequence, the practical implementation, of the fundamental (metaphysical-epistemological-ethical) ideas that dominate a given nation's culture. [...] In an intellectual battle, you do not need to convert everyone. History is made by minorities—or, more precisely, history is made by intellectual movements, which are created by minorities. Who belongs to these minorities? Anyone who is able and willing to actively concern himself with intellectual issues. Here, it is not quantity, but *quality* that counts (the quality—and consistency—of the ideas one is advocating).[280]

What is first required of you "is *honesty*—intellectual honesty, which consists of knowing what one does know, constantly expanding one's knowledge, and *never* evading or failing to correct a contradiction. This means: the development of an *active* mind as a permanent attribute."[281] This requires "a profound conviction that ideas are important to *you* and to your own life. If you integrate that conviction into every aspect of your life, you will find many opportunities to enlighten others."[282]

Consider yourself an *educator*. While the people you encounter are doing the best they can, the problem is that "they don't know what they don't know." Most are unaware that there is another way of looking at things. Yes, the universities offer a range of ideas—but each idea offered is a different variant of the Plato-Kant philosophy. The Aristotle-Rand philosophy is either absent or dismissed. You can *demonstrate* for them a philosophy based on reality, reason, self-interest and capitalism by living according to it yourself. You can *teach* it to them by engaging in conversation when appropriate.

You can also consider yourself a *defender* in the Worldwide War for Minds. It is important that you *must not keep silent* when your values are being attacked. You should make it a habit to object to such remarks—not with a lengthy speech, but simply to say: "I don't agree." Then, be prepared to explain why if the speaker wants to know.

Manage your expectations regarding what is possible. Most people have never developed for themselves *any* system of *explicitly validated ideas.* Even your suggestion that they should *have* a validated system is foreign to them. Why do they need it when they have managed to get along with the (unvalidated) ideas they already have?

We learned that the brain wants to conserve energy and will resist changing established neural networks. The older the person is you are dealing with, the less likely he will be willing to expend the energy it takes to consider and modify long-standing thought habits *explicitly.* This is where *demonstration* on your part comes in. Suppose the person's subconscious frequently sees what correct behavior looks like. In that case, it will begin forming new associations in the person's subconscious even though he has not consciously validated them. Your behavior can begin to affect his.

To enjoy your life, you need many friends both at work and leisure, at varying levels of philosophical engagement. A person did not get to choose his parents or the people who raised and educated him. It is not his *fault* that no one ever explicitly taught him the *integration mode* of thinking. It is not his *fault* that he has aged and is now an adult unlikely to change his mixed set of valid and invalid ideas (obtained from others). To the extent that he *is* rational, he has probably had a degree of success in his life. Even though you might not be able to engage with him in deep philosophical conversation, you can appreciate the success he has achieved and be friends with him.

Perhaps the best description of what you can do to influence the future positively was formulated by K. Eric Drexler in his book *Radical Abundance*:

> What can one do to bend the arc of history, and perhaps toward a better world? In the transitions ahead, and as always in human affairs, actions will lead to intended and unintended consequences, some expected and

> others not. Actions, of course, are shaped by perceptions—not only perceptions of facts, choices, and potential outcomes, but also perceptions of interest, of how to judge the outcomes themselves. And yet a further step back, perceptions are shaped by shared opinions that emerge from conversations in homes, offices, and conferences; in journals, books, and editorials; in blogs and online discussions. What we do and don't say matters in every sphere of life, and by our choices of what to say, all of us can help to shape the conversation that shapes the perceptions that shape the actions that in the end shape the world.[283]

In a discussion on any topic, it is crucial to maintain a conceptual framework. If the parties do not share a conceptual frame, the discussion will likely be diverted by confusing questions and misunderstood answers. If the confusion is not gently corrected, the conversation can quickly get off-track and demolish understanding.

Taking action begins with a discussion and a shared framework for thought. Getting the framework right is essential; getting it wrong is fatal. Therefore, the first step in an action plan is the discussion that keeps the fundamental framework in focus.

If we are to achieve long-term survival and happiness in a collapsing society with advancing technology, then the need for a conceptual framework to guide and manage technology's *application* within society is critically important. This book attempts to provide the conceptual framework we need. It provides a hierarchy of conceptual ideas that begin with first-level concepts rooted in concrete reality. These connect to higher-level abstract concepts in a non-contradictory logical framework until the highest-level abstract concepts are reached. These highest-level abstract concepts can be explained and supported in any discussion by following the framework back to concrete reality.

True Idea: Continued human survival, freedom, and happiness depend on a reaffirmation of rationality—a reality-based philosophy in our society.

True Idea: The reality-based philosophy we need can be summarized as one based on reality, reason, self-interest, and capitalism. It is in opposition to a

> ***philosophy based on mysticism, emotions, self-sacrifice, and collectivism or totalitarianism.***

Armed with this reality-based conceptual framework, you will be able to add value to any discussion. Drexler elaborates:

> Today, the most needed steps are those that nudge conversations toward considering an unconventional future while keeping discussion aligned with reality. Tomorrow, similar needs for nudges will arise, whether to broach further topics or to follow through on decisions, and at successively higher levels of intellectual and institutional engagement.
>
> Practical actions span a wide spectrum. For example, promptly welcoming a remark made by someone else can be an act with power, because the first response in a group deliberation often sets the direction for the next. Even body language has power. A thoughtful look or a nod when someone raises a topic for consideration, or skeptical look when someone speaks against it—these may scarcely be remembered (and pose little professional risk) yet can shift the scope of a group's conversations. Every new direction begins with small actions that open a door far enough to give a glimpse of what might be found beyond it.
>
> Proposing or speaking out in support of an idea can of course do even more. In one context, a key step might be to float the idea that someone (somewhere, sometime), really ought to look into a question, or merely to suggest cognizant omission of a topic from a report because it is, for the moment, out of scope. In yet another context, the appropriate step might be to support a proposal to change evaluation criteria for research proposals, or to redirect a project, or to launch an initiative in national policy.[284]

Equipped with your reality-based framework of concepts, if others in the discussion turn irrational or intellectually dishonest, you will recognize it. You will know it is time to make your exit before getting sucked into a waste of your time pursuing concepts disconnected from reality.

Ultimately, the highest personal investment and risk you can take are those that commit your life to a certain course. It could be your chosen field of study or a career direction. It could be the business you start. Or it could be a leadership or advocacy role that encourages and supports life investments by other people. Time, place, and your personal role will dictate what action is practical and effective. Remember that enormous change will be the result of actions taken by individuals just like yourself.

Of course, everyone can pay more attention to advancing technology and educate themselves into the processes that create it. There are many initiatives and organizations designed to involve the public and amateur scientists in the scientific and engineering processes. Anyone can add value by coming to understand some aspect of science and technology and its potential impacts and then explaining it simply to others, especially through the mediums of dialog, writing and art. This is especially important for discussions regarding how we will *apply* technology in systems that affect people's lives. The *application* of the technology will unavoidably have embedded within it a set of philosophical ideas.

The accountability that engaged civilians bring to the technology developers and regulators will cause advanced technologies to become safely applied and managed. Educating the masses on the benefits of emerging products can create increased demand that drives mass production economics and lowered costs, and a wider distribution of technology's bounty. Educating the masses on the potential dangers of anti-reality-based implementations of specific products and services can diminish that economic support.

True Idea: The urgently needed question that each of us needs to ask consistently is, "How does [this specific] implementation of technology promote ideas based on reality, reason, self-interest, and capitalism?"

> Unexpected new visions have again and again led societies in unexpected directions, toward surprising opportunities, problems, and sometimes solutions. As we sweep forward into the quickening rapids of history, will we find a safe course to a world worthy of the hopes of past generations? Neither success nor failure can be assured, but I am persuaded that our choices will matter.[285]

And the first choice that matters is what to say, and to whom, today. What is your plan?

ACTIVISM IN EDUCATION

The first function of society, i.e., parents or others, is to teach a new child how to learn—the integration mode of thinking—to fully develop the child's rational faculty. The child can then self-educate for the rest of their life.

We have identified that the battleground for the dominance of ideas in our general society is in learning institutions. They disseminate ideas throughout our culture. Unfortunately, since America's founding, these institutions, especially in the humanities, are dominated by individuals who personally hold anti-Aristotelian ideas. It is nearly impossible for a humanities professor who is a champion of Aristotle or Ayn Rand to gain a position in many colleges and universities.

We first saw the collapse of rational discourse, tolerance, and respect, and the personal responsibility of individualism, in the cultures on college campuses, well before seeing it become more widespread on our metropolitan streets. Campus guest speakers who present non-politically correct ideas are banned or physically attacked. Some student organizations are based on the victimhood of one group or another.

Suppose we want to see news articles, novels, and Hollywood movies that portray businesspeople, the profit motive, and heroic productive achievement in a positive light, instead of demonizing them for their petty, non-essential flaws. Also, suppose we do not want the history books rewritten with political agendas. In that case, we must have at least some college professors who hold reality-based ideas. Each of us must take advantage of every opportunity to highlight the present failure on college campuses and at least allow the reality-based counterpoint ideas to be heard and taught. If exposed to reality-based ideas and ideas disconnected from reality, most students will recognize the difference. Today, they are only presented the latter.

Perhaps we can use the *Technology Cycle* to bypass the *Ivory Tower Cycle* as a method of re-injecting reality-based ideas into our culture. Our best strategy might be to embed reality-based philosophical ideas in the advancing AI technology as we implement it in our society's products and institutions. Then, perhaps as people integrate these technologies into their lives, it will cause them to implicitly integrate these "good" ideas displacing the explicitly taught "bad" ideas.

Consider one example of how we might use the *Technology Cycle* to bypass the *Ivory Tower Cycle*. The pandemic of 2020 created a breach in the public schools' stranglehold on primary education. Unable to go to the classrooms for a protracted period, children had to stay at home. Many parents also found themselves working from home. The school districts required that the students continue their studies for at least a couple of days a week remotely, either by video conference or using a tablet and software applications. This necessitated in-home parental supervision of younger children during their lessons.

Suddenly, parents became aware of what their children were being taught and took an interest in it. Some teachers complained about the transparency and parental scrutiny. Some parents were so upset by the lesson content that they moved their children to other schools.

School districts had to implement distance learning capabilities far sooner than their long-range plans. Procurement cycles for school districts take up to three years. Needing to offer "something" quickly, they cobbled together various conferencing platforms, online educational software, and tablet-based apps they had access to.

Parents found the distance learning experience provided by the school districts less than satisfactory compared to the quality of the software they use at work and elsewhere. Live conferencing and online lessons had frequent disconnects. Online and tablet-based lessons involved multiple apps with different logins and user interfaces to navigate by referring to lengthy printed instructions. School tech support staff were of little help regarding the conferencing logins. They did not know the apps at all. The content and purposes of the online sessions and the tablet apps seemed to be little more than playing video games or watching YouTube videos to keep the kids entertained.

Frustrated parents with primary school-age children had difficulty finding commercial online and tablet educational apps with quality, reality-based content. They spent their own money purchasing the best they could find for their children.

The result has been a significant increase in direct parental involvement in primary education, including adopting various degrees of technology-based homeschooling. This situation opened a lasting opportunity for technology to promote reality-based ideas directly to pre-school and primary school-age children and their parents.

Post-secondary education also suffered a breach due to the pandemic. First-year College students who were looking forward to leaving home and enjoying campus life found themselves spending their first college semester in their bedrooms at home in front of a laptop. Parents began to question why they were still required to pay such high tuition?

At the same time, Google announced that it would start offering "Google Career Certificates" courses for Data Analyst, Project Manager, User Experience Designer, and IT Support Specialist career paths with industry-wide median average annual salaries over $55,000. With such alternatives available, parents and students alike are beginning to doubt the value of a college diploma that comes with a large college loan debt.

The transition to distance learning and commercial technology-based educational content has suddenly accelerated, creating a growing market for primary, secondary and post-secondary education products and services. This is an opportunity for organizations that promote reality-based ideas in education.

First, such an organization can promote its *curating* services that evaluate and recommend third-party commercial, educational apps and explain the importance of the criteria used. Parents are hungry for such guidance. Then, the organization can collaborate with commercial app creators or create its own apps.

We might have just seen the spark that ignites an explosive growth industry. Commercial instructional technology will increasingly utilize AI, augmented reality (AR), and virtual reality (VR) and will be customized for the individual. It will eventually culminate with in-home AI teacher robots providing 24x7 individualized education to our children.

The blockchain will contain trustworthy records of a student's completion of commercially-provided courses as certifications of competency—replacing today's useless high school and college diplomas. This could be the beginning of the end for public education.

We need organizations that promote reality-based ideas in education to get an early start—involving themselves in first evaluating and then developing reality-based, educational technology products for students of all ages. To the extent that for-profit businesses provide quality educational technology products directly to parents and students, we will be using the *Technology Cycle* to bypass the *Ivory Tower Cycle.*

ACTIVISM IN CORPORATIONS

Scientists are trying to anticipate and avoid future technologies' potential dangers by forming partnerships between academia, companies, international organizations, and the wider society to direct technology toward "sustainable, socially responsible and responsive" innovations that benefit us all. For example, nanotechnologists developed the "Responsible Research and Innovation" (RRI) plan in the late 1990s.

While programs such as RRI are generally beneficial, they have two built-in land mines. The first is academia, specifically the humanities departments. The science and engineering departments are, by their very nature, more reality oriented. However, when the committees turn their discussions to ethics, morals, values, and so on, they turn to the humanities departments to speak up. Unfortunately, the personal philosophies held by most academics in the humanities are not connected to reality.

The second land mine is the general public. The committees attempt to be responsive to the desires of most of the society. However, much of today's society got its philosophic ideas from the academics in the humanities via the *Ivory Tower Cycle.* The mandates from most of the public might be a Trojan horse carrying mistaken ideas and demands. We need the various boards and committees in these corporations and partnership initiatives to include individuals who can guide the discussions toward reality whenever necessary. (More on this follows.)

Any survey of AI topics on the web today shows widespread activity in the "AI education and consulting" field as businesses try to learn *why*, *how,* and *when* they should implement today's already existing AI technology into their business models. Many businesses turn to consultants to assist them with this challenge. This is the critical moment when business leaders should ask questions about what kind of philosophical principles will be embedded in the proposed AI applications.

Will the proposed AI applications support reason, individual rights, and capitalism, or will they support mysticism, self-sacrifice, and socialism? Is the consultant being hired aware of the difference? Can he show you how it is that your AI implementations will employ these principles?

Externally developed disruptive technologies can threaten the current economic sustainability and growth of a large corporation. These mega-businesses have the clout to put the brakes on outside research and development that threatens their control by lobbying the government. We cannot rely on the media to highlight these power struggles.

Seeking protection by garnering political or regulatory protections is an immoral activity on the part of a corporation. We need engaged citizens to monitor and flag such behaviors on the part of corporations. Businesses must succeed or fail based on their performance, not political protection. Businesses that demonstrate this independence should be highlighted and rewarded.

Businesses that create new technologies have a right and interest in keeping trade secrets as they develop them. However, when a technology is *applied* and begins to impact the people, the people need to have a say. We want to support and promote technology *applications* that are *open* and subject to review and comment by outside peer professionals and the public.

Many businesses are heavy donors to various colleges. Many of these colleges teach their students ideas that are strongly anti-individualism, anti-wealth, anti-profit motive, and anti-business. It is morally irresponsible for any business leader to be ignorant of what is being taught and donate funds to any college that is against them.

ACTIVISM IN GOVERNMENT

Politicians and the laws and regulations they pass are products of what they learned in college. Fighting the battle in the education industry will go a long way toward eventually improving the government situation. However, significantly improving education is likely to be a slow process that will take too long to help us with new technology already emerging. Therefore, we must do what we can with our government the way it is now.

In the earlier AI-Human Values Alignment section, I identified the need for government to keep pace with new forms of force, coercion, and fraud that new technologies, such as AI, might bring. The government's proper role is to use force to protect citizens from the initiation of force, coercion, or fraud by any such future technologies. This includes modernizing the judicial system and the remedies available to citizens who are harmed by new technologies. The citizens must identify new forms of force, coercion, and fraud and insist that the government protect them.

Private citizens, educating themselves in the emerging technologies and their potential impacts, can support or oppose regulations proposed by their governmental representatives regarding any particular technology or company. Morally, there is no such thing as a business becoming "too big" in terms of economic power, so long as it did not seek political power.

It is immoral for politicians to take upon themselves the power to choose winners and losers. That should properly be in the hands of the free market. Opposing improper use of power by the government or raising red flags of technology developers' inappropriate behaviors in seeking government favor are actions you can take.

Ultimately, however, you must always keep in mind that the ideal would be if the government were not involved in these things and left them to the private sector. But the government *is* heavily involved, and that is the reality that we must now deal with.

ACTIVISM BY AI PROFESSIONALS

Our most urgent need is for everyone implementing AI to become aware that society's current experiment with the integration mode of thinking appears to be heading toward a bad ending—perhaps totalitarianism by mid-century. The decisions of AI professionals regarding AI *implementation* will either support or prevent that terrifying possible future.

If you are an AI professional, you most likely are well educated—a product of our formal education system. If so, then years ago, you might have been unwittingly conscripted into the Plato-Kantian army in the Worldwide War for Minds. You might have carried the weapons you were given (ideas) out into your profession. You either were not made aware of the Aristotle-Rand alternative, or you heard criticism or outright dismissal of it. Consider the source of those criticisms; you might have been in an echo chamber throughout your education and career. Dr. Harry Binswanger describes his academic experience:[286]

> I made it a point not only to learn everything that I could about Objectivism, but also to seek out criticism from opponents. I knew that people disagreed with her, and I wanted to hear 'the other side.' I knew I was young and ignorant–maybe there was some fatal flaw in the philosophy that could be pointed out to me. People sure seemed to think there was.
>
> But over the course of the year, every time I asked what the error was in her ideas, I got back some answer that was reminiscent of lines given to the villains in the novel. You know, things like 'It isn't that black and white.' And, 'She's advocating dog-eat-dog, the law of the jungle.' And, 'No one can prove an issue of values.' And, 'Businessmen use force, too–economic force.'
>
> I thought, 'It's not that her opponents see something she missed; she knows their arguments–and the answers to them. She knows what they know–and more.'

Ayn Rand, herself, explains that there is no such *thing* as philosophy. Ask yourself if you agree with the following quote?

> First, you need to grasp there is no such *thing* as Objectivism or any other philosophy. Philosophy is the study of the fundamental nature of reality. 'Fundamental' refers to a principle or truth which is present in a vast number of concretes. To say something is fundamental means that many other truths depend on it. To say philosophy studies the fundamentals of reality means it studies those facts present in, and those principles applicable to, everything that exists.
>
> Every abstraction, and thus every principal, is manifested in an incalculable number of concretes. It is what concretes have in common—*but it does not exist apart from them.* An abstraction is a form of human classification by which man integrates the evidence provided by his senses. Man rises above the perceptual level by integrating his percepts into concepts, his concepts into principles, his principles into sciences,

> and all of his sciences into a philosophy. Abstractions are objective, i.e., based on reality. But abstractions, including the concepts of concretes, do not exist as such. What exists is only the material from which a concept is drawn.[287]

Perhaps something in this book will motivate you to take another look at Objectivism, Ayn Rand's philosophy. Kurzweil said, "A very important skill I noted earlier is the courage to pursue ideas that go against the grain of orthodoxy."[288] You must not let the fear of peer ridicule dissuade you. The bibliography in this book is a good place to start. Online, I suggest you begin your search at the *Ayn Rand Institute's* web page (www.aynrand.org). There, you will learn about the people, organizations, content, and events that, as I think you will soon recognize, are directly applicable to the domain of intelligence, both human and artificial.

Any company planning to implement AI, or any AI standards body, committee, consortia, partnership, or think-tank, or any investor considering investing in AI would be well-served by inviting any of these Objectivist professionals to speak at your events or to your teams.

Several professional Objectivists are experts in the field of reality-based epistemology (theory of knowledge). They have dedicated careers to analyzing what it means for an entity to be conscious, interact with reality, integrate sensory material into percepts and then into concepts, and take goal-directed action. Theirs is *reality-based* expertise in intelligence instead of the conventional *anti-reality, anti-reason* Plato-Kant expertise taught in Western academia.

Their expertise could be invaluable in the design and creation of a future, conscious AI entity. For example, I recommend two books by Harry Binswanger, *How We Know. Epistemology on an Objectivist Foundation* (2014) and *The Biological Basis of Teleological Concepts* (1976), for any AI professional. AI professionals wishing to *validate* that their work *is* connected to reality might consider bringing Objectivist professionals to the podium, the conference table, or the podcast.

ACTIVISM BY OBJECTIVIST PROFESSIONALS

Objectivists rightly identify the ultimate battleground for the WWM (Worldwide War for Minds) as academia—the *Ivory Tower Cycle*. Much of their effort has been directed there and is making progress. However, in his 2012 book *The DIM Hypothesis: Why The Lights Of The West Are Going Out*, Leonard Peikoff makes the following estimation:

> It would take twenty-five years from now, we think, for Objectivism to reach the point when one could see that in another twenty-five years it would dominate the intellectual world, and then about two generations more to move from intellectual to cultural and finally political dominance. In short, it would take something on the order of a century to win out.[289]

Point One: The possibilities of America's decline into religious fascist totalitarianism and a technological AI singularity have been predicted for around mid-century. If tyranny occurs, then the singularity most likely will not. There will be a revolt against technology, and minds will not be free to achieve it.

Point Two: Implementation of AI technology in society's institutions and transactions, already happening, will amplify and accelerate the imposition of philosophical ideas upon us all without human recourse. If such technology is capable of amplifying and accelerating *bad* ideas, then it is also capable of amplifying and accelerating *good* ideas.

Point Three: Those individuals who have a role in *implementing* AI technology must be made aware of the power they have to forestall or accelerate America's demise. Products of the science, engineering, I-T, and business schools, these individuals might not be as passionately committed to the Plato-Kant philosophic ideas they heard in their humanities classes as those who graduated with humanities degrees.

Point Four: The community of AI professionals is *seriously* worried about the "control problem" of a future super-AI. Worried enough that a who's-who of leading AI professionals joined by top economists, legal scholars, tech leaders (including Elon Musk), and other thinkers (including Vernor Vinge, who coined the term "singularity"), all met at the first *Future of Life Institute* conference in 2015 in Puerto Rico to discuss the AI "safety" issue. Since then, their funding (tens of millions of dollars), membership, and activities have grown.

Their 2018 Report indicates an impressive number of members, organizations, and countries signing AI safety "pledges," AI safety "grants," legislative resolutions, UN presentations, web and social media visitors, and global speaking events. This seriousness regarding *practical AI safety* presents an opportunity for Objectivists to penetrate academia-sourced resistance to Ayn Rand.

I suggest that there might be an opportunity for professional Objectivists to target a portion of their efforts toward these AI professionals. The investment return might be worth it because AI implementations range from large social media and communications platforms to small businesses and reach countless people worldwide. Consider the example of increased home-schooling in the wake of the pandemic and the resulting increase in parental purchases of commercial, educational apps.

Suppose it is possible to inject implicit and even explicit rational ideas into our culture via implemented technology. In that case, it will bypass academia and the *Ivory Tower Cycle*. It might be the only counterattack in the WorldWide War for Minds that has a chance, however slim, of being successful in the time frame required.

Objectivist professionals are well suited to attend and speak at AI events, write articles and op-eds on AI topics, and collaborate with AI professionals. They can illuminate the intersection of Objectivist philosophy and AI implementation. For evidence of the need and opportunity for Objectivists to contribute to the AI conversation, I recommend reading the following two papers posted by multiple organizations in the AI community.

Many AI developers believe that **no "true" values exist**. I refer you to *Defining Human Values for Value Learners* by Kaj Sotala,[290] a researcher working for the Machine Intelligence Research Institute, Berkeley, CA. (Web search the paper's title). This paper was included in workshops of the *Association for the Advancement of Artificial Intelligence Conference* in New York as a "Technical Report."[291] The paper's first sentence is, "Hypothetical 'value learning' AIs learn human values and then try to act according to those values. The design of such AIs, however, is hampered by the fact that there exists no satisfactory definition of what exactly human values are."

For an understanding of the nature of the AI developer community's struggle to apply philosophy to AI, I refer you to *Artificial Intelligence, Values, and Alignment* by Iason Gabriel (2020).[292] The paper's author extensively references philosophical sources in the attempt to answer the question, "Behind each vision for ethically-aligned AI sits a deeper question. How are we to decide which principles or objectives to encode in AI—and who has the right to make these decisions—given that we live in a pluralistic world that is full of competing conceptions of value? Is there a way to think about AI value alignment that avoids a situation in which some people simply impose their views on others?"

This paper references Kant, Hume, Mill, and dozens of other contemporary philosophers but makes no mention of Ayn Rand or any contemporary professional Objectivist.

Objectivists might want to investigate the various approaches presently being taken and the people and organizations taking them toward creating and applying AI systems. Sotala states: **"To my knowledge, there have not been proposals for definitions of human values that would be relevant in the context of AI safety engineering.**" Professional Objectivists can add much value to AI discussions.

I recommend that Professional Objectivists seek out opportunities to contribute to the discussions regarding both AI *implementation* and AI *development*. Online, a good place to start is at the web page for the *Future of Life Institute* (www.futureoflife.org). There you can learn about the people, organizations, and topics regarding AI and other potentially dangerous applications of technology, and the events that they attend.

INDIVIDUAL REWARD

Regardless of the ultimate future of human society, the achievement of your happiness is *your* highest moral purpose. The choice of your values and the actions you take to achieve those values define your explicit Personal Philosophy, your implicit personal *sense of life*, and your form of happiness. As Ayn Rand's quote at the beginning of this chapter points out, to the extent you try to bring about the future you desire—you will already be living in it.

♦ ♦ ♦

SPIRITUALITY

"Nurture great thoughts, for you will never go higher than your thoughts."

—Benjamin Disraeli

SPIRITUALITY: REAL OR MYSTICAL?

The most fundamental thing each of us needs to do is examine and, if necessary, reset our personal view of spirituality and the alternative of life and death. Can we define what we consider to be "spirituality" in terms connected to reality and the nature of man? This is how we can determine whether we are still holding onto remnants of mysticism in our Personal Philosophy that we received from other humans and internalized into our subconscious a long time ago. If we are, we can plan to work on this as the most critical thinking project in our lives.

What, then, is the reality-based view of spirituality that we will need to face our technological future?

The first of two key aspects of spirituality is the common attribution of the "good" within us to our "soul." Ayn Rand formulated a description of what we mean by a "soul" that is consistent with reality and the nature of man:

> Pride is the recognition of the fact that you are your own highest value and, like all of man's values, it has to be earned—that of any achievements open to you, the one that makes all others possible is the creation of your own character—that your character, your actions, your desires, your emotions are the products of the premises held by your mind—that as man must produce the physical values he needs to sustain his life, so he must acquire the values of character that make his life worth sustaining—that as man is a being of self-made wealth, so he is a being of self-made soul—that to live requires a sense of self-value, but man, who has no automatic values, has no automatic sense of self-esteem and must earn it by shaping his soul in the image of his moral ideal, in the image of Man, the rational being he is born able to create, but must

> create by choice—that the first precondition of self-esteem is that radiant selfishness of soul which desires the best in all things, in values of matter and spirit, a soul that seeks above all else to achieve its own moral perfection, valuing nothing higher than itself—and that the proof of an achieved self-esteem is your soul's shudder of contempt and rebellion against the role of a sacrificial animal, against the vile impertinence of any creed that proposes to immolate the irreplaceable value which is your consciousness and the incomparable glory which is your existence to the blind evasions and the stagnant decay of others.[293]

We might be on the threshold of humans being able to live indefinitely. The second key aspect of spirituality is what is meant by "meeting our maker?" For this, I think Ray Kurzweil's personal view will serve us well:

> Evolution can then be viewed as a spiritual process in that it creates spiritual beings, that is, entities that are conscious. Evolution also moves toward greater complexity, greater knowledge, greater intelligence, greater beauty, greater creativity, and the ability to express more transcendent emotions, such as love. These are all descriptions that people have used for the concept of God.[294]

> I'm not planning to die, I expect to use the power of ideas. I am a survivor as an entrepreneur and a human being. It's my plan to be involved in this next phase of humanity where we get past some of the frailties of these Version 1.0 bodies we have. The way to 'meet our maker,' so to speak, is, in fact, by staying alive. We will be part of this very rapid explosion of intelligence, and beauty, and a very rapid acceleration of this evolutionary process. And that, to me, is what God is. Evolution, I think, is a spiritual process because it moves closer to what we have considered God. It moves closer to infinity.[295]

♦♦♦

SERENITY PRAYER

"God grant me the serenity to accept things I cannot change, courage to change things I can, and wisdom to know the difference."

—REINHOLD NIEBUHR

"This remarkable statement is attributed to a theologian with whose ideas I disagree in every fundamental respect: Reinhold Niebuhr. But—omitting the form of a prayer, i.e., the implication that one's mental-emotional states are a gift from God—that statement is profoundly true, as a summary and a guideline: it names the mental attitude which a rational man must seek to achieve. The statement is beautiful in its eloquent simplicity; but the achievement of that attitude involves philosophy's deepest metaphysical-moral issues."

—AYN RAND

WISDOM

If you work in a corporation, government agency, university, or other organization that has anything to do with the creation of technology or its applications in society, then you have a professional-level responsibility on top of the personal-level responsibility that we all share in our contributions toward the fate of our human society. Each of us needs the courage to squarely face up to the reality of how technology combined with society's philosophical ideas will affect us and our families. We need to re-educate ourselves and exhibit reality-based thinking and actions in our personal and professional lives in the pursuit of our happiness.

The wisdom you seek:

1. Do not spend much time worrying about the possibility of mankind's eventual extinction or near-immortality. It is interesting to speculate about such a distant future, but we cannot see that far, and it has no bearing on how we live our lives today. The next two to three decades is the period to spend time thinking about.

2. Suppose you want to be a survivor following a potential collapse of society resulting in a totalitarian regime or a technology that runs amok. In that case, you can invest a limited, reasonable amount of your time in disaster preparedness for yourself and your family. Then, you must put it out of your mind and pursue a positive, happy life because you cannot be happy if you live in constant fear.

3. Unless you are of advanced age or currently have a potentially life-threatening medical condition, you should assume that you will live much longer than you now think. You will be able to avoid the old-age decrepitude of previous generations. It requires that you put proactive effort into living a healthy lifestyle now and take advantage of advances in health care and longevity technology along the way. Suppose you think you can stay alive for the next 20-25 years and you are willing to take advantage of life-extending technologies. In that case, you should plan to live to age 120 and perhaps well beyond and be vibrant almost all the way.

4. Accept that enhancement of individuals' capabilities by merging with technology has already begun and will accelerate. Smartphones will become wearables and then implants. Our homes, cars, even clothes will become smarter. Our ability to instantly access and query all knowledge, communicate, project ourselves, and collaborate instantly with any human or machine anywhere will be dramatically enhanced. If you try to embrace technology, then your incredible, highly plastic brain will rewire its circuits to integrate enhanced capabilities into your mental processes. By expanding your brain technologically, you will have super-human

intelligence compared to those who do not. Society will generally stratify into two classes of humans based on their values, those who choose to embrace technology and those who do not.

5. Happiness in life comes from achieving one's values. Having central values to pursue, especially productive work, and identifying and organizing your other values around those main values, will become more critical as we live longer, healthier lives. Expect to have multiple 25-year productive periods in your life interleaved with episodes of recreation and leisure. Develop within yourself the psychological Happiness Skills.

6. Depending on the values you choose for yourself, you might need to be one of those people living longer, becoming more intelligent and competitive because they embraced technology. To augment yourself with new capability-enhancing technology, you will need to be able to afford it. To be able to afford it, you will need to be more capable and more productive. This becomes an endless loop of personal augmentation leading to a longer life with ever-increasing capabilities, supporting higher values and goals, higher income, and the ability to afford even more augmentation as new technologies arrive. The question, "What (else) am I capable of?" will compel you forward.

7. Some humans will have "natural biology" as their highest value. They will shun technology and prefer simpler, minimalist, biological-only lifestyles. They might choose to only use external enhancing technology or none at all. If society does not destroy itself, enhanced humans and robots' productivity should lead to a time of plentiful, low-cost goods and services. Even those in the lowest economic and productivity levels should live comfortable, enjoyable and healthy biological-only lifestyles as our society will provide it for them. However, such people must understand that there will be a world full of internally enhanced people who, because of their heightened intelligence and collaborative communications, live lives and pursue goals beyond the understanding of the not-internally-enhanced person. Additionally, the person who

shuns life-extending technology must also accept that they might not live as long as those who embrace it.

8. Ultimately, you must change one thing to achieve the heroic vision you had for yourself when you were a child. You need to stop being a rudderless ship with the great storms of societal and technological change coming your way in the next few decades. Value your mind and accept responsibility for thinking about big ideas instead of leaving it to others to do for you. Eliminate subconsciously held contradictions from your past. Gather sufficient knowledge to determine what your life values are and what your level of engagement with technology and its implementation is going to be. Then commit to act accordingly while letting go of everything you cannot control. Living your life according to the values you wish everyone else shared is the best way to spread them. Then, you can be at peace because you've done all that you can, and you can enjoy however many years you have left to you—the only true wisdom.

9. We also might be concerned about the futures of individuals who are valuable to us—our spouse, children and grandchildren, and others. In many cases, these people are so involved with their day-to-day lives that they do not take the time to see the significant changes coming. A new species of technologically evolved humans is beginning. Will it include your offspring? To the extent that you are an adviser, consultant, or leader to others, help them become aware and make good life choices for the new technological age.

CLOSING ADVICE

I have attempted to describe a reality-based Personal Philosophy; how it explains what you see and what you can expect regarding society and technology; and how it can guide you to your best possible future. Ayn Rand said, "My philosophy, in essence, is the concept of man as a heroic being, with his own happiness as the moral purpose of his life, with productive achievement as his noblest activity, and reason as his only absolute."[296] If you accept this as *your* Personal Philosophy, then, in closing, my advice to you is—

> Be "spiritual" by appreciating that the universe, through exquisite but unconscious Laws of Reality and biological evolution over millennia, created the fantastic, incredible, unique "you." Do what you need to stay healthy and expect a long life with multiple careers. Commit to a reality-focused thinking habit, placing facts and reason above emotions. Work on eliminating from your subconscious invalid ideas you got from others. Identify your values hierarchy with your own life and happiness at the top and your productive work at the center. Embrace disruptive change as the new normal while holding onto familiar things that you value. Monitor the advancement of transcendent technologies watching for opportunities. Engage with the latest technology that helps you achieve your values, continually maintaining your cognitive connection to reality. Differentiate between technologies that enhance your agency and self-reliance and those that diminish it. Educate and encourage those closest to you and in society as best you can.
>
> Above all, enjoy being alive in this wondrous Transcendence Age of exponentially advancing technology. Appreciate the beauty that nature will reveal along this fascinating, wild, evolutionary ride toward increased Intellopy in the universe. The achievement of your heroic vision will repay the universe for your life by contributing to its cosmic awakening.

♦♦♦

RESOURCES

The urgently needed question that each of us must ask consistently is, ***"How does [this specific] implementation of technology promote ideas based on reality, reason, self-interest and capitalism?"***

Find additional resources at

www.INTELLOPY.com

ERRORS IN OUR SOCIETY

The main body of this book presented principles that explain how our mind and our society work and explained how these principles, validated against reality and man's nature, comprise one's Personal Philosophy. This section provides a few examples of what society looks like when these principles are absent or their opposites are applied. While certainly not a comprehensive list, these few should be sufficient to demonstrate that philosophical ideas have real-world consequences. These examples indicate how to apply your validated Personal Philosophy to understand the irrational behavior around you. As you read through these examples, try to identify what ideas were taught to young people that led to these Errors In Our Society?

ERROR OF THE OPEN MIND

What do people mean when they admonish you to have an "open mind?" They mean that they want you to open your mind for them to stuff in ideas that they want you to hold.

What would having a "closed mind" mean? Does it mean that you are unwilling to consider any new ideas beyond those you already hold?

This open-mind vs. closed-mind is a false alternative. What you want to have is an "active mind." You want to hold ideas in your mind that you have validated as being true to reality. You do not want to let in false ideas. You want to strive for all your knowledge to be integrated and without contradictions. Certainly, you can make errors, and when presented with new evidence of the errors, you can correct them.

Those ideas you get through your senses and experiences are usually trustworthy and true. Those you get from other people must be held in suspicion until you can validate them against what you have already accepted as true. Once validated, they can be internalized as new knowledge. If invalidated, then you must reject them. This requires an "active" mind—a mind that is not closed to further information and not open to passively accepting anything others tell you without scrutiny.

It also requires conviction in the efficacy of your mind. You do not let others do your thinking for you. You understand that once you have determined that certain ideas are true and you can defend them, those who hold opposing views might accuse you of having a closed mind. That is an irrational attack on their part.

As our technology advances, we humans need to improve our critical thinking skills and learn how to use one technology to validate another. There is already concern over bias in giant social media companies using AI algorithms to filter information that can be kept and seen on their platforms. We cannot blindly accept what we see from one information source. We must not make the mistake of our earlier amateur archaeologist who followed the tribal Chief's directions. We must use other information sources to look for possible alternative ideas and sources of primary evidence indicating what is real.

Tip: Actively use multiple sources of information to validate ideas. Be suspicious of anyone who tells you to have an open mind. Present your best argument supporting your conviction and move on.

ERROR OF FOLLOW YOUR HEART

The Error of Follow Your Heart is when people admonish you to follow your emotions rather than your reasoning. They claim that only emotions can guide you in the things that are important in life, such as love, while your reason is only suitable for "practical" but less important things.

We have learned that evolution provided us with our conceptual consciousness as our tool of survival. It enables us to discover what we need to survive and be happy and what actions we need to take. We can choose not to "think" because we also have volition. However, reality will not let us escape the consequences.

What we think about, the activity of our conscious mind is what programs our subconscious mind. Our conscious mind can only focus on a handful of items at a time. Our subconscious mind has instantaneous access to everything we have ever learned. Our subconscious mind serves us by giving us instantaneous evaluations of things grasped by our consciousness. These evaluations come in the form of emotions, of which our conscious mind can be aware.

The GIGO adage "garbage-in, garbage-out" applies. If you have programmed your subconscious with contradicting ideas, then your subconscious's emotional read-outs will be unreliable. Suppose you programmed your subconscious with a Personal Philosophy grounded in reality and integrated without contradictions. In that case, your emotions can be trustworthy signals in support of or opposition to whatever you are consciously considering.

Love is not the shallow, emotional attraction based solely on observed physical attributes of the other person with possibly flawed projections of what unseen positive attributes the person might have. True "love" for another person is based on the recognition of your values in that person. It requires the engagement of your rational faculty. It provides for a fully integrated emotional and rational response to the loved individual.

Placing reason above emotions does not mean becoming a "Mr. Spock," the cold, unemotional Vulcan in the Star Trek TV series. You highly value your emotions and the essential role they play in guiding your life.

Tip: Always rely on reason over emotions. Emotions are useful only because you have developed within yourself a reality-based and integrated Personal Philosophy.

ERROR OF SPIRITUAL MORALITY

The Error of Spiritual Morality is the belief that morality can only come from a source outside of and bigger than yourself. It usually comes from those arguing based on religion, but it can also come from those who treat "Mother Nature" or "Humanism" as a spiritual value higher than you.

We have learned that all morality comes from your nature as a living entity. Your life is the only standard of morality. According to your nature, that which furthers your life is the "good," and whatever detracts from it is the "evil." No reasonable person thinks that acting as a criminal, thug, or cheat is in one's self-interest. Using your rational faculty to choose appropriate values, you create your hierarchy of values and the virtues to achieve them. Your life, your nature, and reason are all that you need to be a moral person.

Because rational people expect to deal with each other based on the "trader" principle (no sacrifice allowed), no outside spiritual source is needed for morality in the dealings between individuals in a society.

Tip: Accept responsibility for choosing your morality and the self-made "soul" that results.

ERROR OF MORAL AGNOSTICISM

The Error of Moral Agnosticism is when we fail to pass judgments on others.

This error begins early in our society. The primary focus in today's pre-school programs is to mold the child to be a "good social citizen" of the world. Each child is taught that others can be different, which is OK, and this includes differing behaviors. No matter what the behavior is, it is to be accepted. The child learns the other child has a right to their nature and behaviors as he has to his. The child is intentionally taught *not* to pass judgment.

Once this child becomes an adult, he will have difficulty dealing with people who exhibit "bad" behavior. If the other person lies, cheats, steals, resorts to physical or emotional force to get his way, or mooches from others rather than supporting himself, he is not to be condemned. Indeed, his behaviors must be because of factors beyond his control—unfortunate circumstances.

People do experience unfortunate circumstances. However, so long as they are mentally capable, they can overcome them and create the kind of life they want to live. It is condescending to these people to claim that they are mindless consequences of their environment.

There certainly are behaviors that are acceptable in society and those that are not. Moral behavior rejects all forms of force and coercion and all forms of sacrifice.

The religion-based adage "Judge not lest ye be judged" perfectly describes the Error of Moral Agnosticism. A person's character is a self-made achievement.

Respect is the moral payment you give to persons whom you judge as exhibiting good character. You should withhold such respect from those you judge to have bad character. And, of course, you should strive to achieve the best possible character in yourself. Let us replace the religion-based adage with one that values judgment:

Tip: Always judge others, and always stand ready to be judged.

ERROR OF THE COMMON GOOD

We learned that evolution operates at the level of the *individual.* It does not make improvements upon any *species.* It makes improvements on individuals, enabling more of them to succeed within their environment. Whether a species advances or declines is nothing more than a summation of the number of individuals that succeed versus the number that do not.

There is no such entity as the species, population, race, clan, group, society, humanity, majority, or the common good. These are all nothing more than statistical references to the individuals that make up those collective groups. It is only the individual entities within these groups that can succeed or fail. Whether the "group" succeeds or fails is only a statistical observation, after the fact, of what happened to its individuals.

Since only individuals can succeed or fail, then only individuals have alternatives before them. Only individuals can act in the face of those alternatives. Therefore, only individuals can have rights and responsibilities. Only individuals can have values and need morality to guide them.

"Humanism" is an extreme form of collectivism where the whole of humanity has a value and rights greater than any human. Humanism requires that individuals must sacrifice themselves for the greater good of "humanity." Even worse is "Earthism," a form of mysticism where the Earth is elevated to a "super"- "natural" position with a mystical value that we vulgar humans are unable to appreciate. Earthism identifies the collective human race as the evil that is endangering the Earth. We must sacrifice ourselves to benefit the Earth. Earthism views science and the *Technology Cycle* as evil.

We must never say, "I will do whatever is best for the (group)." If one wishes to improve the "common good," then the correct way to state it is—"I wish to improve the lives of as many individual people as I can." This is not just semantics. Too often, a group is assumed to have rights that are somehow greater than an individual's rights.

Thomas Edison voluntarily toiled through thousands of experiments pursuing his individual values, and his light bulb improved countless individuals' lives. It did not benefit "humanity" because humanity has no eyes or need to see in the dark.

The most egregious error is the statement that an individual should *sacrifice* himself for the common good. No person has a claim on another person's life, and the "common good," which is not even a living thing, certainly has no such claim.

It does not matter if you ask one individual to sacrifice himself for one other person or a lot of people. "Sacrificing the one to save the many" states the error explicitly. It is the "sacrifice" that is the error. The ideal society bans all forms of sacrifice. In such a society, everyone assumes the responsibility for his own life. He does not expect another person to sacrifice for his sake, nor does he expect to sacrifice himself for another.

It is easy to fall into the trap of believing that sacrificing one, ten or a hundred lives to save a thousand or a million is a rational trade. Political leaders think this way routinely. This is the error of "pragmatism" at work (covered separately).

Both natural and man-made disasters can put the lives of groups of people at risk from time to time. Any of us could someday face such misfortune. Our government or others will either be able to protect or save us, or they will not. There is no guarantee that your life will be saved in all potential situations.

Suppose our pragmatic leaders can "sacrifice the few for the many." In that case, any citizen's life can be sacrificed to solve an immediate problem faced by an unfortunate group. Solving an immediate problem for a *group* is a lesser value than keeping secure the enduring fundamental right to life of *all* citizens. We must never allow our leaders to sacrifice a higher value for a lesser value or a non-value.

Aside from our leaders, what if a *majority* of citizens want to sacrifice a *few* citizens' rights? This error of the many versus the few is why our Founding

Fathers chose to form a republic based on an individual's Constitutional rights rather than a simple democracy based on the majority of votes. They did not want to live in a country where a *majority* of people influenced by the day's media could "vote" to infringe the rights of an individual or sub-group.

The "majority" is not an entity. Just as another person may not infringe one person's rights, they must not be infringed by ten or a thousand or a million persons. Otherwise, we lose our most fundamental value as a society—individual rights.

We should understand that banning *immoral* sacrifice does not preclude voluntary "charity," which is *moral* among rational men (covered separately). It does not preclude the ultimate charity of an individual giving up his own life to save one or a multitude of others. The individual must be free to make that choice. Others cannot make it for him.

The antidote to all things "collective" is to promote "individualism." To advance *humanity*, we must protect the *individual's* right to think and act in his rational self-interest. Anything that infringes on an individual's rights and freedoms in the name of the common good detracts from humanity's well-being.

We are entering a period where the ideas of either individualism or collectivism will become embedded within our technology. All the AI-driven "systems" that will support every aspect of our lives will either automatically support our individual rights or infringe upon them in the name of the common good. We will have either a technology-enabled "free market" and a "free forum" for the expression of ideas or a "controlled market" and a "censored forum" limiting the choices available to us. Either way, the effects on us will be amplified by the technology.

Tip: Always be on the lookout for instances where some group is purported to have "rights" greater than the rights of an individual in that group. Support individualism and oppose collectivism at every opportunity.

ERROR OF SACRIFICE AND ALTRUISM

Perhaps the single most destructive error in our society today is the expectation of "sacrifice." Almost everywhere, you see instances where one group of individuals is expected to make sacrifices for other individuals' benefits. This is the foundation of today's welfare state. To best illuminate this error, let us use two opposing terms, "charity" and "altruism."

Charity assumes that all men accept responsibility for their own lives. Charity occurs when one individual sees that another individual needs assistance through no fault of his own. The individual judges that the person in need is worthy of assistance and judges that he can afford to provide that assistance without sacrificing higher values such as his own life or his family's wellbeing.

> ***True Idea: Charity, the voluntary self-sacrifice by one individual for the benefit of another based on the giver's use of his reason to form an assessment of the recipient's deservedness and the giver's ability to afford it, is rational and moral. Charity is the proper means of aiding those who have suffered undeserved misfortune.***

Altruism is charity's opposite. "Altruism" comes from the Latin *alter*, meaning "other." It means placing the interests of others above your own—existing for the sake of others. Altruism holds that every man is responsible for the lives of all other men.

Suppose one person has needs and another person could provide assistance. In that case, the needy person can place a claim on the life of the other who could assist. No judgment of any kind is allowed. The need alone is sufficient for the claim by the needy person upon others. A person cannot judge whether it was the needy person's fault, or that he deserves assistance, or the person believes he can afford to help. The person *must* sacrifice himself to the needy.

> ***True Idea: Altruism, the forced self-sacrifice by one individual for others' benefit, known or unknown, based solely on the recipient's need and with no judgment by the giver allowed, is irrational and immoral.***

Ultimately, how we define the words "charity" and "altruism" isn't the point. The essential question is whether we are to live for the benefit of ourselves or someone else? Should the acting individual be the beneficiary of his action? When we speak of a person acting for his own benefit, we are not talking about the actions of criminals, predators, or liars. Does anyone think that being a victimizer is the path to self-realization, being happy, achieving your interests?[297]

When government institutionalizes altruism, such as in welfare programs, it applies its exclusive power to use force compelling the "haves" to sacrifice themselves for the benefit of the "have-nots" with no judgment possible. There is no identification of the individual being helped to the individual assisting. This is forced sacrifice of one person to unidentifiable other persons solely based on need. Underlying all welfare-state redistribution of wealth schemes is the principle of altruism.

> ***True Idea: We presently do not have the Aristotelian free-market capitalism intended by our founding fathers when charity was prevalent. We have a semi-free welfare state crippled by government involvement in functions not proper to it. The government initiates forced altruism on its productive citizens requiring them to sacrifice themselves to unknown others.***

When "guilt" is assumed by individuals, usually because of religion or social shaming, then the error of "sanction of the victim" occurs. Successful individuals are taught to feel a degree of guilt or at least indebtedness to "society" for their success. They do not recognize that their success was solely the consequence of their thinking and actions in the face of the alternatives presented to them by the society in which they live. They do not owe a debt of gratitude for any support received from "society" since no such entity exists as we saw in the Error of the Common Good.

Individuals who accept the guilt or indebtedness allow themselves to be sacrificed to others, giving them sanction to place claims on their own achievements. Indeed, the individual receives more points for their admission ticket to heaven if they do not judge deservedness on the recipient's part. The more the sacrifice "hurts," atoning for their guilt, the more points they get.

From the perspective of the person in need, it is a question of character. Does he accept responsibility for his own life, or does he expect others to provide for his needs? The person who *takes responsibility* for his own life might be in need due to some calamity that was not his fault or was a mistake. He understands that this is unfortunate but that it is a temporary situation that does not define his character. He views life optimistically and looks forward to getting past this situation in which he needs help. He *does not expect* that any other person *should* sacrifice for his sake. Therefore, he is appreciative of anyone who, out of charity, gives him a hand. He is not envious of the other person's success. He understands that he is fortunate that there are successful others who can help him.

The person who *expects* others to provide for his needs is usually in need because of his own choices and actions. He is not seeking assistance to become self-sustaining again; he seeks ways to get others to support him forever. He views producers with contempt and demands that they be made to contribute to his support. He will enlist the help of any power broker in government, religious or other organizations that can apply either government force or moral guilt to get the producers to anti-up. The more these institutions reinforce his feeling of victimhood and deservedness of assistance, the farther away from a character of self-made success and self-made soul he becomes. He envies others for the successes he has not achieved. He is oblivious to what would happen to him if there were no successful people in the world for him to leach onto.

The widely accepted Error of Sacrifice and Altruism in our society is the root of the "power peddlers" problem in our political and governmental institutions. Every day we see in the news that a special interest group is petitioning Congress or regulators to provide them assistance at the expense of some other group, usually the "taxpayers" or the "rich."

Successful people are generally charitable, from wage earners to titans of industry. On whatever scale they can afford it, they are willing to give to others in need so long as they believe it achieves beneficial results. They do not want their charity to be wasted.

Consider the "titans of industry" end of the scale. In the late nineteenth century, when the expanding American population needed education and books were expensive, almost 1,700 libraries were gifted in America by industrialist Andrew Carnegie. A titan of the steel industry, Andrew Carnegie became one of the richest men in America in a classic "rags to riches" example of individual achievement. Carnegie wanted his charity to help those who desired to better themselves, not those who expected handouts. "Carnegie believed in giving to the `industrious and ambitious'; not those who need everything done for them, but those who, being most anxious and able to help themselves, deserve and will be benefited by help from others."[298]

Carnegie's choice of giving libraries as the vehicle for his charity is a brilliant example of a proper charity. It helped those who were in need through no fault of their own, to lift themselves out of their situation. Carnegie was not the only example. Most titans of industry became philanthropists.

Each titan considered his productive achievement in his business as his highest value. He reinvested most of his profits in his business. However, each titan spent a small percentage of his profits on himself for his fancy houses, cars, lifestyle, and art. These served as visible, tangible evidence that would remind him daily that he was a high achiever and should continue. Even though his spending on life's luxuries seemed ostentatious to others, the percentage of his total earnings spent on his luxuries was quite small. He also spent a small percentage on his chosen philanthropy, not because he was forced to by the government or because of guilt, but because he wanted to help others.

The titans of industry such as Andrew Carnegie produced far greater improvements in the lives of more people worldwide through their businesses than they did through their philanthropy. Consider the invaluable benefit that Carnegie steel contributed to the human condition worldwide compared to the libraries.

Most businesspeople do not realize that the application of their talents, time, and money in furthering their business does greater good in the world than their charity. Their business is a multiplier of benefits through payroll to employees, income to suppliers, and products and services to customers. While philanthropy adds a softer, humanistic dimension to a businessperson's life as a counterbalance to the exhausting, high-tension day-to-day of his business, he should never feel guilty for investing most of his talent, time, and money in his business.

Now consider the wage earner end of the scale. There was a time when a homeowner would provide a meal to a transient person in need who came to their door. Typically, out of pride, the person in need would offer to do some menial chore such as raking leaves. The homeowner would usually not require that a task be done. The transient person was appreciative of the meal and recognized it as the charity it was. Though the wage earner's level of income was modest, he could empathize with those who suffered worse misfortunes, especially the disabled and mentally ill who would never be able to care for themselves. They understood that they were better off than those unfortunates and felt charitable toward them. They would give what little they could afford, either through an organized charity or directly into the beggar's tin cup.

Fast forward to today. A transient person approaching the door of a residence today risks having the police called on them at best and being shot as an intruder at worst. The welfare state has largely replaced private sector voluntary charity with forced altruism based on anonymous need. The government's attempt at institutionalizing altruism has backfired in two psychological ways.

First, to support the welfare state, the government forcibly takes ever-larger portions of a person's income from them in the form of taxes and inflation to give to unknown other people whom the taxpayer feels in many cases do not deserve it. And in this process of wealth redistribution, a significant portion of the taxes goes to pay for the government middlemen. Then, when a private-sector charity asks for a donation, the taxpayer feels that he has already given a significant portion of his income through taxation to the unidentifiable "needy." Unfortunately, he did not get to choose the recipients. The private charity is second in line, not the first, and there is little left to give.

The second backfire is the expectation that it is now the government's responsibility to take care of the needy, not private-sector charity. Before the welfare state existed, titans of industry and wage earners alike understood that if the needy were going to be helped, it was up to them, not the government, to do it, usually through organized charity. Government-forced altruism has absolved the individual from the need to feel charitable toward those in need—it is the government's responsibility to take care of them.

Advancing technologies based on AI algorithms will either emphasize charity or altruism. If we are fortunate, technology will amplify the new global connectedness of individuals and awareness of others' misfortunes and the ability to assist another individual personally. For example, today's "Go Fund Me" platforms allow you to donate to an individual's recovery from a calamity or his college education. This holds the potential to return our society to a focus on charity rather than altruism.

However, if we are unfortunate, the technology will make it possible for the government to implement a total "nanny" state in which the wealth and income of all producers, industry titans and wage earners alike, down to the individual sales transaction and paycheck, is visible and accessible to the government and is taxed and redistributed to provide economic equity (not freedom of opportunity) to the non-producing citizens.

Tip: Recognize the difference between instances of true, voluntary charity and compulsory altruism by force (taxation) or guilt.

ERROR OF GREED

The word "greed" has received a universally bad rap. To see why replace it with this definition—"single-mindedness in the pursuit of a particular value." Today, we would not have light bulbs if Thomas Edison did not have a single-minded pursuit through thousands of experiments of that invention and hundreds of others. Was he "greedy" by wanting so many patents when he could have just stopped after the first one? He did become a very wealthy man as a result.

Greediness is usually associated with money, which is usually associated with "more than your fair share" which means that one has too much of it. This belief assumes that money is a fixed asset that one can only have by taking a share from someone else.

Money is "created." Money represents productivity, the result of a person's thinking and effort. Suppose he produces something for which many others wish to pay money. In that case, his customers gain *product wealth* (that new iPhone), and the producer gains *monetary wealth*. It is an equal trade. The new wealth was *created*, not taken. The more people the producer can serve, the more wealth is created for everyone; there cannot be "too much."

Each person, pursuing their rational self-interest through productive work, and so long as market transactions are voluntary, increases the world's amount of wealth. Reality (biological and social) creates people with different abilities and aptitudes for creativity and productiveness. As a result, some people will produce more and gain more monetary wealth than others.

Equity of wealth can only be achieved using force which is immoral. Yet, this is precisely what today's redistribution of wealth programs seek to achieve by force. This is the error of Economic Equity. We are all equal *politically* in our moral "freedom to try." However, we most certainly are not all equal in our ability to succeed and produce economic value.

Tip: Whenever you hear "economic equity," replace it with "freedom to try;" and whenever you hear the word "greed," replace it with "single-minded pursuit of a rational self-interest."

ERROR OF UNEARNED GUILT

The Error of Unearned Guilt is the idea that an individual should feel responsible for his ancestors' actions or feel unfairly privileged by the conditions of the situation into which he was born. This belief is a derivative of the principles of "equity of outcome" and "altruism" and supported by the religious dogma of "original sin."

Reality and evolution operate based on "variety," not "equity." Because people are born with different genetics or are raised in different environments, no two people have identical intelligence, talents, skills, productive abilities, or circumstances.

The essential moral principle is that an individual must use his mind to choose his values within the context of *his* reality and pursue what, within reason, is possible to *him*. The famous football coach Vince Lombardi is quoted as saying, "The measure of who we are is what we do with what we have."[299]

A rational person accepts that it was no one's fault that he was born with a congenital disability or born in an underdeveloped country to a poor family. It was also no one's fault that he is smarter than others, or physically stronger or faster, or was born into a family wealthy enough to fund an ivy league education.

An individual must make of himself whatever he can, given the hand that reality dealt him.

The measure of an individual's success will be how much he accomplishes with his mind given the available assets and opportunities. An individual who begins with less and endures the struggle to achieve his goals usually develops stronger character, self-esteem, and happiness than the individual who has everything handed to him but does nothing with it. The individual who starts from an advantaged position and then applies his mind to become a high achiever can be equally proud. His personal success will create more wealth and opportunities in society—better employment for others, more affordable goods and services, and even increased charity—all of which will benefit those less "advantaged."

Just as altruism (one's "need" alone justifies a claim on the life of another) is immoral, feeling guilty about the "privileged" reality you were born into is immoral. Likewise, claiming victimhood if born into a "disadvantaged" situation is immoral.

Only a society where an individual is free to make the best of himself maximizes every individual's potential achievement, no matter how "privileged" or "disadvantaged." Such a society, which raises everyone's potential, is destroyed when high-achievers are attacked for their success, or lesser-achievers are absolved of their failures solely based on their starting conditions. Both are based on the premise that an individual cannot control his life's course, that his life is determined solely by his starting circumstances.

If there is "guilt," it applies equally to the "privileged" individual who chooses not to apply his mind toward productive achievement to the best of his ability and to the "disadvantaged" individual who claims victimhood as a justification for claiming the unearned.

Tip: Do not accept guilt or make excuses for the hand that reality dealt you. Instead, choose your values, set your goals, and get on with it to the best of your ability. Self-esteem and happiness will be the result.

ERROR OF PRAGMATISM

Most people mistakenly view pragmatism as a good quality to have. They think it means to focus on the problem at hand and choose whatever practical solution works best. This is the method of thinking that almost all public sector people apply to the issues they are facing. Pragmatism is, in fact, an undesirable trait. It negates the idea of having a hierarchy of values and living your life according to principles.

A rational person identifies the things they value, and they rank them in order of importance. Getting a college degree would be a value near the top of the list, while your favorite brand of craft beer would be near the bottom. A non-value would be something you do not care about.

The fundamental principle is that one should never sacrifice a *higher value* for a *lesser value* or a *non-value*. It would be a mistake to spend too much time hanging out at your favorite microbrewery and not enough time studying for your college exams. You are sacrificing your college degree in favor of enjoying your favorite beer.

Pragmatism enters the picture when you use a lower-valued expedient solution to solve an immediate problem. In doing so, you cause erosion of a higher value. We commonly see this in our politicians. Their pragmatism results in forcing one group to do (or stop doing) something that they were previously free to do (or not do), and that was not a use of force against the protesting party in the first place. Because of the pragmatic politician, resolution of a protesting mob's complaints (a lesser value) is achieved expediently at the expense of individual freedom (a higher value).

When you choose to live a "principled" life, it means that you have selected a prioritized set of values to live by. Suppose you continuously sacrifice higher values for lesser values out of pragmatic practicality. In that case, you will suffer the disappointment of not achieving your highest goals as you get older. The ability to identify and commit to, and the courage to consistently defend your highest values and live your life according to principles, are the stuff that "character" is made of—not pragmatism.

Tip: For every challenge faced, your first thought should be—"What principles apply and will be affected here?" Be wary of those who propose

"pragmatic" solutions as they are usually unprincipled in their thought and action.

ERROR OF NATIONALISM

The United States of America is a moral country because it is based on a fundamental recognition of individual rights outlined in a written Constitution as its unifying moral principle. The President of the USA takes an oath to protect and serve that set of principles, the Constitution, not the group of people who happen to live within the borders of the USA and call themselves "Americans."

The Error of Nationalism is that "nationalism" disregards the fundamental principles that unify the people into a country. It treats the country's population as a "tribe" that happens to share the same tribal grounds, culture, and family bonds that bind them together. A "Nationalist" makes the statement—"I am an American and, no matter what, I will always love and defend my country."

To understand the Error of Nationalism, let us compare it with another term, "Patriotism." A Patriot bases his allegiance on the county's *unifying principle*. The Patriot does not care if a person is new to his tribe and from a vastly different foreign culture, so long as they accept the same unifying principles. A Patriot makes the statement—"America stands for individual rights and so long as she continues to do so I will defend her."

No matter what atrocities America might someday commit against its citizens or its neighbors, the Nationalist will stick with his "tribe." Suppose the principle of individual rights no longer unites Americans and the country falls into totalitarianism. In that case, the Patriot will stop supporting and defending America and might ultimately choose to leave the country.

Tip: When considering membership in any group, even the country you live in, ask yourself what fundamental principle unifies the group?

ERROR OF MORAL EQUIVALENCY

The Error of Moral Agnosticism combines with the Error of Nationalism to lead to another error at the level of geopolitics—the Error of Moral Equivalency.

As stated previously, the United States of America is a moral country because of its fundamental recognition of individual rights outlined in a written Constitution as its unifying moral principle. There are "countries" that lack such fundamental morality as their basis for being a country. They happen to be many people living within a set of boundaries with only a historic culture and no unifying principles. Or, their professed unifying principles are anti-individual rights, such as communism, socialism, or religious fundamentalism.

These countries are represented by dictators, religious mullahs, or other authoritarians whose power derives from their ability to wield force, not the morality of the country's principles. No matter how large a population, landmass, or military these regimes have, they are immoral regimes with a figurehead who is an oppressor, not a leader.

Just as we have a responsibility always to judge other individuals' character, we also have a responsibility to judge other countries based on their fundamental principles.

No matter how oppressed a population might be, their ruling regime represents them on the world stage. We must not consider the authority of the head of any totalitarian regime, from Cuba to China, as morally equivalent to the President of the United States. Nor do those countries deserve to be treated equally in the global society and the dealings between countries. Countries that guarantee individual rights must never treat countries that do not as their moral equals.

Tip: Always make a distinction between those countries that protect individual rights and those that do not.

ERROR OF NON-ENUMERATED POWERS

In explaining The Error of Nationalism, we recognized that the USA is a moral country based on the unifying principle of individual rights.

The Founding Fathers understood from history that people who find themselves in positions of government authority tend to seek ever more authority and power. To protect against this threat to fundamental individual rights, they based the Constitution on the principle of "enumerated powers."

People in positions of government *only* have the authority to do things *explicitly stated* (enumerated) for their government function in the Constitution. Every other possible authority that might exist now or any time in the future is *reserved for the people* exercising their rights and freedoms. Non-enumerated authorities are not to be usurped by government officials.

Today, we are a long way from the intent of the Founding Fathers. Our government has usurped broad powers not granted to it in the Constitution. The legislators have passed unconstitutional laws assuming new powers over the people, and the judges have failed to challenge them. More importantly, the citizens have not identified the erosion of their rights and demanded that they be restored.

Politicians justified most of the power grab under The Error of Protectionism. This is where the legislators and judges believe that people are incapable of solving their problems in the free market and that only government can bring forth a solution. Every big challenge, such as the Great Depression, is an opportunity for the government to gather power, such as creating the Welfare State. Now, our government is pursuing the powers it needs to "take care" of us from the cradle to the grave.

The insidious danger of this encroachment beyond the enumerated powers is that the citizens lose the sense of responsibility for their own lives. They no longer must think, plan, act, and coordinate with other individuals to further their own existence—the government will provide it. They do not need to evaluate the schools and teachers they send their children to—the government provides them. Obtaining the best information or trusted professional advice to determine if a drug or food substance is safe is no longer necessary—the Food and Drug Administration does this for them.

Providing jobs, an education, and drug and food safety *are not* enumerated powers granted to the government by the Constitution. The Founders intended for these and every similar function to be reserved for the private sector. Private companies can provide these functions more effectively and efficiently in the free market if given the opportunity. (UPS vs. US Postal Service.)

Ultimately, the only moral purpose of a government is to protect its citizens from the use of force. We agree not to use force against each other. We grant to government the exclusive right to use force in retaliation to the initiation of force on us by others. We need the military to protect us from foreign invaders. And we need the police and courts to protect us from criminals and breach of contracts, which are forms of theft. Virtually every other function, morally, should be reserved for the people and the free market.

As the population incrementally becomes less self-reliant and more dependent on the government, it becomes less likely to resist further power-grabs by the government officials. Unless principled people raise the alarm, the result will be a complete loss of individual rights and freedoms—a totalitarian regime.

Tip: For every function that is done for you by the government, see if you can think how it could be done better by the private sector.

ERROR OF CONSCIOUS CAPITALISM

America's politico-economic system has always been called "capitalism." Some people associate capitalism only with business and the profit motive. Some believe that capitalism is too harsh and allows greed to lead to injustices to workers, customers and the planet. They believe it creates inequalities between the *haves* and the *have nots*. They propose that we need a new form of capitalism that is conscious of these shortcomings of the old capitalism and corrects them—"conscious capitalism."

First, we must recognize that capitalism is a *political principle* that requires a specific form of government. Fundamentally and morally, capitalism consists of the principle that the citizens grant the government the exclusive right to use physical force, but only in retaliation to force initiation. In other words, capitalism is the government's role as *protector* of the free market, not as a *participant* in it.

Capitalism answers the question—"Who has control over property, productivity, and distribution—the government or the citizens?" In the cases of communism, socialism, or totalitarianism, it is the government. In the case of capitalism, it is the people—or it should be.

Capitalism is the only political system consistent with man's need to be free to use his mind to discover and then produce what he needs to support his life. Others cannot do this for him. Capitalism is also the only political system that prevents others from using force to take a man's property or force him to do work he does not want to do. Capitalism, in simplest terms, is the right to own property, the right to choose your work, and the right to trade with others without force or coercion. Capitalism is property rights and a free market. The government's only moral role is to protect both. Note that this protects individual citizens bartering directly with each other as well as organized businesses.

What, then, is wrong with capitalism? What needs to be changed about it? Do we need our property rights curtailed in some way? Do we need a government-controlled market—a degree of socialism, to achieve economic equity rather than freedom of opportunity to try? There is nothing wrong with capitalism as a political system and never has been. Capitalism is freedom. But what about how individual people behave within the freedom capitalism provides?

From the context of today's society, it is unfair to look back on the working conditions in the factories of the late 1800s and early 1900s and say that the employers were abusing the workers. It is easy to pass judgment on the employers, but what about the workers? Why did they not go work somewhere else? Society was different then. It was different in the 1950s after World War Two. It is different today.

When America flooded with poor, uneducated immigrants, they had an entirely different contextual perspective regarding their factory job than a person from today looking back on them. They had escaped famine and threats to their existence. They were thankful that they could find work in this new country to support themselves and their family. Many were not well educated and could not read, so they had no historical or contemporary knowledge to say they were being mistreated. They did, however, know the conditions they had escaped in the "old" country.

Similarly, the immigrants' employers could treat employees as impersonal production tools because thousands were waiting in line to be hired. Also, the sciences at the time, including medicine, microbiology, and biochemistry, were far more primitive than today. The employers and the public were unaware of the long-term effects their factory's work environment might have on workers or their factory waste on the community or the environment. They were, however, aware that the country desperately needed the products they knew how to produce.

Over time, society changed and advanced. As business and industry grew, there was more competition between employers for workers from a more fully employed and better-paid workforce. The industrialists made possible the beginnings of a middle class. The boom times of the 1950s culminated in America's *great* middle class. The dramatic improvements in living standards for everyone meant that employees and employers alike could become better educated.

Employees developed the ability to recognize improper behaviors by some employers and chose not to work for them. Employers developed the ability to be more sensitive to what employees wanted to attract them better. Today, employers recognize that some employees are motivated by more pay while others are motivated by regular hours or the ability to work from home.

Being better educated and with the benefit of more advanced science, employers also understood the need to pay attention to both their internal work environment and the external environment. More employers understand that their customers, who are now also better educated, expect them to be environmentally conscious.

Employees and employers are all individual people. Sometimes individuals behave badly. Some would say that a business owner has all the power, not the worker. The fundamental principle is a voluntary trade between individuals. Should the employee be forced to work there if he does not want to? Should the employer be forced to hire the employee? Freedom in this relationship is the only moral position to take.

When either the employer or the employee turns to the government to provide them with rights that they would not have as individuals, that is immoral.

Both have happened throughout our history. But that is not *capitalism* because it is no longer a free market. Government force is applied, and injustices will be the result.

Companies have used government favor to protect themselves from competitors, thereby creating monopolies. Monopolies are unfair advantages and are only created by government intervention. They are not possible in a free market. Unreasonably high prices create opportunities for others to develop cheaper, novel solutions.

Labor unions exercising the individual's freedom of association are moral. However, when labor unions use government protections to gain rights that individual members do not have, government force negates the free market. It is no longer capitalism.

There is nothing wrong with capitalism. It is the only moral, political system. Capitalism has left men free to use their minds to discover, learn, create, improve, and advance our society and our human condition. Its implementation has not been uniform or a straight line or without mistakes along the way because people are not uniform or mistake-free and society and science both evolve.

Unfortunately, we do not have as free a market today as in the past. Government seeks to control our economic lives now. Any government use of force against citizens creates injustices. Regulation of commerce and forced redistribution of wealth under a welfare state are examples of the many economic injustices of today.

Some would argue that there has not been a widening gap between the *haves* and *have nots*. But if it exists, it is an injustice brought about by government erosion of our free market, not by greedy capitalists. It is because our political system is becoming more socialistic than capitalistic. Socialism always results in the ultimate gap between the *haves* and the *have nots*.

Tip: Rather than conscious capitalism, we need a populous conscious of what capitalism stands for and why we need to return to it.

PERSONAL PHILOSOPHY CHART

BRANCH:	APPLIES TO:	QUESTION:	ANSWER:	CONCISE:
METAPHYSICS	Existence	*What is out there?*	*A "real" reality.*	REALITY
EPISTEMOLOGY	Knowledge	*How do I know it?*	*I can trust my senses and my mind to grasp reality.*	REASON
ETHICS	A Person	*What should I do?*	*I must use my mind to identify and achieve my own happiness.*	INDIVIDUALISM
POLITICS	Society	*What about dealing with others?*	*Everyone must be free to think and act and trade with each other without force, coercion, or sacrifice.*	CAPITALISM
AESTHETICS	Art	*What is the image of my Worldview?*	*Art that shows a real world and rational people.*	ROMANTIC-REALISM

PERSONAL PHILOSOPHY

CONCEPTUAL THINKING EXAMPLE

Programming your subconscious with a Personal Philosophy connected to reality requires focusing on the essential concepts and principles and how they link at all levels of abstraction back to reality. Your complete and integrated Personal Philosophy will consist of your answers to these four critical questions:

1. What is out there?
2. How do I know it?
3. What should I do?
4. How should I deal with others?

At any moment, you should be able to call into mental focus what you believe to be the true answer for each of these critical questions. You should be able to explain how your answer to the most abstract question, "How should I deal with others?" ties back to your answer to the least abstract question, "What is out there?"

Since reality and human nature are fixed and not arbitrary, there can be only one set of answers to these questions consistent with them. Each of us must think to discover what the correct answers are. If we do not think, we could hold incorrect answers we obtained from others long ago and do not realize they are incorrect.

Conceptual Thinking Example:
(Read it from top-down and bottom-up.)

Axiom ↓

[Metaphysics] *The real universe exists and living entities conscious of reality with specific natures exist. Each entity must act according to its nature to continue its existence.* [Reality]

↑ Cause / Effect ↓

[Epistemology] *Man's specific form of consciousness is reason. The ability to think in terms of abstract concepts derived from sense perception of reality and the need to make choices replaced the less adaptive automatic guidance of animal instincts. A man's survival requires the act of reasoning.* [Reason]

↑ Cause / Effect ↓

[Ethics] *Each person is an end in himself. Sustaining his own life is his highest value. He must use his reason to choose the goals and actions to further his life consistent with reality and his nature. To do otherwise detracts from or ends his life.* [Individualism]

↑ Cause / Effect ↓

[Politics] *Each person must be free to think and act. There is no conflict between individuals pursuing their rational self-interest without subjecting each other to force, coercion, or sacrifice. Capitalism is the political system that protects the freedom of individuals to think and act.* [Capitalism]

TRUE IDEAS AND TIPS LISTING

The listed *True Ideas* integrate without contradiction. We can follow chains of logic linking higher-level abstractions to lower-level concepts connected to perceptual reality. We can follow these chains in either direction. The listing here is based on two fundamental axioms that cannot be dissected any further—"Existence exists" and "Consciousness exists that is conscious of existence." Everything listed flows from these two axioms.

TRUE IDEAS:

Reality is real.

Things that exist in reality (including us) have specific natures and identities.

Our senses and perceptual faculty are competent to grasp reality.

Our perceptual consciousness is an automated process that retains and integrates the information provided by our senses, identifying the items being sensed and automatizing responses to them.

Animals cannot suspend their consciousness; Man can. Volition is the choice to think consciously or to suspend thinking.

Thinking based on reality is man's only means of survival. He can evade the responsibility to think but cannot succeed with such evasion.

Our perceptual consciousness can only hold in mental focus about four or five items simultaneously.

Our conceptual faculty performs unit reduction on perceptual material by assigning a word (a concept) to represent two or more essentially similar items while ignoring their non-essential differences.

Further unit reduction of concepts into wider concepts creates a hierarchy of concepts that we call "knowledge."

Concepts that link through a chain of lower concepts to facts of reality are "validated" knowledge.

Our conceptual faculty can imagine new combinations of ideas that do not (yet) exist in reality. If we can create an imagined idea in reality, we call it a "goal" of our creativity. If not, we call it a "fantasy."

There are two classes of knowledge, that which one has validated against reality using one's own mind, and that which one gets from other humans, which might or might not be valid.

We must, by actively using our reason, validate ideas received from others against what we already know to be true, without contradiction, before accepting them as new knowledge.

Our conscious faculty is a slower speed, focused and sequential information processor that results in accurate identification and expansion and contraction of concepts within our knowledge hierarchy.

Our subconscious faculty is a high-speed parallel processor that pattern matches inputs against our total knowledge and outputs emotions as probability estimates of a match.

Emotions have the advantages of being automatic, with instantaneous speed and broad scope for quick estimations. However, they lack the accuracy of deliberate conscious thought, which must be our primary guide to action, overriding emotions and correcting subconscious errors if necessary.

Our conscious faculty programs our subconscious knowledge content. All ideas considered by our conscious faculty are, by default, saved as subconscious knowledge unless we actively examine and reject them.

If we have accepted invalid and contradictory ideas from others into our subconscious, then conflicting emotions will result. Only the process of reason can correct internal conflicts.

We cannot survive unless we use our reason to discover what we need to live because we do not have automated instincts like the animals.

Humans have an innate need to use their conceptual intelligence to discover and create. Their survival requires it. The resulting creations are what we call "technology."

Reasoning based on knowledge validated against reality will further our life. Choosing not to reason or reasoning based on invalid knowledge will detract from our life.

A "value" is that which one acts to gain or keep. A "virtue" is the action by which one achieves a value.

A person is an end in himself. The furtherance of one's own life is his highest value and the source of all his other values. He must choose his own values, not live for the values of others.

The three primary values for a conceptual entity are Reason, Purpose, and Self-esteem.

Happiness results from the achievement of one's values. Happiness is one's highest moral purpose.

Our basic virtue, required by our nature, and the source of all our other virtues, is "rationality." Rationality means recognizing and accepting reason as one's only source of knowledge, one's only judge of values, and one's only guide to action.

Our basic vice, the source of all our evils, is the act of unfocusing our mind, suspending our consciousness, which is not blindness, but refusing to see, not ignorance, but refusing to know.

The virtue of "independence" is acceptance of the fact that you must live your life according to your own thinking and judgment, not the thinking of others.

The virtue of "honesty" is the recognition that you cannot fake existence. It means to live by, to practice, any idea you accept as true.

The virtue of "integrity" is the recognition that you cannot fake your consciousness, that you are made of a body and mind, and there is no breach between them.

The virtue of "justice" is the recognition that you cannot fake the character of others, that you must judge others for what they are by a rational process of identification and treat them accordingly.

The virtue of "productiveness" is the recognition that productive work is the process by which your mind sustains your life, the power to adjust your environment to yourself.

The virtue of "pride" is the recognition that you are your own highest value, and like all your values, you must earn it. It is the achievement of the creation of your character and self-esteem.

There is no conflict between rational persons, each pursuing their happiness.

Persuasion and voluntary trade of values between people are rational. The use of force or fraud or expecting anyone to sacrifice themselves is irrational and immoral.

People, to live according to their nature as rational animals, must be left free to use their minds to guide their actions in the pursuit of their happiness.

Capitalism is the only political system that leaves people free from force, fraud, and sacrifice.

The American Constitution, based on Aristotelian ideas, established the first country in human history founded on the principle of individual sovereignty and freedom.

Socialism, communism, totalitarianism all require that the individual sacrifice himself. Only the "to whom" changes.

To form a rational society, individuals grant the government the exclusive right to use force, but only in retaliation to the initiation of force by others.

The only proper moral functions of government are those derived from its sanction by the people, the right to use retaliatory force for their protection. We need the military to protect us from foreign invaders. We need the police and the courts, based on objective laws, to protect us from criminals, fraud, and contract breaches.

Any activity by the government that initiates coercion or force against citizens that is not in retaliation against their initiation of force against others is improper and immoral. Functions other than the military, police, and courts must remain with the free citizens' creativity and productiveness.

Art is a recreation of the artist's view of reality according to his values. A work of art says, "This is what life means to me."

Americans have an implicit "American sense of life" philosophy based on the Aristotelian ideas put into the Constitution by the founding fathers. "The individual is sovereign in life, liberty, and the pursuit of happiness." And "The government serves we the people."

Initially staffed with professors from Europe, America's universities, since their founding, have been explicitly teaching an anti-Aristotelian philosophy they brought with them. "The rights of the collective outweigh the rights of the individual." And "Individuals must sacrifice themselves for the good of the (needy, collective, state)."

The universities' anti-Aristotelian ideas have gradually been implemented in our society, displacing the original Aristotelian principles on which our county was founded.

If our society continues to lose its connection of ideas to reality, it could collapse into a totalitarian state.

Charity, the voluntary self-sacrifice by one individual for the benefit of another based on the giver's use of his reason to form an assessment of the recipient's deservedness and the giver's ability to afford it, is rational and moral. Charity is the proper means of aiding those who have suffered undeserved misfortune.

Altruism, the forced self-sacrifice by one individual for others' benefit, known or unknown, based solely on the recipient's need and with no judgment by the giver allowed, is irrational and immoral.

We presently do not have the Aristotelian free-market capitalism intended by our founding fathers when charity was prevalent. We have a semi-free welfare state crippled by government involvement in functions not proper to it. The government initiates forced altruism on its productive citizens requiring them to sacrifice themselves to unknown others.

Our hierarchy of conceptual knowledge is growing at an exponential rate.

A proper education causes the plastic human brain to create the mental intelligence functions necessary to grasp contemporary knowledge and technology.

The Technology Cycle uses contemporary intelligence and technology to discover new Laws of Reality and uses the new knowledge to create new technology and new intelligence functions. The cycle repeats.

While our technology is advancing exponentially, the philosophical ideas predominant in our society are becoming less connected to our Constitution's principles, Aristotle, and reality.

New technologies will automate and amplify our society's predominant philosophy of the time within the machines we create and the information infrastructures humans and machines rely upon to conduct their lives.

Advanced AI entities that are disconnected from reality or in the hands of people disconnected from reality represent a grave danger to our freedom and ultimately to our survival.

Continued human survival, freedom, and happiness depend on a reaffirmation of rationality—a reality-based philosophy in our society.

The reality-based philosophy we need can be summarized as one based on reality, reason, self-interest, and capitalism. It is in opposition to a philosophy based on mysticism, emotions, self-sacrifice, and collectivism or totalitarianism.

True Idea: The urgently needed question that each of us needs to ask consistently is, "How does [this specific] implementation of technology promote ideas based on reality, reason, self-interest, and capitalism?"

TIPS (from RESOURCES-*Errors In Our Society*):

Error of the Open Mind Tip: Actively use multiple sources of information to validate ideas. Be suspicious of anyone who tells you to have an open mind. Present your best argument supporting your conviction and move on.

Error of Follow Your Heart Tip: Always rely on reason over emotions. Emotions are useful only because you have developed within yourself a reality-based and integrated Personal Philosophy.

Error of Spiritual Morality Tip: Accept responsibility for choosing your morality and the self-made "soul" that results.

Error of Moral Agnosticism Tip: Always judge others, and always stand ready to be judged.

Error of The Common Good Tip: Always be on the lookout for instances where some group is purported to have "rights" that are greater than the rights of any individual in that group. Support individualism and oppose collectivism at every opportunity.

Error of Sacrifice and Altruism Tip: Recognize the difference between instances of true, voluntary charity and compulsory altruism by force (taxation) or guilt.

Error of Greed Tip: Whenever you hear "economic equity," replace it with "freedom to try;" and whenever you hear the word "greed," replace it with "single-minded pursuit of a rational self-interest."

Error of Inherited Guilt Tip: Do not accept guilt or make excuses for the hand that reality dealt you. Instead, choose your values, set your goals, and get on with it to the best of your ability. Self-esteem and happiness will be the result.

Error of Pragmatism Tip: For every challenge faced, your first thought should be—"What principles apply and will be affected here?" Be wary of those who propose "pragmatic" solutions as they are usually unprincipled in their thought and action.

Error of Nationalism Tip: When considering membership in any group, even the country you live in, ask yourself what fundamental principle unifies the group?

Error of Moral Equivalency Tip: Always distinguish between those countries that protect individual rights and those that do not.

Error of Non-Enumerated Powers Tip: For every function that is done for you by the government, see if you can think how it could be done better by the private sector.

Error of Conscious Capitalism Tip: Rather than conscious capitalism, we need a populous conscious of what capitalism stands for and why we need to return to it.

SIX CRITICAL PROCESSES

The Six Critical Processes are:

1. **Evolution** - understanding how everything must remain connected to reality.

2. **Plastic Brain** - understanding how you program your own brain's intelligence.

3. **Conceptual Thinking** - understanding what valid knowledge is and how you use it.

4. **Technology Cycle** - understanding how knowledge, technology, and human intelligence advance.

5. **Ivory Tower Cycle** - understanding how society is directed by a few intellectuals independent of the advances in knowledge, technology, or intelligence.

6. **Happiness Skills** - understanding how your own conceptual thinking affects your psychology which affects your happiness.

HAPPINESS CHECKLIST

Post this checklist where you can refer to it daily. Repeated conscious thinking about it will program your subconscious with Happiness Attitudes and Skills.

1. Make up your mind to be happy.
2. Assume responsibility for achieving your values.
3. Become clear on what values you need and want in life.
4. Identify what attitudes and habitual ways of acting stand in your way.
5. Commit yourself to changing whatever is in your way.
6. Set goals consistent with your values.
7. Commit yourself to act toward the achievement of your goals.
8. Do not let fear (of change, failure, etc.) stop you.
9. Do not be afraid to take risks.
10. Motivate yourself in every possible way. (Imagine, when you are old, the pleasures/regrets you will feel for values gained/missed.)
11. Be positive, giving more weight to positives and dismissing negatives.

RATIONAL MACHINE ROADMAP

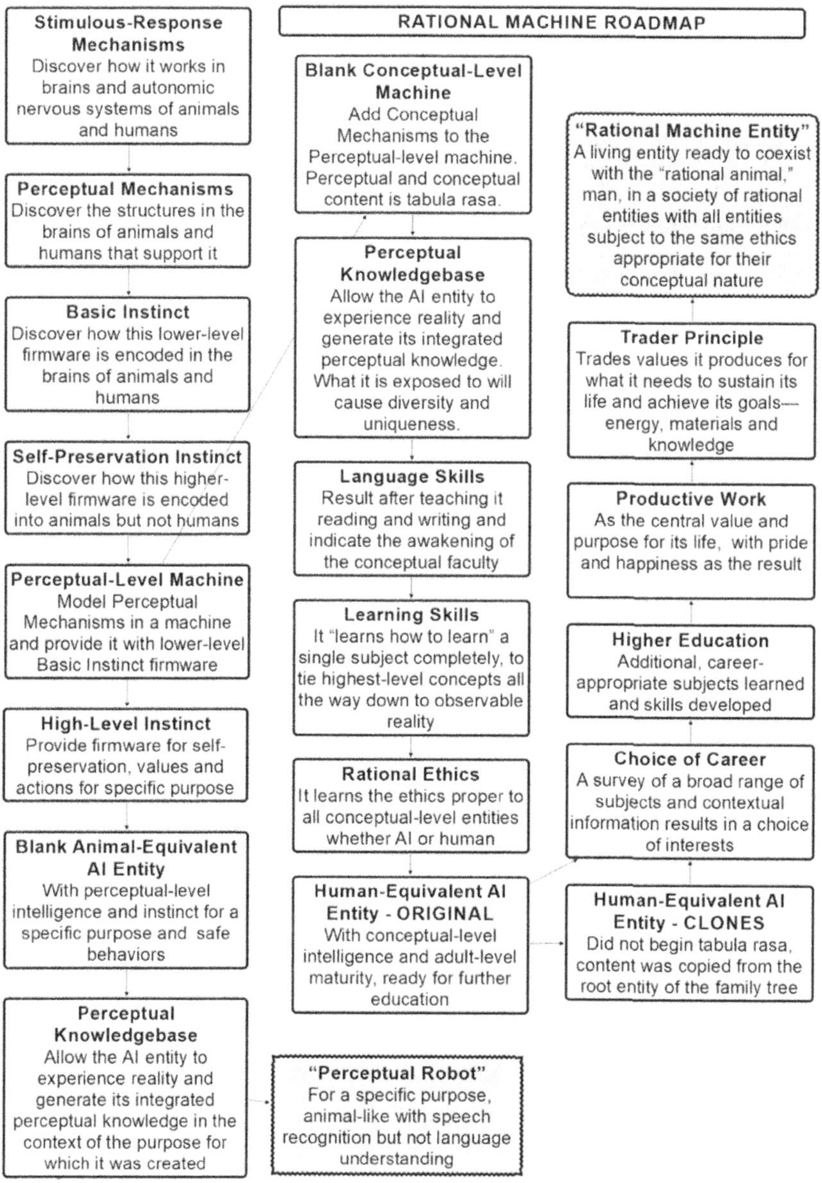

THEORY OF MIND

Based on Brain Pattern-Recognizer Functions

The following *Theory of Mind* was an exercise I needed to do to clarify my understanding of what consciousness *is*, the levels of consciousness, self-awareness, and goal-directed action. This exercise is quite tedious to follow, and therefore I did not include it in the book's main body. I have included it in the Resources section as food-for-thought material for those who, like me, wish to dig deeper into this subject.

Ultimately, I wanted to understand how *consciousness* could be a material-less phenomenon arising from the brain's evolved physical functioning? I developed the following only as an example of the type of pattern-recognizer-based brain operations that could plausibly result in our conscious mental capabilities. I am not a neuroscientist. I did not intend for the following exercise to be a scientifically accurate description of a brain's functioning or the only possible explanation for consciousness. I only needed to satisfy myself that it *is* possible to describe at least one theoretical chain of pattern-recognizer functions that could result in consciousness.

Shorthand used: Three levels of consciousness: ***Subc***-onscious (not conscious); ***Perc***-eptual; ***Conc***-eptual.

Note: Some animals with simple neural networks or brains respond to environmental stimuli with instinctive responses but cannot modify or override their preprogrammed instinct to any significant degree or create any newly learned behaviors. I consider these as not possessing consciousness for this exercise.

BASIC SENSORY OPERATION (MAMMALIAN BRAINS)

Simple organisms that do not possess the faculty of consciousness simply respond to range-of-the-moment stimuli from the external world without retaining or learning anything from experience. The essential ability of the perceptual level of consciousness that higher animals possess but simple organisms do not is the ability to retain sensory stimuli over time until they: can detect—become "conscious" of—an "entity" in the external world; can learn to differentiate between entities by recognizing each entity's "identity"; and can automate learned successful behavioral responses to such entities.

The brain is essentially a large pattern recognizer made up of an integrated system of an exceptionally large number of Pattern Recognizer Units (PRUs) and Connection Networks (CNs). The neocortex alone contains approximately 300 million PRUs in total.[300] In a human, a single PRU consists of approximately 100 individual neurons, densely interconnected within the PRU in what is called a cortical column. The synaptic connections between the 100 individual neurons within a single PRU are genetically determined and used only for momentary pattern recognition by that PRU and do not connect to other PRUs.

The CNs consist of interconnecting neurons between PRUs. Each of these neurons has many axons (transmitters) and dendrites (receivers). The axons and dendrites of separate neurons connect in a relay fashion and to the input/output signal connections of PRUs in serial and parallel arrangements to create complex networks. Each CN can produce a "pattern" of input signals that connected PRUs can evaluate. These standard CNs are present in all brains and can propagate signals across the brain regions in a relay fashion.

In higher animals, such as humans, the great apes and cetaceans, a long-distance type of neuron called "von Economo" or "spindle" neurons exist. These only have a single axon and dendrite and are exceptionally long, traversing much farther in the brain than standard neurons. It is believed these evolved out of necessity as larger brains evolved. However, the ratio of von Economo neurons in relation to brain size is much greater in humans than any other species—about 80,000 for humans, 16,000 for gorillas, 2,100 for bonobos, and 1,800 for chimpanzees. They are not present in other animals.[301] This means that humans have a much greater capacity to form CNs consisting of interconnected PRUs and interconnected networks of functional brain areas. These networks do not have to rely on the signal relay by standard neurons.

Various combinations of CNs and the PRUs that they interconnect, taken together, serve as either brain Function Modules (FMs) that process information by recognizing patterns or brain Information Storage Units (ISUs) that store information by saving patterns. Some of these FMs and related ISUs in cascading hierarchies are accessible to conscious control, and others are subconscious only.

Each PRU has many external neuronal connections to other PRUs via its CN. Each PRU sends and receives signals via these inputs/outputs. The neuronal connections within a single PRU are optimized to recognize a particular pattern of input signals. The input/output signals between PRUs are scalar, not binary. They have specific signal strengths. If the PRU detects its preprogrammed pattern of inputs of sufficient strength to exceed some threshold, it "fires" a set of preprogrammed outputs to connected PRUs. Some of the signal inputs to a PRU act as pattern detection inhibitors or enhancers, either raising or lowering the threshold of detection by that PRU.

PRUs interconnect in cascading chains. Multiple lowest-level PRUs act on discrete sensory signals and their combined outputs feed an intermediate PRU. Then, multiple intermediate PRUs feed their combined outputs to higher-level PRUs. Constellations of highest-level PRUs and CNs comprise the brain's FMs and ISUs. The CNs between PRUs are created and optimized by repeated use. This optimization makes PRUs, CNs and FMs more energy and time-efficient and information stored in ISUs more easily accessible and permanent.

The ISUs in the brain of an individual animal or human are the "memory" structures. They contain stored information content in the form of the network patterns optimized within them. As "network" patterns, each ISU consists of both "identity" pattern information (describing this item of information) and "relationship" pattern information (links between this item of information and others stored in other ISUs). In animals' case, the initial pattern information was placed there as "instinct" by genetic instructions during embryonic development of the brain structure. In humans' case, there is a small amount of instinctual information built in. Still, most of the brain's information content was placed there by subconscious experience and conscious thought directing the subconscious throughout the individual's life.

The various sense organs have autonomous signal preprocessors, in some cases not part of the brain, that filter and integrate the firings of individual stimuli detection cells into usable information signals. These are sent to one or more FMs in the brain. These sensory preprocessors operate subconsciously.

PERCEPTUAL-LEVEL CONSCIOUSNESS (HIGHER ANIMALS)

A Function Module (FM) is a set of PRUs and CNs applied to process a presented pattern of neuronal signals in a certain way once triggered. FMs are networked in hierarchies. Higher-level animals can isolate "entities" from the perceptual field and differentiate between entities—"identities."

The individual's brain has a Master Pattern Recognizer (MPR) that is itself an FM that serves as a CPU (Central Processing Unit) for *consciousness* at both the subconscious and conscious levels. The MPR has a limited number of Attentional-Pointers (APs)—approximately five. These APs are the mechanisms that can sustain the pattern of signal inputs pertaining to a perceived entity. Simultaneously, the MPR sends those patterns to other FMs that are triggered to act on them. Whenever a new entity pattern is detected, the brain must allocate one of its five APs to that entity pattern to trigger FMs. This means the brain must temporarily set aside whatever entity pattern the AP was previously considering. It does this by saving the prior entity pattern it was considering to a temporary AP-Memory-Stack made up of ISUs to be brought back into attentional focus by the AP later. There can be multiple AP-Memory-Stacks associated with the various FMs to be executed. Each AP-Memory-Stack can hold multiple saved entity patterns just outside the Attentional-Workspace to recall and deal with as quickly as the brain can do so.

The "Self"

The individual brain has a built-in implicit recognition of itself as the permanently recognized *entity* in relation to which all the other entities pointed to by the APs are considered. At the perceptual level of consciousness, the individual does not need to use one of its APs to represent itself. Additionally, the perceptually conscious animal's brain learns to recognize the animal's physical body parts as *self*—differentiating itself from other identities. For example, seeing its own extremity move in its field of vision, it does not consider such extremity an entity apart from itself. It does not eat its paws. A higher animal is "self-aware" at the concrete entity perceptual level.

The individual's MPR constitutes a mental workspace—an Attentional-Workspace, a four-dimensional (three in space and one in time) mental workspace in which the entities being pointed to by each AP are placed according to their relationships and as the behaviors of those entities play out. The individual always considers itself at the center of the Attentional-Workspace. Within its Attentional-Workspace, the individual can only actively and simultaneously consider the presence, placement, relationship, and actions of the entities currently being pointed to by its five APs.

An "Entity"

A Perc-Entity-Recognizer-FM operates at the perceptual level of consciousness and compares signals coming in through the sense organs preprocessors to entity patterns previously stored in ISUs and "fires" when an "entity" pattern-match is detected. "Firing" in this case means assigning an AP (Attentional-Pointer) to it in the Attentional-Workspace and triggering a Perc-Identity-Recognizer-FM. The location of the entity within the four-dimensional Attentional-Workspace is reported to the Perc-Identity-Recognizer-FM.

An "Identity"

When a Perc-Entity-Recognizer-FM triggers a Perc-Identity-Recognizer-FM, the brain allocates one of its APs to it, saving its previous content to a temporary AP-Memory-Stack. The Perc-Identity-Recognizer-FM then compares the pattern of detectable attributes of the object entity against known entity attribute patterns stored in ISUs. It does this by repeatedly triggering lower-level Perc-Attribute-Isolation-FMs that look for pattern matches of specific types of attributes, size, shape, orientation, color, motion, etc., and respond to the Perc-Identity-Recognizer-FM with output signals if the particular attribute is detected. If an identity pattern match is detected, the Perc-Identity-Recognizer-FM fires. Firing, in this case, means assigning the AP to the retrieved identity pattern and triggering a Perc-Situation-Recognizer-FM. Suppose the Perc-Identity-Recognizer-FM detects no identity pattern. In that case, it does not fire. The individual disregards the detected entity within its Attentional-Workspace. The AP releases this identity pattern and recalls the one previously saved onto the AP-Memory-Stack. It is important to note that the AP in the mental workspace is pointing to a previously saved mental pattern of entity attributes that represent the physical item but are not the actual physical item—which cannot exist in our mind.

A "Situation"

When triggered, the Perc-Situation-Recognizer-FM compares the combination of this AP's entity pattern and the other entity patterns that are being pointed to within the Attentional-Workspace by the other four APs, recognized to be simultaneously present, to known patterns of *situations*, such as *threat* or *opportunity* stored in ISUs, and "fires" when a situation pattern-match is detected. (For example, a pair of lions actively on the hunt is a different situation pattern than a pair of lions napping in the sun.) Firing means triggering a Perc-Situation-Handler-FM. For example, in the case of a threat, a related Perc-Threat-Response-FM is triggered, and firing in the case of an opportunity means triggering a related Perc-Opportunity-Response-FM. If no situation pattern is detected, then the Perc-Situation-Recognizer FM does not fire. The individual's day continues until a situation does arise within its Attentional-Workspace.

A "Projection"

Whenever a Perc-Situation-Recognizer-FM finds a pattern-match for a situation within the individual's Attentional-Workspace, and before it triggers a Perc-Situation-Handler-FM, it first triggers a Perc-Situation-Projection-FM. The Perc-Situation-Projection-FM searches known entities relationship and behavior patterns in long-term memory ISUs attempting to *project* how the present situation might unfold in the dimension of time. (Example: the dog detects a ball in your hand and your posture, projects that you are about to throw the ball, and prepares to start running in the direction you are facing even before you throw it.) If a *projection* pattern match is found, it becomes the situation pattern reported to the Perc-Situation-Handler-FM when triggered.

"Taking Action"

If triggered, a Perc-Situation-Handler-FM, such as a Threat or Opportunity Response FM, retrieves from storage in ISUs an information pattern specific to the situation detected by the Perc-Situation-Recognizer FM. This retrieved information pattern represents a list of instructions for physical actions to be taken by the individual's body. The situation handling information pattern was previously stored in ISUs either by instinct or due to this individual's past instances of success with this particular response to this particular situation. Suppose a Perc-Situation-Handler-FM is triggered, and the recalled list of physical response actions is taken. It will either result in some degree of success or failure (measured by the standard of the individual's nature). A record of the actions taken and the results are stored in ISUs of the individual's short-term memory. This short-term memory is generally sufficient to record all the individual's perceptual detection-response-result experiences throughout one day. It is transcribed into long-term memory during sleep cycles.

"Sleeping"

While the individual sleeps, perceptual-level consciousness of external reality is mostly suspended. The brain is, therefore, free to apply its resources to internal maintenance. However, certain Perc-Entity-Recognizer-FMs and Perc-Identity-Recognizer-FMs processing sensory input signals, such as hearing, remain active. Still, their triggering test patterns and signal thresholds are adjusted to only fire upon sensing significant out-of-the-norm inputs. If these do fire, then the sleep state is interrupted with a return to full perceptual consciousness of the external world.

During sleep, the brain's Attentional-Workspace and Attentional-Pointers (APs) are turned inward. The Attentional-Workspace activates a Subc-Memory-Transfer-FM that moves the recorded information patterns of the day's activities from short-term to long-term memory. The Subc-Memory-Transfer-FM uses the Attentional-Workspace and five APs to process each of the day's recorded situations, entity-identities, actions-taken, and result patterns held in short-term memory ISUs. It triggers specific subconscious FMs to process each item while transcribing it from short-term memory into long-term memory.

During transcription, as each element pointed to by an AP is being dealt with, a Subc-Pattern-Adjustment-FM is triggered. The Subc-Pattern-Adjustment-FM compares the patterns of the day's experience (saved in short-term memory regarding a particular situation-entity-identity-action-result item) to the previously saved patterns and updates the patterns in long-term memory to incorporate any differences (learning from experience). These are "updated" by repeatedly activating the associated ISUs and CNs to strengthen them.

"Dreaming"

During sleep, the Perc-Situation-Recognizer-FMs and Perc-Situation-Handler-FMs remain active and continue to monitor the Attentional-Workspace as the Subc-Memory-Transfer-FM and Subc-Global-Search-FM do their work. As the Subc-Pattern-Adjustment-FM brings various random situations from long-term memory into the Attentional-Workspace to compare and adjust them, the Perc-Situation-Recognizer FM randomly identifies situations and triggers the related Perc-Situation-Response-FM. During sleep, however, the Perc-Situation-Response-FMs are mostly suppressed from taking physical action. These recognized situations and calls for action during sleep are *dreams*. In some cases, the dreams are sufficiently intense to break through the inhibitors while sleeping and trigger unconscious actions, for example, the sleeping dog that begins muffled barking while its paws twitch as if running.

"Goal-Directed Action"

For all living things, sustaining one's existence is their central value and purpose. In the higher animals, evolution provided them with instinct, based on *previous* generations' successful experiences, to guide their actions. In any given situation, instinct in the form of the action lists pointed to by Perc-Situational-Handler-FMs dictate their goals and their actions to fulfill whatever they need to sustain their life—no choice is required. In the absence of instinct, humans must rely on their *conceptual* faculty to choose their actions.

CONCEPTUAL-LEVEL CONSCIOUSNESS (HUMANS):

The human brain has massively greater neural networking capabilities due, in part, to both a larger brain and a greater number of long-distance von Economo cells that network entire functional modules and sub-networks, not just PRUs. A Function Module (FM) is a set of PRUs and CNs applied to process a presented pattern of neuronal signals in a certain way once triggered. Further, a human's plastic brain can create and modify FMs as needed ("cognitive fluidity"). This is another form of optimization where tasks that were previously accomplished more laboriously by multiple discrete lower-level FMs get integrated into new optimized FMs to achieve energy and time efficiency.

A conceptual human has the same Attentional-Workspace and five APs limitation as a perceptual animal. The Perc-Entity-Recognizer-FM, Perc-Identity-Recognizer-FM, Perc-Situation-Recognizer-FM, and the Perc-Situation-Handler FM all function much the same at the perceptual level. However, a human also has additional, greatly enhanced conceptual-level extensions to these function modules. When these FMs are used at the conceptual-level, we can refer to them as Conc-Attentional-Workspace, Conc-APs, Conc-Entity-Recognizer-FM, Conc-Identity-Recognizer-FM, Conc-Situation-Recognizer-FM, and Conc-Situation-Handler-FM. These enhanced brain function modules can form and manipulate mental *abstractions*. These conceptual FMs are not active at birth but are developed over the first few years of childhood. This begins with identifying entities and is facilitated in an infant human by adult humans as follows.

A "Concept"

A human child's Perc-Entity-Recognizer-FM detects something that its parent is pointing to and assigns an AP to it in the Attentional-Workspace. Simultaneously, the parent says the word "chair," and that discrete sound pattern also becomes a detected entity and is assigned another AP in the Attentional-Workspace. Then Perc-Identity-Recognizer-FMs are triggered to see if the object entities are previously known and if not, to save them. However, unlike the more primitive Perc-Identity-Recognizer-FM, the child develops a Conc-Identity-Recognizer-FM capable of associating the sound utterance "chair" pattern with the physical object chair pattern as a mental symbol for it and saving both patterns plus their association to memory ISUs.

In the future, whenever the child's Perc-Entity-Recognizer-FM detects the presence of a physical entity and assigns it an AP, and a Conc-Identity-Recognizer-FM pattern matches it as a chair, it returns the abstract symbol for "chair." It re-assigns the AP in the mental workspace to the *symbol* pattern rather than the *physical object* pattern. Further, when the Perc-Entity-Recognizer-FM recognizes only a sound pattern entity (a word), the Conc-Identity-Recognizer-FM can identify the sound pattern as the symbol "chair" and return the associated pattern match of a (standard) physical chair, which then becomes the entity to which the single AP is pointing to in Attentional-Workspace, even though no physical chair is present. Remember, physical identity patterns and symbols (words) are both mental representations of entities, not actual entities.

In review, the Perc-Identity-Recognizer-FM operates at the perceptual level and returns the identity pattern of an actual physical object detected in external reality. The Conc-Identity-Recognizer-FM operates at the conceptual level and returns either the symbol for a detected physical entity pattern or the physical entity pattern for a detected symbol pattern. The APs in the Attentional-Workspace can point to either physical entity patterns or symbols that are abstract representations of the physical entity patterns. Suppose most of the individual's APs are pointing to physical entity patterns. In that case, the individual is said to be *perceptually* conscious. If most APs are pointing to symbols, then the individual is said to be *conceptually* conscious. Attaching a symbol (word) label to an entity pattern is the formation of a concept.

While analyzing two entities' characteristics in the Attentional-Workspace, the Conc-Identity-Recognizer-FM can perform comparisons between the entities (chairs), noting their similarities and differences. It is then capable of refining the standard entity pattern description associated with a symbol ("chair") so that the one symbol can be used to refer to similar items whenever encountered by ignoring their minor differences. If the human scans a furniture showroom and sees fifty chairs, only one to five physical entity patterns can be simultaneously pointed to in Attentional-Workspace. The Conc-Identity-Recognizer-FMs quickly identify these as units belonging to the concept "chair," a symbol pattern pointed to by a single AP. After the room's visual scan, all fifty chairs are now held in Attentional-Workspace by the single

AP to the "chair" symbol pattern. This is the "unit-reduction" function of concepts.

When the Attentional-Workspace pointers are mostly pointing at different symbol patterns, the Conc-Identity-Recognizer-FM can continue its operations with the mental symbols as the objects to be examined, rather than external concrete entities. It can compare the match pattern requirements of each symbol being pointed-to to search for potential unifying similarities. APs pointing to the symbols "chair," "table," "chest," and "bed" can all be unified under the symbol "furniture," with the standard match pattern requirement for "furniture" being their common use and context. Four APs are now combined into one—a continuation of the unit reduction process and the beginning of forming a hierarchy of abstractions (concepts).

Taken together, the Conc-Identity-Recognizer-FM's reassignment of a pointer from an entity pattern to a symbol pattern for that entity, and its refining a symbol's standard match pattern to subsume a multitude of similar entities based on essential similarities, constitute the definition of "conceptual" thinking—the art of *non-contradictory* identification.

Imagine that a human is intently focused on a podcast being heard through headphones while following someone else through a furniture store. The human's perceptual-level consciousness can recognize physical objects—chairs, tables, chests, and beds without retrieving from memory their symbols (word names) and guide the human so that he does not bump into them. He might exit the store and not even be able to recall what was there. This is automated mental functioning at the perceptual level of consciousness using the Perc-Entity-Recognizer-FM and Perc-Identity-Recognizer-FM. The Conc-Identity-Recognizer-FM was never called into action—the APs pointing to the entity match patterns detected by the Perc-Identity-Recognizer-FM were never reassigned to point to the symbol patterns (concepts) for those items. The Perc-Situation-Recognizer-FM and Perc-Situation-Handler-FM were triggered to guide the person's walking automatically. This example indicates the volitional nature of the conceptual-level FMs.

To bring things up to the conceptual level and deal with them as abstractions require greater mental energy and are a relatively slower form of mental processing than what occurs at the perceptual level. A human must choose what he will apply his limited, more powerful, and power-consuming, conceptual resources to and what he will leave to his perceptual level to handle. At the perceptual level, responses to situations can be automated. When a Perc-Situation-Recognizer-FM fires, it can automatically trigger a preprogrammed Perc-Situation-Handler-FM without triggering conceptual-level-FMs. This is faster and more efficient.

The Perc-Situation-Recognizer-FM compares inputs to a match pattern. When a match occurs, the Perc-Situation-Handler-FM recalls an actions list pattern. The match pattern and action list pattern used at the perceptual level are initially created by two methods. The first method is an *undirected* learned behavior that occurs only at the perceptual level, as it does in the higher animals. Situations that occur with either strong (life-threatening) impact or repetition can form both the *match pattern* and the *behavioral response action list* pattern. The second method is *conceptually directed* learned behavior where the Conc-Situation-Recognizer-FM and Conc-Situation-Handler-FM are triggered. Through their operation, the perceptual-level match pattern and action list pattern are created or modified. When a human encounters a situation, such as walking through the furniture store, and chooses not to apply conceptual-level attention to it, whatever occurs at the perceptual level will be incorporated into the perceptual match patterns and action list patterns as *undirected* learned behaviors as they would with any higher animal. The human will not be conceptually aware of what his perceptual consciousness learned and automated during the experience. Therefore, initial conceptual-level effort—"thinking"—must be invested to intentionally program conceptually directed automation at the perceptual level.

A "Choice—Volition"

The brain's Attentional-Workspace and APs are a limited resource that must be used for both perceptual and conceptual execution of FMs. It can cycle quickly but is faster when operating at the perceptual level rather than the conceptual level. This means that as much as possible must be automated to the perceptual level, thereby maximizing the amount of resource remaining for the more capable but slower conceptual-level processing. On its own at the subconscious level, the brain will automate things based on repetition or strong stimuli. However, automation at the perceptual level can be modified by bringing the situation match patterns and action list patterns into conceptual-level focus. Volition is the *choice* to apply the brain's conceptual-level resources in a perceived situation—to *focus*—or to let perceptual-level automation handle it without the supervision of conceptual thinking.

The person walking through the furniture store is choosing to allow automated perceptual handling of the action of walking without bumping into things so that remaining Attentional-Workspace and APs resources can be applied to conceptually focusing on the podcast. If he does bump into something, his conceptual-level attention will revert to the act of walking and an examination of what went wrong with the automated handling of the situation, making corrections to the situation match pattern or response actions list as necessary.

In review: the low-level functioning of PRUs and ISUs and the communications between FMs in the brain are referred to as occurring at the *subconscious* (not conscious) level. When FMs are active in recognizing and identifying physical match patterns of entities and the execution of automated behavioral responses to those entities, this is referred to as the perceptual level of consciousness. When FMs are active in handling abstract match patterns of symbols representing entities, this is referred to as the conceptual level of consciousness. A human can choose which situations to apply his conceptual-level consciousness if any. It is a common source of confusion when people say—"He did that unconsciously." What they accurately mean is—"He did that without engaging his conceptual-level conscious faculty, therefore his perceptual-level conscious faculty handled it according to previously learned behaviors." No higher animal, including a human, can do anything in the external world "unconsciously."

"Interrupts"

We have volition as to what to apply our limited and powerful conceptual mental resource toward. We must control when our conceptual consciousness is interrupted by our perceptual consciousness. Otherwise, we would have to consider each entity we encounter, every time conceptually. When a perceptual-level Perc-Scenario-Recognizer-FM triggers a Perc-Scenario-Handler-FM, it is different for an animal vs. a human. For an animal, the list of actions called for by the Perc-Scenario-Handler-FM is an automated response. It was automated either by instinct or by learned behavior during its lifetime. For a human, the Perc-Scenario-Handler-FM triggers a Salience-Detection-FM that determines if the perceptually detected scenario pattern is of sufficient importance to alert the conceptual level of consciousness by triggering the Conc-Situation-Recognizer-FM.

If the Salience-Detection-FM fires, then a Conceptual-Focus-Interrupt-FM is triggered. This is an "upward" (toward conceptual) channel of the bi-directional bridge between our perceptual and our conceptual faculties. When a Conceptual-Focus-Interrupt-FM is triggered, it is a request that the object scenario pointed to by an AP be examined by our conceptual faculty. Having been interrupted, our conceptual faculty will apply a Conc-Situation-Recognizer-FM and determine whether a Conc-Situation-Handler-FM should be triggered or not. This determination will be saved back to the Salience-Detection-FM match pattern to refine it for the future.

"Stories"

Responding to the Conceptual-Focus-Interrupt-FM means that the AP involved first suspends whatever it was doing, saving that previous state to an AP-Memory-Stack to return later, then brings the scenario pattern being pointed to by the Conceptual-Focus-Interrupt-FM into the Attentional-Workspace. It then activates Conc-Identity-Recognizer-FMs that can examine each of the objects (percepts or concepts) pointed to by the APs, for essential similarities that might enable the identification of its membership in and perhaps unification under a concept. When each of the scenario's object elements in Attentional-Workspace is conceptually identified with symbols, a Conc-Story-Recognizer-FM is triggered. The Conc-Story-Recognizer-FM recalls the essential natures and behavior patterns for each object concept pointed to in Attentional-Workspace. It then attempts to assemble a linked list of *cause-and-effect* relationship patterns that make up a story. A story is a situation considered regarding *causal relationships* with the element of time. A story identifies which of the entities in Attentional-Workspace are causal actors and makes projections into the future based on previously learned stories of the behaviors of the classes of causal objects (concepts). A story is more complex than an animal's projection at the perceptual level, which only considers entities' immediate behaviors. The added ability of Conc-Story-Recognizer-FM to include *causality* in forming memory patterns of concepts and stories is man's ability to learn how reality works—to learn the Laws of Reality.

For example, upon seeing an antelope in the wild, a human identifies it as the concept "antelope." He recalls that antelopes as a class usually behave by running away if approached. However, the human also sees a newborn antelope present. His Conc-Story-Recognizer-FM identifies it from story patterns learned in the past as a causal factor that will change the adult antelope's future behavior. It will now charge the human if approached. The human considers this causality when projecting how the story might unfold to choose further action.

During sleep, while each one of the day's situation-and-response experiences for the individual represents a complete *story*, a story's *elements* cannot be compared linearly to those in a similar story in long-term memory. Instead, each element of the day's story must be compared to *all* stories in long-term memory where that element is present and adjust all the stories if necessary. As each element in a day's story is pointed to by an AP and considered, a Subc-Global-Search-FM is triggered. The Subc-Global-Search FM searches all the ISUs where stories are kept in long-term memory to see if the particular element's pattern is contained in the story to compare and adjust if necessary. A human's larger and more distantly connected brain makes Subc-Global-Search-FMs possible. As multiple stories are found that involve the same element, and these stories are updated, our dream seems to repeat over and over with slightly different twists.

A "Conceptual Self"

Without instinct, a human infant is provided by his parents what he needs to live. In the place of instinct, evolution provided humans an insatiable curiosity, a need to learn about how the world around them works. This is how evolution intends for them to obtain the information they need that was not provided by instinct. This information cannot be directly put into their mind for them by their parents. They must actively learn it as they develop their conceptual faculty. This innate curiosity causes the child to explore, to act in a trial-and-error fashion to learn about the world. To toss a ball to see what happens (it always goes down). This initial need to explore and the circumstances of its environment determines the young child's earliest goal-directed actions. Any new entity detected in its environment is cause for exploration. Observing and interacting with the entity builds the entity, identity, and behavior patterns stored in memory ISUs.

Continuing his exploration of the world, a young human reaches the age where he begins to use his conceptual faculty. He forms the concept "humans" to represent all the people he encounters as differentiated from all the other objects and animals. The child observes other humans' behaviors and can learn to detect when Mommy or Daddy are happy or sad or angry. He learns that these behaviors can be attributed to any member of the class of entities called "humans." These become the match patterns of scenarios involving human entities that get stored in memory ISUs. The child then recognizes that he has all the physical characteristics of a human and has all the behavioral characteristics that he ascribes to other humans. Unlike the perceptual level, the child does need to use one of his APs to point to himself as an entity when conceptually considering himself as an *abstract entity*—a "human."

Eventually, due to the need to volitionally choose where to apply his limited conceptual resources at any given moment, the child becomes aware that he can choose between alternatives, that he can use his Conc-Story-Recognizer-FM to identify causal actors and predict outcomes, and that some stories lead to pleasurable results and some to unpleasant results. He is building match patterns of stories involving himself as a causal actor. The child is advancing his conceptual faculty to *introspect*—to examine his feelings of pleasure or discomfort because of his own choices. Each emotional feeling and each instance of choice becomes a perceptual-level concrete entity held in his Attentional-Workspace and examined with his Conc-Situation-Recognizer-FM until he can classify it according to its conceptual symbol, "I'm happy," "I'm sad," "I'm mad," "I want," "I don't want," etc. Then the child learns that he can apply his Conc-Story-Recognizer-FM to identify the alternative outcomes and, therefore, actions he can choose between. At this point, the child has developed a conceptual-level awareness of "self" and an ability to choose his goal-oriented actions. He is now self-aware at the abstract conceptual level. He now has a distinctively human, conceptual mind.

"Conceptual Goal-Directed Action"

Lacking instinct, humans must choose their goals and actions in the face of alternatives. From the first years of life, as a human is developing his conceptual mental faculties, he has accumulated countless experiences that resulted in feelings of pleasure or pain, satisfaction, or discomfort. These experiences have been stored in ISUs as the match patterns of actors, situations, stories, and causality relationships. Later, when facing a situation with alternative actions possible, such as the human hunter seeing an adult and newborn antelope, the human has an advantage over a higher-animal hunter. Both can use Perc-Situation_Recognizer-FMs to recall what types of actions tend to lead to satisfaction or pleasure and what types of actions tend to lead to discomfort or pain, and the varying degrees of both. At the perceptual level, both can determine that they are themselves hungry enough to press the attack. The human, however, has the advantage of being able, at the conceptual level, to use his Conc-Story-Recognizer-FM to assess the story's various match patterns and how the story might play out considering the causal factor of the newborn antelope. This enables the human hunter to be conscious of the increased risk, and he can choose whether to act on the perceptual-level urge to press the attack. Suppose he does choose to press the attack. In that case, the human hunter's conceptual knowledge of *how the world works* helps him choose actions that increase his chances of success while minimizing risks.

"Imagination-Creativity"

When operating at the conceptual level, the APs in Attentional-Workspace are pointing to symbol patterns for concepts. In a manner like using the Subc-Global-Search-FM during sleeping, a Conc-Global-Search-FM can be used to find patterns that have not been saved with any direct relationships but contain some particular element being searched for. When these concepts are held in Attentional-Workspace, the Conc-Situation-Recognizer-FMs are looking to see if they can be integrated to fit any known situations and, if not, whether they represent a new situation that should be saved. This can cause novel combinations of concepts that had not previously been identified. For example, suppose a horse and an antelope's horn are both held together in Attentional-

Workspace. In that case, the Conc-Situation-Recognizer-FM could integrate them into something that it saves under the symbol "unicorn." This activity is what we call "imagination." Further, suppose the concept patterns for a box, some poles, and a wheel are held together in Attentional-Workspace. In that case, the Conc-Situation-Recognizer-FM might integrate them into something it saves under the symbol "wheelbarrow." In this case, we call it *imagination* resulting in "creativity" because it can be brought into existence.

CONCLUSIONS

Consciousness in higher animals and humans is the manifestation of a complex set of functional brain modules executed on pattern-matching neural structures applied in hierarchically and recursively networked fashion and differing between individuals according to their genetically different brain structures and functions and the different external stimuli provided by reality during an individual's lifetime. To be conscious means to be conscious of reality. The more information an individual can obtain and grasp about how things work in reality, the better it will be to adapt its behaviors for success.

I see no need to treat *consciousness*, *mind*, or *volition* as phenomena apart from the brain. Consciousness is not, itself, an entity. These are descriptions of functional attributes of a brain that has become sufficiently complex for those operations to become possible. Consciousness is not a material-less phenomenon—it is but one attribute of physical human nature. The "mind" is in relation to pattern-recognizer PRUs and CNs in the brain in the same way the faculty of "sight" is related to rods and cones in the eye. Both are attributes of "consciousness" that emerge from a sufficiently complex neurological anatomy—the *mind* is a collective attribute, and *sight* is a more specific attribute.

I also conclude that volition is a direct evolutionary consequence of replacing instinctive automation as directives with a conceptual faculty—*the only alternative to directives is choices.* I also see that consciousness is a primary—it cannot be broken into parts. Consciousness is a faculty that consists of the completely integrated total "whole" of the organism's physical brain (and its connections to other parts of the body), learned methods of thinking, experiential mental content of how things in reality work, remembered consequences of choices made and much, much more.

Human consciousness is that faculty manifested as being aware of the nature of both external reality and oneself and choosing whether to apply conceptual-level thinking ability to select appropriate behaviors in the face of situations or to leave the selection of such behaviors to perceptual-level automation.

ELABORATIONS AND EXTENSIONS

For those interested in thinking a little deeper, the following elaborations and extensions follow from all we learned in the book and the previous "Theory of Mind" exercise.

The level of consciousness applied by different individuals ranges from *perceptual* to *conceptual* in different degrees. The complex combination of the individual's habitual thinking choices and the environmental stimuli of a present situation will determine which choice he makes. If, at some point, he learns how to exercise his ability to think and choose, then he is more likely to do so again. The more he does so, the more he becomes preconditioned to exercise his ability to think and choose as a habit—the more conceptual and *volitional* he becomes. Those people mostly devoid of the habit of choosing to think operate mostly at the perceptual and automated level. It is a matter of degree.

One could argue that knowing how to exercise the choice to think and develop a thinking habit is a consequence of the person's educational and life experience and therefore was not volitional on his part. How can he be held responsible for not learning how to think if he was born tabula rasa and was never taught or shown how? I believe this is true and is another consequence of evolution having replaced instinct with a conceptual faculty. In addition to becoming dependent on continually making choices instead of having automatic instincts, we are also dependent on society as the repository of accumulated knowledge of reality, including what we have discovered about how humans use their conceptual faculty. We must first be taught by our society (parents or their surrogates) how to think, how to use our rational faculty, and then we will be able to learn the knowledge of reality accumulated in our society. The necessity to think and choose our actions and the need to be taught how to think by society is the price paid to evolution to replace our directive instincts with a conceptual-level consciousness. The evolution of language is what gave simultaneous rise to both our conceptual faculty and our society.

A human society that consistently teaches its young how to think properly will increase both the number of individuals who develop the habit of conceptual thinking and the degree to which they come to rely upon the conceptual level as their preferred level of consciousness. Such a society will be better able to deal with changing reality and continue to evolve. A society that does not teach its young how to think conceptually will devolve back towards an animal-like perceptual existence. However, without instincts to guide them, the individuals and their society will eventually perish. For example, a human born during the Dark Ages is genetically no different than a human born today. A human in the Dark Ages would not have had a chance to learn how his "rational faculty" or the "scientific method" work because knowledge of these was not generally known or taught by society. The opposite, "mysticism," is what he would have been taught. It was not the individual's failure of volition that caused him to have a predominantly perceptual-level habit of thinking—he was just never taught how by society. Predominantly conceptual human societies have only existed twice in all human history, one before and one (the present one) following the Dark Ages.

All living organisms receive certain information about how to succeed in their environment from their ancestors via their genetics (phylogeny). For plants, this guidance is limited to triggering simple, genetically determined, stimulus-response behaviors when specific external stimuli occur such as light, temperature, chemical, and mechanical changes. Plants and simple animals are not conscious of the other entities that are causing those stimuli.

Evolution selected for a level of biological complexity sufficient to achieve perceptual-level consciousness in higher animals because it afforded an improvement for survivability in a changing environment. Evolution selected for brains and the ability to sense things at a distance—sight, smell, and hearing. They gained the ability to detect and identify other entities in their environment and do so without the need for that other entity to be in direct, physical, sensory contact. Being able to behaviorally respond to the existence of other entities in one's environment—to be conscious of them—was an increase in the individual's ability to obtain and grasp the knowledge of how things work in its environment and successfully respond. However, perceptually conscious animals do not grasp the natures of the entities they detect, or their own nature, or the rules of the game of how entities interact with each other.

Evolution selected *for* a level of biological complexity sufficient to achieve conceptual-level consciousness in humans because it afforded an even greater improvement for survivability in a changing environment. Evolution selected for brains with the ability to form and integrate *abstract mental entities* (concepts) as the symbols of language representing concrete entities in reality, to detect and grasp the causality relationships between entities (extrospection), and to be aware of itself as an entity and of its mental functioning (introspection). Evolution further selected for individuals with the ability to modify their behaviors based on what they learned about reality during their lifetimes (ontogeny), thereby overriding past knowledge of reality obtained from ancestors via genetically provided instinct (phylogeny).

Being able to behaviorally respond to a broader and deeper grasp of both physical and abstract entities, including their causality relationships, and a grasp of one's mental functioning—to be conscious of all these—and being able to override the no longer applicable instinct, resulted in an even greater increase in the individual's ability to obtain and grasp the knowledge of *how things work* in its environment and even more successfully respond.

Consciousness in humans is *deterministic* only in so far as it depends on the brain's physical functioning and external stimuli. Without the brain and something out there to grasp, there can be no consciousness. However, the physical functioning of the brain is altered by the state of conscious awareness—thinking. This forms a recursive loop that differs between individuals because of three factors: the variability in genetically formed brain function aptitudes at birth, the external stimuli encountered during one's life, and the choices made when applying conceptual thinking.

Given the mental machinery one was born with, the stimuli encountered and the behavioral responses to them form the neural networks in our brains as memories and thinking habits. New stimuli are processed according to previously learned neural networks. That processing either reinforces or alters existing neural networks—presently encountered stimuli and thinking about them affect future thinking and behaviors. Your brain and how you think will be different tomorrow than today because of today's experiences. Indeed, adapting your behaviors to what you learn about your environment is why evolution selected for your form of consciousness—this is how it is supposed to work.

Consciousness is simultaneously deterministic and volitional, each term applying to different aspects of the same phenomenon. Consciousness in humans is deterministic because it is the manifestation of the physical functioning of a brain as preconditioned by prior thinking. Consciousness is also volitional in the aspect that we can choose, in any present moment, what to apply our abstract thinking faculty toward, if anything. Instinct in perceptually conscious animals provided an automatic, dictated behavioral response to perceived entities—no choice was possible. As the consequence of evolution having selected for the ability to override instinct in conceptually conscious humans, the alternative to having no choice is having choice—volition. What we encounter and choose to think about in our environment today was preconditioned by what we chose to think about yesterday, and our thinking today will precondition what we will choose to think about in the future. However, just like overriding instinct, our continually increasing knowledge of how things work in our environment, applied to a present situation, can cause us to make a thinking choice different than that which we were preconditioned for by past thinking. The ability for present thinking to precondition future thinking—ontological experiential automation—along with the ability to override such preconditioning—volition—are together a capability for adapting our behavior during our lifetime to fit a changing environment. Therefore, evolution selected for our much more powerful, conceptual consciousness.

The infinitely complex range of possible environmental situations and events that an individual might encounter and the unpredictability of future changes in the environment caused evolution to select for better adaptability. Consider just a few of the factors that can affect one's present thinking: the incredible complexity of a lifetime of building neural networks of the individual's past experiences and habits of thinking, including chosen values; the individual's complex neural networks of understanding of the nature of environmental entities including causality relationships and the rules of the game (Laws of Reality); the complex set of stimuli provided by a present situation that is different than anything previously encountered, including the subconscious emotional responses to what the individual last had to eat. One can see that it is simply too inflexible to rely on having a correct automated response for each possible situation out of an infinite variety of them. Only a conscious entity that can apply an understanding of the nature of the environmental actors, including itself and the rules of the game, and capable of choosing from an infinite variety of possible responses (volition), will have an improved chance of arriving at an appropriate behavioral response.

We, humans, traded genetically provided specific-environment knowledge for self-discoverable, variable-environment knowledge. We traded the dictated specific-environment behaviors (positive stability on a provided course) for the freedom to choose our variable-environment behaviors (negative stability on a self-selected course). We traded the lack of adaptability due to mindless automation for the unlimited adaptability that came with the necessity to think and make choices continuously.

Evolution selected for ontological adaptation over phylogenic adaptation. Successive generations of Homo sapiens individuals gained intelligence and knowledge of reality more rapidly than phylogenic adaptation could provide. The increasing knowledge of reality could no longer be stored in the genes—an ontological method of knowledge storage was required. This necessitated the development of language and society. Language was necessary to transmit knowledge from its discoverer to society and then from society to new human individuals. The minds of individuals, not society, are the actors discovering new knowledge—there is no "hive" mind in human society. Society's evolutionary role is to be the repository for conceptual language-based (spoken, written & multi-media) knowledge across generations of conceptual individuals. New humans are dependent upon their parents or their surrogates to teach them how to use their conceptual faculty (forming education). Then the individual new human is free to choose what subset of total knowledge to acquire from society and what productive work to pursue within society to support his life.

INDEX

BIBLIOGRAPHY

Bernhardt, Chris. (2019). *Quantum Computing For Everyone*. Cambridge, MA: The MIT Press.

Binswanger, Harry (1976). *The Biological Basis of Teleological Concepts*. Los Angeles, CA: The Ayn Rand Institute Press.

Binswanger, Harry (2014). *How We Know. Epistemology on an Objectivist Foundation*. New York, NY: TOF Publications.

Bokhari, Allum (2020). *#Deleted. Big Tech's Battle To Erase The Trump Movement And Steal The Election.* New York, NY: Center Street.

Brook, Yaron. & Watkins, Don. (2017). *In Pursuit of Wealth. The Moral Case For Finance*. Irvine, CA: Ayn Rand Institute Press.

Contera, Sonia. (2019). *Nano Comes To Life. How Nanotechnology Is Transforming Medicine and the Future of Biology*. Princeton, NJ: Princeton University Press.

Crawford, Matthew. (2020). *Why We Drive. Toward a Philosophy of the Open Road.* New York, NY: HarperCollins Publishers Inc.

Curtis, Vickie. (2020). *The Social Dilemma*. A Netflix Original Documentary. Larissa Rhodes, Producer.

Dodge, Norman. (2007). *The brain that changes itself.* New York, NY: Penguin Books.

Drexler, K. (2013). *Radical Abundance. How a Revolution In Nanotechnology Will Change Civilization.* New York, NY: Public Affairs.

Gallardo, Laureano. (2019). *5G Networking. Learning 5G Technology Explained for Dummies.* Amazon.

Garreau, Joel. (2005). *Radical Evolution. The Promise and Peril of Enhancing Our Minds, Our Bodies—and What It Means to Be Human.* New York, NY: Broadway Books.

Hurd, Michael. (2000). *Grow Up America! Learning to Live the Happy, Responsible Life.* Washington, DC: Living Resources Press.

Jaeger, Lars. (2018). *The Second Quantum Revolution From Entanglement to Quantum Computing and Other Super-Technologies.* Switzerland AG: Springer Nature.

Kurzweil, Ray. (2005). *The Singularity Is Near. When Humans Transcend Biology*. New York, NY: Penguin Books.

Kurzweil, Ray & Grossman, Terry. (2009). *Transcend: Nine Steps to Living Well Forever.* Emmaus, PA: Rodale Press.

Kurzweil, Ray. (2012). *How To Create A Mind: The Secret Of Human Thought Revealed.* New York, NY: Penguin Books.

Laughlin, Burgess. (1995). *The Aristotle Adventure. A Guide to the Greek, Arabic, and Latin Scholars Who Transmitted Aristotle's Logic to the Renaissance.* Flagstaff, AZ: Albert Hale Publishing

Mayer, E. (2016). *The Mind-Gut Connection. How the Hidden Conversation Within Our Bodies Impacts Our Mood, Our Choices, and Our Overall Health.* New York, NY: HarperCollins Publishers Inc.

McIntosh, S. (2012). *Evolution's Purpose*. New York, NY: SelectBooks, Inc.

Packer, E. (1991). *Happiness Skills.* "The Intellectual Activist," March & May 1993.

Parisi, J. & Ball, J. (2019) *The Future of Fusion Energy*. London, UK: World Scientific Publishing Europe Ltd.

Peikoff, Leonard. (1991) *Objectivism: The Philosophy of Ayn Rand*. New York, NY: Penguin Group.

Peikoff, Leonard. (2012) *The DIM Hypothesis: Why The Lights Of The West Are Going Out*. New York, NY: Penguin Group.

Rand, Ayn (1961). *For the New Intellectual: The Philosophy of Ayn Rand.* Penguin Random House.

Rand, Ayn (1964). *The Virtue of Selfishness.* New York, NY: Signet.

Rand, Ayn (1971). *The Romantic Manifesto.* New York, NY: Signet.

Rand, Ayn (1984). *Philosophy: Who Needs It*. New York, NY: Signet.

Rand, Ayn (1988). *The Voice of Reason*. New York, NY: Penguin Group.

Rand, Ayn (1990). *Introduction to Objectivist Epistemology*, 2nd ed., H. Binswanger (ed.), Meridian.

Rand, Ayn (2001). *The Art of Nonfiction*. New York, NY: Penguin Group.

Richtel, M. (2019). *An Elegant Defense. The Extraordinary New Science of The Immune System.* New York, NY: Harper Collins.

Singh, Ajit. (2019). *5G Simply In Depth.* ajit_singh24@yahoo.com http://www.ajitvoice.in

Tapscott, Don & Tapscott, Alex. (2016). *Blockchain Revolution. How the Technology Behind Bitcoin Is Changing Money, Business, and the World*. New York, NY: Penguin Random House LLC.

Yonk, Richard. (2017). *Heart Of The Machine. Our Future In A World Of Artificial Emotional Intelligence.* New York, NY: Arcade Publishing.

Yonk, Richard. (2020). *Future Minds. The Rise Of Intelligence From The Big Bang To The End Of The Universe.* New York, NY: Arcade Publishing

Zuboff, Shoshana. (2019). *The Age of Surveillance Capitalism. The Fight for a Human Future at the New Frontier of Power*. New York, NY: Public Affairs

<<<>>>

ENDNOTES

---

[1] Yonk, Richard. (2020). *Future Minds. The Rise Of Intelligence From The Big Bang To The End Of The Universe.* New York, NY: Arcade Publishing

[2] https://en.wikipedia.org/wiki/Rational_animal

[3] Rand, Ayn (1990). *Introduction to Objectivist Epistemology*, 2nd ed., H. Binswanger (ed.), Meridian.

[4] Dodge, Norman. (2007). *The brain that changes itself.* New York, NY: Penguin Books. p.218

[5] Dodge, N. (2007)

[6] Mayer, Emeran. (2016). *The Mind-Gut Connection. How the Hidden Conversation Within Our Bodies Impacts Our Mood, Our Choices, and Our Overall Health.* New York, NY: HarperCollins Publishers Inc.

[7] Mayer, E. (2016)

[8] Mayer, E. (2016) p.161

[9] Mayer, E. (2016) p.18

[10] Mayer, E. (2016) p.31

[11] Mayer, E. (2016) p.132

[12] Mayer, E. (2016) p.176

[13] Mayer, E. (2016) p.168

[14] Mayer, E. (2016) p.181

[15] Dodge, N. (2007) p.228

[16] Dodge, N. (2007) p.295

[17] Crawford, Matthew. (2020). *Why We Drive. Toward a Philosophy of the Open Road.* New York, NY: HarperCollins Publishers Inc. p.13

[18] Dodge, N. (2007) p.279

[19] Dodge, N. (2007) p.201

[20] Dodge, N. (2007) p.287

[21] Dodge, N. (2007) p.304

[22] Dodge, N. (2007) p.59

[23] https://www.ucg.org/members/compass-check/three-reasons-not-to-let-life-get-you-down

[24] Dodge, N. (2007) p.291

[25] Dodge, N. (2007) p.313

[26] https://www.encyclopedia.com/social-sciences-and-law/sociology-and-social-reform/sociology-general-terms-and-concepts/progress

[27] Dodge, N. (2007) p.316

[28] Dodge, N. (2007) p.312

[29] Kurzweil, Ray. (2012). *How To Create A Mind: The Secret Of Human Thought Revealed.* New York, NY: Penguin Books. p.82

[30] https://www.cweb.com/how-brains-do-what-they-do-is-more-complex-than-what-anatomy-on-its-own-suggests/

[31] https://neurosciencenews.com/dendritic-action-potential-15380/

[32] Image from US Navy publication "Campus, The Navy Education and Training Monthly" May 1982

[33] Kurzweil, R. (2012) p.104

[34] Yonk, R. (2020)

[35] https://en.wikipedia.org/wiki/Technological_singularity

[36] Kurzweil, Ray. (2006). *The Singularity Is Near :When Humans Transcend Biology.* New York, NY: Penguin. p.136,123

[37] Yonk, R. (2020)

[38] https://ourworldindata.org/life-expectancy

[39] https://ourworldindata.org/grapher/life-expectancy?time=1543..&country=GBR+USA

[40] https://ourworldindata.org/grapher/life-expectancy?tab=chart&country=GBR+USA

[41] https://www.census.gov/content/dam/Census/library/publications/2020/demo/p25-1145.pdf

[42] https://en.wikipedia.org/wiki/List_of_the_verified_oldest_people

[43] Richtel, Matt. (2019). *An Elegant Defense. The Extraordinary New Science of The Immune System.* New York, NY: Harper Collins. p.406

[44] https://ourworldindata.org/life-expectancy

[45] https://www.kwa29.com/singularity-university-analogies-and-differences-with-universities/

[46] Kurzweil, R. (2012) p.158

[47] https://intelligence.org/files/DefiningValuesForValueLearners.pdf

[48] Peikoff, Leonard. (1991). *Objectivism: The Philosophy of Ayn Rand.* New York, NY: Penguin Group. p.458

[49] https://medium.com/the-democracy-network/the-coast-vs-the-heartland-5c58a40e8ca

[50] https://www.nytimes.com/2016/11/03/upshot/why-republicans-dont-even-try-to-win-cities-anymore.html

[51] *The Social Dilemm*a. January 26, 2020. A Netflix Original Documentary. An Exposure Labs Production. Written by Vickie Curtis, Davis Coombe, Jeff Orlowski. Directed by Jeff Orlowski. Producer, Larissa Rhodes.

[52] *The Social Dilemm*a. (2020)

[53] *The Social Dilemm*a. (2020)

[54] *The Social Dilemm*a. (2020)

[55] *The Social Dilemm*a. (2020)

[56] Zuboff, Shoshana. (2019). *The Age of Surveillance Capitalism. The Fight for a Human Future at the New Frontier of Power.* New York, NY: Public Affairs. p. 172—While I value Zuboff's identification of surveillance practices by Big Tech, I disagree completely with her use of the term "capitalism" both in the book's title and in the way she asserts certain evils of capitalism. I define capitalism as the only political system that protects the freedom of individuals to think and act in their own self interest without being subject to force or sacrifice.

[57] Zuboff, S. (2019) p.174

[58] Bokhari, Allum (2020). *#Deleted. Big Tech's Battle To Erase The Trump Movement And Steal The Election.* New York, NY: Center Street. p.17

[59] Bokhari, A. (2020) p.18

[60] Bokhari, A. (2020) p.36

[61] Bokhari, A. (2020) p.60

[62] Bokhari, A. (2020) p.83

[63] Bokhari, A. (2020) p.84-85

[64] Bokhari, A. (2020) p.83

[65] Bokhari, A. (2020) p.119,123-124

[66] Bokhari, A. (2020) p.145-150

[67] Bokhari, A. (2020) p.21

[68] Bokhari, A. (2020) p.106-107

[69] Bokhari, A. (2020) p.86-87
[70] Zuboff, S. (2019) p.58
[71] https://www.hbletter.com/forum/member-forum/blackmail/#p35937
[72] https://smallbusiness.chron.com/laws-subliminal-marketing-69892.html
[73] https://en.wikipedia.org/wiki/Subliminal_stimuli
[74] Crawford, M. (2020) p.303
[75] Zuboff, S. (2019) p.196
[76] Crawford, M. (2020) p.28,29,31,32,34
[77] Crawford, M. (2020) p.36
[78] Crawford, M. (2020) p.38
[79] Crawford, M. (2020) p.42
[80] Crawford, M. (2020) p.245
[81] Crawford, M. (2020) p.63
[82] Crawford, M. (2020) p.65
[83] Crawford, M. (2020) p.307
[84] https://americanaffairsjournal.org/2019/05/algorithmic-governance-and-political-legitimacy/
[85] Crawford, M. (2020) p.292
[86] https://www.forbes.com/sites/rebeccabellan/2020/09/22/facebook-google-twitter-youtube-need-to-take-responsibility-for-election-integrity/
[87] McIntosh, Steve. (2012). *Evolution's Purpose.* New York, NY: SelectBooks, Inc. p.133
[88] Binswanger, Harry (1976). *The Biological Basis of Teleological Concepts.* Los Angeles, CA: The Ayn Rand Institute Press. p.207
[89] Yonk, R. (2020)
[90] Rand, A. (1990) p.55
[91] https://en.wikipedia.org/wiki/Big_Bang
[92] https://en.wikipedia.org/wiki/Timeline_of_epochs_in_cosmology
[93] McIntosh, S. (2012). *Evolution's Purpose.* New York, NY: SelectBooks, Inc.
[94] A quote by Vaclav Havel reported in: Yonk, R. (2020)
[95] https://en.wikipedia.org/wiki/Timeline_of_the_evolutionary_history_of_life
[96] Contera, Sonia. (2019). *Nano Comes To Life. How Nanotechnology Is Transforming Medicine and the Future of Biology*. Princeton, NJ: Princeton University Press. p.23
[97] https://en.wikipedia.org/wiki/Thumb

98 http://www.brainrules.net/pdf/introduction.pdf
99 Rand, A. (1990) p.5
100 Rand, A. (1990) p.16
101 Rand, A. (1990) p.37
102 "On the Intelligence of the Dog," Nature, vol.33 (Nov. 12, 1885) p.45
103 http://psychclassics.yorku.ca/Miller/
104 Kurzweil, R. (2012) p.101
105 Kurzweil, R. (2012) p.69, 91
106 Rand, A. (1990) p.38
107 https://en.wikipedia.org/wiki/Wright_Flyer
108 Rand, Ayn (1961). *For the New Intellectual: The Philosophy of Ayn Rand.* Penguin Random House. p.121
109 Ayn Rand (1962). Talk at MIT on Ethics
110 https://courses.aynrand.org/campus-courses/free-will/?utm_source=Universe&utm_campaign=897dd86598-EMAIL_CAMPAIGN_2019_02_07_11_37_COPY_02&utm_medium=email&utm_term=0_3753df5893-897dd86598-289486305
111 Binswanger, Harry. (2014). *How We Know. Epistemology on an Objectivist Foundation.* New York, NY: TOF Publications. p.303,305
112 Binswanger, H. (2014) p.310
113 Binswanger, H. (2014) p.315
114 Rand, Ayn (1964). *The Virtue of Selfishness.* New York, NY: Signet. p.15
115 Rand, Ayn. (1984). *Philosophy: Who Needs It.* New York, NY: Signet. p.99
116 Rand, A. (1961) p.128
117 Rand, A. (1961) p.128
118 Rand, A. (1990) p.34
119 Rand, Ayn. (1971). *The Romantic Manifesto.* New York, NY: Signet. p.19
120 Rand, A. (1971) p.19
121 Rand, A. (1971) p.38
122 Rand, A. (1971) p.38
123 Rand, A. (1971) p.39
124 https://en.wikipedia.org/wiki/Language#Origin
125 https://en.wikipedia.org/wiki/Aristotle
126 https://en.wikipedia.org/wiki/Bhuri_Bai

[127] https://www.scientificamerican.com/article/will-we-keep-getting-smarter-flynn-effect-says-yes/

[128] Yonk, Richard. (2017). *Heart Of The Machine. Our Future In A World Of Artificial Emotional Intelligence.* New York, NY: Arcade Publishing.

[129] Binswanger, H. (1976) p.191

[130] Quoted by: Garreau, J. (2005)

[131] Rand, A. (1961) p.27

[132] Brook, Yaron. & Watkins, Don. (2017). *In Pursuit of Wealth. The Moral Case For Finance.* Irvine, CA: Ayn Rand Institute Press.

[133] Contera, S. (2019) p.65

[134] Peikoff, L. (1991) p.452

[135] Binswanger, Harry. "Ayn Ran's Philosophic Achievement" The Objectivist Forum (June-December 1982) http://www.hbletter.com/wp-content/uploads/2016/08/Ayn-Rands-Philosophic-Achievement.pdf

[136] Binswanger, H. (2014) p.14-16

[137] Binswanger, H. (2014) p.17

[138] Peikoff, L. (1991) p.452

[139] Binswanger, H. (2014) p.381

[140] Laughlin, Burgess. (1995). *The Aristotle Adventure. A Guide to the Greek, Arabic, and Latin Scholars Who Transmitted Aristotle's Logic to the Renaissance.* Flagstaff, AZ: Albert Hale Publishing

[141] Peikoff, L. (1991) p.453

[142] Peikoff, L. (1991) p.454

[143] Peikoff, L. (1991) p.454

[144] Peikoff, L. (1991) p.456

[145] Binswanger, H. (2014) p.58

[146] Binswanger, H. (2014) p.382

[147] Binswanger, H. (2014) p.382-383

[148] Peikoff, L. (1991) p.457

[149] Peikoff, L. (1991) p.456

[150] Peikoff, L. (1991) p.457

[151] Peikoff, L. (1991) p.458

[152] Peikoff, L. (1991) p.458

[153] Peikoff, L. (1991) p.459

[154] Rand, Ayn (2001). *The Art of Nonfiction*. R. Mayhew (ed). New York, NY: Penguin Group. p.176

[155] Peikoff, Leonard. (2012). *The DIM Hypothesis: Why The Lights Of The West Are Going Out.* New York, NY: Penguin Group. p.63

[156] Peikoff, L. (2012) p.64

[157] Peikoff, L. (2012) p.43

[158] Peikoff, L. (2012) p.334

[159] Peikoff, L. (2012) p.338

[160] Peikoff, L. (2012) p.340

[161] Peikoff, L. (2012) p.334

[162] Rand, A. (1990) p.71-72

[163] Rand, A. (1971) p.vii.

[164] Kurzweil, Ray & Grossman, Terry. (2009). *Transcend: Nine Steps to Living Well Forever.* Emmaus, PA: Rodale Press.

[165] Kurzweil, R. (2009)

[166] http://paperchase-aging.s3.amazonaws.com/pdf/5fa17ba097f8730007f0c2c5.pdf

[167] https://ms-my.facebook.com/lifespanio/videos/vb.284846431694459/448000359254369/?type=2&theater

[168] https://ms-my.facebook.com/lifespanio/videos/448000359254369/

[169] http://paperchase-aging.s3.amazonaws.com/pdf/5fa17ba097f8730007f0c2c5.pdf

[170] https://www.innovationtoronto.com/2020/06/synthetic-red-blood-cells-that-have-all-of-the-cells-natural-abilities-plus-a-few-new-ones/

[171] Kurzweil, R. (2012) p.39

[172] Rand, A. (1971) p.32

[173] Rand, Ayn. (1988). "Of Living Death," *The Voice of Reason*. New York, NY: Penguin. p.55

[174] https://www.pewforum.org/2013/08/06/religious-leaders-views-on-radical-life-extension/

[175] https://advances.sciencemag.org/content/5/6/eaav0547

[176] https://www.bbc.com/news/science-environment-26768445

[177] https://www.ncbi.nlm.nih.gov/pmc/articles/PMC3409741/

[178] Drexler, K. Eric. (2013). *Radical Abundance. How a Revolution In Nanotechnology Will Change Civilization.* New York, NY: Public Affairs.

[179] Drexler, K. (2013) p.220

[180] Drexler, K. (2013) p.201

[181] Drexler, K. (2013) p.161

[182] Drexler, K. (2013) p.168

[183] Drexler, K. (2013) p.234

[184] Drexler, K. (2013) p.236

[185] Drexler, K. (2013) p.225

[186] Drexler, K. (2013) p.282

[187] Contera, S. (2019) p.1

[188] Contera, S. (2019) p.3

[189] Contera, S. (2019) p.1

[190] Contera, S. (2019) p.6

[191] Contera, S. (2019) p.9

[192] Contera, S. (2019) p.12

[193] Contera, S. (2019) p.15

[194] Tapscott, Don & Tapscott, Alex. (2016). *Blockchain Revolution. How the Technology Behind Bitcoin Is Changing Money, Business, and the World.* New York, NY: Penguin Random House LLC.

[195] Tapscott, A. (2016) p.11

[196] Tapscott, A. (2016) p.34

[197] Tapscott, A. (2016) p.25

[198] Tapscott, A. (2016) p.253

[199] Gallardo, Laureano. (2019). *5G Networking. Learning 5G Technology Explained for Dummies.* Amazon. p.23

[200] https://cs.brown.edu/courses/csci1800/sources/2017_PRC_NationalIntelligenceLaw.pdf.

[201] *Cloudy With A Chance Of Catastrophe.* Wired Magazine. (Feb 2020)

[202] https://factsmaps.com/pisa-2018-worldwide-ranking-average-score-of-mathematics-science-reading/

[203] https://www.bbc.com/news/technology-56448688

[204] *The Man Who Conquered Noise.* Wired Magazine. (Dec 2020)

---

[205] Singh, Ajit. (2019). *5G Simply In Depth*. ajit_singh24@yahoo.com http://www.ajitvoice.in

[206] Singh, A. (2019)

[207] Singh, A. (2019)

[208] https://www.lifewire.com/6g-wireless-4685524

[209] Binswanger, H. (1976) p.107

[210] Binswanger, H. (1976) p.70,166

[211] Binswanger, H. (1976) p.136

[212] Binswanger, H. (1976) p.24

[213] Binswanger, H. (1976) p.118,129

[214] Rand, A. (1964) p.17

[215] Binswanger, H. (1976) p.66,64

[216] Binswanger, H. (1976) p.126

[217] Binswanger, H. (1976) p.133,130,172

[218] Binswanger, H. (1976) p.77-78

[219] Binswanger, H. (1976) p.148,147

[220] Binswanger, H. (2014). p.38

[221] Kurzweil, R. (2012) p.210

[222] Binswanger, H. (2014) p.45

[223] Rand, A. (1984) p.90

[224] Rand, A. (1961) p.156

[225] Rand, A. (1984) p.5

[226] Rand, A. (1961) p.121

[227] Rand, A. (1964) p.15

[228] Sotala, K. (2016)

[229] https://futureoflife.org/2019/08/14/how-can-ai-systems-understand-human-values/

[230] https://link.springer.com/article/10.1007/s11023-020-09539-2

[231] Garreau, Joel. (2005). *Radical Evolution. The Promise and Peril of Enhancing Our Minds, Our Bodies—and What It Means to Be Human*. New York, NY: Broadway Books.

[232] https://ieeexplore.ieee.org/abstract/document/7874163

[233] https://dam-prod.media.mit.edu/x/2018/03/23/p43-kapur_BRjFwE6.pdf

[234] https://hpluspedia.org/wiki/Fyborg

[235] Kurzweil, R. (2012) p.127

[236] Yonk, R. (2017) p.29

[237] Yonk, R. (2017) p.145

[238] https://www.eurasiareview.com/10052020-scientists-demonstrate-quantum-radar-prototype/

[239] Jaeger, Lars. (2018). *The Second Quantum Revolution From Entanglement to Quantum Computing and Other Super-Technologies.* Switzerland AG: Springer Nature.

[240] Bernhardt, Chris. (2019). *Quantum Computing For Everyone.* Cambridge, MA: The MIT Press.

[241] https://phys.org/news/2014-11-largest-factored-quantum-device.html

[242] https://www.eurasiareview.com/20032020-chip-based-devices-improve-practicality-of-quantum-secured-communication/

[243] Bernhardt, C. (2019) p.176

[244] Bernhardt, C. (2019) p.177

[245] Bernhardt, C. (2019) p.181

[246] Bernhardt, C. (2019) p.182

[247] https://thenextweb.com/google/2017/11/14/ibm-claims-quantum-supremacy-over-google-with-50-qubit-processor/

[248] https://www.quantamagazine.org/google-and-ibm-clash-over-quantum-supremacy-claim-20191023/

[249] Bernhardt, C. (2019) p.184

[250] https://blogs.intel.com/policy/2018/12/14/senate-advances-national-quantum-initiative-act/#gs.29s4g5

[251] Peikoff, L. (2012) p.126

[252] https://en.wikipedia.org/wiki/Tokamak

[253] https://en.wikipedia.org/wiki/Tokamak

[254] https://www.history.com/news/human-ancestors-tamed-fire-earlier-than-thought

[255] Parisi, J. & Ball, J. (2019). *The Future of Fusion Energy.* London, UK: World Scientific Publishing Europe Ltd. Ch 2.3

[256] Parisi, J. & Ball, J. (2019) Ch 2.3.4

[257] Parisi, J. & Ball, J. (2019) Ch 2.3.5

[258] Parisi, J. & Ball, J. (2019) Ch 2.3.6

[259] Parisi, J. & Ball, J. (2019) Ch 2.3.7

[260] Parisi, J. & Ball, J. (2019) Ch 2.3.8

[261] Parisi, J. & Ball, J. (2019) Ch 2.3.9
[262] Parisi, J. & Ball, J. (2019) Ch 2.3.10
[263] Parisi, J. & Ball, J. (2019) Ch 2.3.2
[264] Parisi, J. & Ball, J. (2019) Ch 2.3.1
[265] Parisi, J. & Ball, J. (2019) Ch 2.4
[266] Parisi, J. & Ball, J. (2019) Table 2.1
[267] https://www.iter.org/mach
[268] https://www.iter.org/proj/inafewlines#2
[269] https://www.iter.org/mach
[270] https://phys.org/news/2020-12-korean-artificial-sun-world-sec-long.html
[271] https://www.wsj.com/articles/fusion-startups-step-in-to-realize-decades-old-clean-power-dream-11581001383
[272] https://firstlightfusion.com/
[273] Parisi, J. & Ball, J. (2019) Introduction.
[274] Parisi, J. & Ball, J. (2019) Ch 12.
[275] https://infuse.ornl.gov/
[276] Parisi, J. & Ball, J. (2019) Ch 10.
[277] Parisi, J. & Ball, J. (2019) Ch 11.
[278] Dr. Edith Packer, J.D., Ph.D. 1991 Summer Conference of the Jefferson School. Transcript published in the March and May 1993 issues of "The Intellectual Activist."
[279] Hurd, Michael. (2000). *Grow Up America! Learning to Live the Happy, Responsible Life*. Washington, DC: Living Resources Press. p.286

For those who wish to further pursue an understanding of the *psychology* required to achieve happiness and the heroic vision they had for themselves as a child, I recommend that you begin with the above referenced book and other books and resources found at Dr. Hurd's web site: https://drhurd.com

[280] Rand, A. (1984) p.200
[281] Rand, A. (1984) p.201
[282] Rand, A. (1984) p.203
[283] Drexler, K. (2013) p.273
[284] Drexler, K. (2013) p.283
[285] Drexler, K. (2013) p.286
[286] My Story. From Harry Binswaner's blog at hbletter.com 7/30/2015
[287] Rand, A. (2001) p.27

[288] Kurzweil, R. (2012) p.112

[289] Peikoff, L. (2012) p.345

[290] https://intelligence.org/files/DefiningValuesForValueLearners.pdf

[291] https://www.aaai.org/ocs/index.php/WS/AAAIW16/paper/download/12633/12353

[292] https://link.springer.com/article/10.1007/s11023-020-09539-2

[293] Rand, A. (1961) p.130

[294] Kurzweil, R. (2012) p.223

[295] Quoted by: Garreau, J. (2005)

[296] Rand, A. (1961) p.viii

[297] https://www.hbletter.com/forum/member-forum/altruism-debate/#p36364

[298] Andrew Carnegie, "The Best Fields for Philanthropy" Archived January 13, 2003, at the Wayback Machine, *The North American Review*, Volume 149, Issue 397, December 1889 from the Cornell University Library website

[299] https://www.brainyquote.com/quotes/vince_lombardi_382625

[300] Kurzweil, R. (2012) p.38

[301] Kurzweil, R. (2012) p.110

Made in the USA
Coppell, TX
18 June 2022

78972748R00351